S0-BHS-315

# ArcView GIS®/ Avenue™ Programmer's Reference

*Class Hierarchy Quick Reference and 100+ Scripts*

Second Edition

**Amir H. Razavi and Valerie Warwick**

# ArcView®/Avenue™ Programmer's Reference
## Class Hierarchy Quick Reference and 100+ Scripts
**Amir H. Razavi and Valerie Warwick**

Published by:

OnWord Press
2530 Camino Entrada
Santa Fe, NM 87505-4835 USA

All rights reserved. No part of this book may be reproduced or transmitted in any form or by any means, electronic or mechanical, including photocopying, recording, or by any information storage and retrieval system without written permission from the publisher, except for the inclusion of brief quotations in a review.

Copyright © Amir H. Razavi and Valerie Warwick

Second Edition, 1997

SAN 694-0269

10 9 8 7 6 4 3 2 1

Printed in United States of America


Cataloging-in-Publication Data

Razavi, Amir H., 1957-

ArcView/Avenue programmer's reference : class hierarchy quick reference and 101 scripts / Amir H. Razavi, Valerie Warwick. — 2nd ed.

    p.   cm.

Includes index.

ISBN 1-56690-123-5

1.   Avenue  (Computer program language)   2. ArcView.    I.  Warwick, Valerie,

2-  .

QA76.73.A895R39    1997

910'.285'574—dc21                                                    96-29962

                                                                            CIP

## Trademarks

ArcView GIS is a registered trademark and Avenue is a trademark of Environmental Systems Research Institute (ESRI), the world's leading supplier of geographic information systems software. OnWord Press is a registered trademark of High Mountain Press, Inc. All other terms mentioned in this book that are known to be trademarks or service marks have been appropriately capitalized. OnWord Press cannot attest to the accuracy of this information. Use of a term in this book should not be regarded as affecting the validity of any trademark or service mark.

## Warning and Disclaimer

This book is designed to provide quick reference information and scripts for the purpose of customizing the ArcView GIS program through Avenue, ESRI's object oriented programming language for ArcView GIS. Every effort has been made to make the book as complete, accurate, and up to date as possible; however, no warranty or fitness is implied.

The information is provided on an "as is" basis. The authors and OnWord Press shall have neither liability nor responsibility to any person or entity with respect to any loss or damages in connection with or arising from the information contained in this book.

## About the Authors

**Amir H. Razavi** is a professional engineer registered in the states of Maryland and Virginia. He has a B.A. degree in civil engineering, and a Master's of information systems management from George Washington University. Amir has been developing software since 1982, and has served as GIS manager for the Civil Rights Division at the Department of Justice. In 1994, he founded Razavi Application Developers (RAD) to provide quality software development services to businesses and government agencies. RAD specializes in ArcView, MapObjects, and Oracle applications. (Amir can be contacted at *razavi@compuserve.com* or through *http://ourworld.compuserve.com/homepages/razavi* on the Internet.)

**Valerie Warwick** is an ESRI-certified instructor for PC ARC/INFO and ArcCAD. She has written commercial applications in ArcView GIS and Avenue for oil and gas companies in Texas and Louisiana. Valerie also teaches and designs classes on ArcView GIS and Avenue for wptc (we provide technical consulting). Based in Austin, Texas, wptc specializes in technical education and consulting for ArcView GIS, Avenue, and the World Wide Web.

## Acknowledgments

This book would not have been a reality without the efforts and guidance of editors Daril Bentley and Barbara Kohl at High Mountain Press. I am very grateful to both. I also appreciate the cooperation of ESRI and technical reviewers throughout the development of this book.

I dedicate this book to my wife, Tema, who has always supported me in my activities.

*Amir H. Razavi*

I would like to thank my parents, Oliver and Genevieve Berglund, for their guidance and support over the years, and my children, Brandy, Bonnie, and Charlotte Warwick, for their never-ending patience and love.

*Valerie Warwick*

## OnWord Press...

Dan Raker, President
David Talbott, Acquisitions and Special Projects Director
Barbara Kohl, Acquisitions Editor
Daril Bentley, Project Editor
Carol Leyba, Senior Production Manager
Cynthia Welch, Production Editor
Michelle Mann, Production Editor
Lauri Hogan, Marketing Services Manager
Kristie Reilly, Assistant Editor
Lynne Egensteiner, Cover designer, Illustrator

# Contents

**INTRODUCTION** . . . . . . . . . . . . . . . . . . . . . . . . . . . vii

   How to Find Classes and Scripts . . . . . . . . . . . . . . . . . . . . . . . . . vii

   How to Read Class Entries . . . . . . . . . . . . . . . . . . . . . . . . . viii

   How to Read Script Entries . . . . . . . . . . . . . . . . . . . . . . . . x

**TOPICAL INDEX OF AVENUE CLASSES** . . . . . . . . . . . . . . . . . . . . xiii

**TOPICAL INDEX OF AVENUE SCRIPTS** . . . . . . . . . . . . . . . . . . . . xix

**CLASS HIERARCHY QUICK REFERENCE** . . . . . . . . . . . . . . . . . . . . 1

**101 SCRIPTS** . . . . . . . . . . . . . . . . . . . . . . . . . . . 279

   Applications . . . . . . . . . . . . . . . . . . . . . . . . . . . 281

   DocGUIs . . . . . . . . . . . . . . . . . . . . . . . . . . . 283

   Documentation . . . . . . . . . . . . . . . . . . . . . . . . . 306

   Files . . . . . . . . . . . . . . . . . . . . . . . . . . . 358

   Graphics . . . . . . . . . . . . . . . . . . . . . . . . . . . 368

   Message Boxes . . . . . . . . . . . . . . . . . . . . . . . . . 383

   Miscellaneous . . . . . . . . . . . . . . . . . . . . . . . . . . 389

   Projects . . . . . . . . . . . . . . . . . . . . . . . . . . . 402

   SEds . . . . . . . . . . . . . . . . . . . . . . . . . . . 412

   Tables . . . . . . . . . . . . . . . . . . . . . . . . . . . 442

   Themes . . . . . . . . . . . . . . . . . . . . . . . . . . . 482

   Views . . . . . . . . . . . . . . . . . . . . . . . . . . . 502

**INDEX** . . . . . . . . . . . . . . . . . . . . . . . . . . . 519

# Introduction

Welcome to the *ArcView GIS/Avenue Programmer's Reference: Class Hierarchy Quick Reference and 100+ Scripts*. The first part of the book is the class hierarchy quick reference, a complement to ArcView GIS/Avenue's online help capabilities. The latter half of the book presents scripts ready for use in your projects. All material in the book is written to 3.x of ArcView GIS/Avenue.

The quick reference portion of the book provides nuts and bolts descriptions of Avenue classes, class requests, instance requests, and enumerations. Thus, the quick reference does not pretend to be a comprehensive guide to Avenue. The scripts portion of the book assumes that you possess a working knowledge of Avenue, including syntax and other rules of usage. In brief, it is assumed that readers are knowledgeable about their operating systems, and are somewhat experienced in writing Avenue code.

The scripts appearing in this book are available on diskette as a separate product from OnWord Press. Ask for *100+ ArcView GIS/Avenue Scripts: The Disk* by Valerie Warwick, 1997, ISBN 1-56690-133-2.

## ◆ *How to Find Classes and Scripts*

***Alphabetically by class or script name***. Class and script names can be found alphabetically in respective sections of the book.

***Topical listing***. Class names are listed by topic in an index preceding the class hierarchy section of the book. In turn, a topic listing of script names precedes the script section of the book.

***Class cross-references***. At the end of many class entries, the code example line references scripts that demonstrate usage of a given class.

***General alphabetical index***. Class enumerations appear in the general index. Topics and search keys pertaining to each script are cross-referenced in the general index.

# ◆ How to Read Class Entries

All entries in the class hierarchy section of the book begin with the class name in large boldface type at the far left. The class name is followed by a brief description and one or two lines indicating the parent class the class inherits from, and the child class(es) the class is inherited by. Classes that do not exist in version 3.0 are flagged with a Note.

If you cannot locate a specific request for a given class, check the requests of the class's parent or ancestors (parent's parents). Through inheritance, you can use requests of the parent or ancestor class with a child class.

Child classes are important because they provide task specialization. For instance, it is easier to use the NorthArrow class to place a north arrow on a layout than using the parent GraphicGroup class. In brief, NorthArrow is specialized to handle north arrows. The format and organization of class entries are illustrated and described in the material that follows.

## ClassName

 This icon to the right of a classs name indicates that the class is a new one.

 This icon to the right of a class name indicates that thie class has changed significantly.

A brief description of the class appears immediately below the class name. If the class is not available in ArcView 3.0, a note appears immediately below the **Inherits from** and/or **Inherited by** lines.

**Inherits from** ParentClass

**Inherited by** ChildClass

> ⊷ **NOTE:** *This class is not available in ArcView GIS (v. 3.0).*

Class and instance requests (if applicable) appear next in a tabular format. The function or action of the request is described in the right column. Requests that are *not* applicable to ArcView GIS only are flagged with a brief statement in parentheses in the right column.

The request, parameters (if applicable), and returned object (if applicable) appear in the left column. If the request requires parameters, the parameters appear in parentheses. Multiple parameters inside parentheses are separated by commas. If the request returns an object, the returned object appears after a colon ( : ). Do not type the colon in an Avenue script. In a script, you use the equal sign ( = ) as the assignment operator to store the returned object.

| Class Requests | |
| --- | --- |
| MakeDefault (UpdateSubclass, aField, anFTab) : anAttrUpdate | Makes a default single or range update rule for aField and anFTab. |
| AddMissingRules (aRuleList, anFTab) | Applies the list of rule enumerations to the fields or paired fields of a feature table. |
| ReturnDefaultRules (anFTab) : aRuleList | Returns the default rule enumerations assigned to a feature table. |

| Instance Requests | |
| --- | --- |
| GetField : Field | Gets the field in AttrUpdate object. |
| HasField (aField) : aBoolean | Returns true if a field is assigned to an object of AttrUpdate. |

Appearing below are examples of how requests for the Date class are presented in this book, and how they are used in an Avenue script.

| Class Requests | |
| --- | --- |
| Make (dateString, formatString) : aDate | Creates a date object from a date string and format. |

| Instance Requests | |
| --- | --- |
| GetDay : aString | Gets the full day name. |

```
' Create a data object.

myBirthDate = Date.Make ( "08/08/1970:","MM/dd/yyyy")

MsgBox.Info ( "I was born on a "++myBirthDate.GetDay,"")
```

The message box displays the following: "I was born on a Saturday." The parameter and returned object names begin with lower-case letters. Where meaningful, the class name was appended to the object name. For instance, *formatString* in the Make request of the Date class is a string object.

Applicable enumerations are presented under the class entry that uses them. All enumeration names end with Enum.

### Enumerations

| RasterFillStyleEnum | |
| --- | --- |
| #RASTERFILL_STYLE_EMPTY | No shading. |
| #RASTERFILL_STYLE_OPAQUESTIPPLE | Stipple shading with solid background. |

Next, at the end of every entry is a cross-reference line to help you locate similar or related classes.

**See also**: Duration; String

The last line for many entries is a cross-reference to a script in the second half of the book that demonstrates usage of the class.

**Code example**: ButtonAdd

## ◆  *How to Read Script Entries*

All script entries in the second half of the book begin with the name of the script in large boldface type to the far left. The next line is a description of the script.

### ScriptName

A description of the script and topics and search keys related to the script appear immediately after the script name.

***Topics***: SEd

***Search Keys***: SEd, GetSource, SetInsertPos, Insert

If applicable, additional information following the Requirements, Comments, and Test Drive headings appears after the Search Keys line(s).

The script is reproduced in a monospaced font. Long lines of code "wrap" to the next line, but do *not* signify that you should insert a hard return at the end of the line as it appears on the page. In these instances, the space (leading) between lines is quite narrow. An example of a wrapped line of code appears below. When inputting this statement, you would hit <Enter> or <Return> only after typing *Meadows").*

firstQueryValue = msgBox.Input("Value in" ++ theAField.GetName ++ "Field equals:","First Query","Rose Meadows")

Lines of code that end with a hard return are separated by larger spaces. An example of the spacing between lines of code ending with hard returns follows. In this instance, you would input a hard return after *exit end* and *theAField.GetType.*

```
if (theAField = nil) then exit end

theAFieldType = theAField.GetType
```

A sample code segment containing wrapped code lines and lines of code ending with hard returns follows.

```
' Build query string.

' Query A.

fldList = theVTab.GetFields

theAField = MsgBox.List(fldList,"Choose a field to Build First
Query","FIRST QUERY FIELD-Subname")

if (theAField = nil) then exit end

theAFieldType = theAField.GetType

firstQueryValue = msgBox.Input("Value in" ++ theAField.GetName ++
"Field equals:","First Query","Rose Meadows")

if (theAField.IsTypeString) then
```

# Topical Index
# of Avenue Classes

## ◆ Address Matching

| | | | |
|---|---|---|---|
| AddressStyle | 3 | MatchKey | 146 |
| EventDialog | 67 | MatchPref | 147 |
| GeoNa | 90 | MatchPrefDialog | 148 |
| LocateDialog | 141 | MatchSource | 149 |
| MatchCand | 145 | RematchDialog | 198 |
| MatchCase | 146 | StanEditDialog | 232 |
| MatchField | 146 | XYName | 276 |

## ◆ Application Development

| | | | |
|---|---|---|---|
| EncryptedScript | 62 | Printer | 188 |
| File | 71 | Project | 192 |
| FileDialog | 73 | Script | 203 |
| FileName | 73 | ScriptMgr | 204 |
| LineFile | 137 | SEd | 221 |
| LockMgr | 141 | SourceDialog | 227 |
| ODB | 173 | Textfile | 245 |

## ◆ Application Interface

| | | | |
|---|---|---|---|
| AppleEvent | 6 | DLL | 54 |
| AppleScript | 7 | DLLProc | 55 |
| Application | 8 | Mac | 142 |
| Clipboard | 30 | RPCClient | 199 |
| DDEClient | 46 | RPCServer | 200 |
| DDEServer | 47 | System | 239 |

# ◆ Chart Document

| | | | |
|---|---|---|---|
| Axis | 13 | ColumnChartSymbol | 34 |
| Chart | 21 | Display | 51 |
| ChartDisplay | 23 | PieChartSymbol | 182 |
| ChartLegend | 24 | Title | 254 |
| ChartPart | 25 | XAxis | 274 |
| ChartSymbol | 26 | YAxis | 276 |

# ◆ Database Themes

| | | | |
|---|---|---|---|
| AddDBTheme | 3 | SDEFeature | 210 |
| DBTheme | 42 | SDELayer | 213 |
| DBTQueryWin | 45 | SDELog | 219 |
| SDEDataSet | 204 | | |

# ◆ Digitizer

| | |
|---|---|
| Digit | 48 |

# ◆ Extension

| | | | |
|---|---|---|---|
| DocumentExtension | 60 | LegendExtension | 134 |
| Extension | 67 | SrcExtension | 230 |
| ExtensionObject | 68 | ThemeExtension | 253 |
| ExtensionWin | 69 | | |

# ◆ Graphics

| | | | |
|---|---|---|---|
| Annotation | 6 | ColorMap | 33 |
| BasicMarker | 15 | CompositeFill | 35 |
| BasicPen | 16 | CompositeMarker | 35 |
| Circle | 27 | CompositePen | 36 |
| Color | 32 | Fill | 75 |

| | | | |
|---|---|---|---|
| Font | 76 | PolyLine | 186 |
| FontManager | 78 | PolyLineTextPositioner | 187 |
| GEdgeRec | 89 | RasterFill | 195 |
| GNodeRec | 91 | Rect | 197 |
| Graphic | 91 | Shape | 224 |
| GraphicGroup | 94 | Stipple | 233 |
| GraphicLabel | 95 | Symbol | 236 |
| GraphicList | 96 | SymbolList | 237 |
| GraphicSet | 98 | SymbolWin | 238 |
| GraphicShape | 99 | TextComposer | 244 |
| GraphicText | 100 | TextPositioner | 246 |
| Icon | 116 | TextSymbol | 247 |
| IconMgr | 116 | VectorFill | 260 |
| Line | 136 | VectorPen | 262 |
| Marker | 144 | VectorPenArrow | 262 |
| MultiPoint | 157 | VectorPenDiamond | 263 |
| Oval | 174 | VectorPenDot | 263 |
| Palette | 179 | VectorPenHash | 263 |
| Pen | 180 | VectorPenHollow | 263 |
| Point | 183 | VectorPenScallop | 264 |
| PointTextPositioner | 184 | VectorPenScrub | 264 |
| Polygon | 184 | VectorPenSlant | 264 |
| PolygonTextPositioner | 185 | VectorPenZigZag | 265 |

## ◆ *Layout*

| | | | |
|---|---|---|---|
| ChartSymbol | 26 | PageManager | 177 |
| ColumnChartSymbol | 34 | PageSetupDialog | 178 |
| Display | 51 | PictureFrame | 181 |
| DocFrame | 58 | PieChartSymbol | 182 |
| Frame | 79 | ScaleBarFrame | 201 |
| Layout | 128 | Template | 244 |
| LegendFrame | 134 | TemplateMgr | 244 |
| NorthArrow | 168 | TGraphic | 249 |
| NorthArrowMgr | 169 | ViewFrame | 268 |
| PageDisplay | 175 | | |

# ◆ *Literals*

| | | | |
|---|---|---|---|
| Boolean | 18 | NameDictionary | 158 |
| Collection | 32 | Nil | 168 |
| Date | 39 | Number | 169 |
| Dictionary | 47 | Pattern | 180 |
| Duration | 61 | Stack | 231 |
| EnumerationElt | 63 | String | 233 |
| Interval | 122 | Units | 259 |
| List | 139 | Value | 260 |

# ◆ *Network Analyst*

| | | | |
|---|---|---|---|
| ClosestFacilityWin | 30 | NetWork | 164 |
| GraphicFlag | 93 | NetWorkSrc | 166 |
| LandMark | 128 | NetworkWin | 166 |
| NetCostField | 160 | NetworkWinSrc | 167 |
| NetDef | 161 | ServiceAreaWin | 223 |
| Net Units | 162 | ShortestPathWin | 225 |

# ◆ *Projection*

| | | | |
|---|---|---|---|
| Albers | 4 | Nearside | 159 |
| Cassini | 20 | NewZealand | 167 |
| CoordSys | 38 | ObliqueMercator | 172 |
| EqualAreaAzimuthal | 63 | Orthographic | 174 |
| EqualAreaCylindrical | 64 | Perspective | 181 |
| EquidistantAzimuthal | 65 | Prj | 189 |
| EquidistantConic | 65 | ProjectionDialog | 194 |
| EquidistantCylindrical | 66 | Robinson | 199 |
| Gnomonic | 91 | RSO | 200 |
| Hammer | 114 | Sinusoid | 226 |
| Lambert | 127 | Spheroid | 228 |
| Mercator | 151 | Stereographic | 232 |
| Miller | 152 | TransverseMercator | 258 |
| Mollweide | 153 | Units | 259 |

# ◆ Spatial Analyst

| | | | |
|---|---|---|---|
| AnalysisEnvironment | 5 | GTheme | 112 |
| AnalysisPropertiesDialog | 6 | Interp | 121 |
| DensitySurfaceDialog | 47 | InterpolationDialog | 122 |
| FocalStatisticsDialog | 76 | NbrHood | 159 |
| Grid | 101 | Radius | 195 |
| GridLegendExtension | 111 | ReclassEditor | 196 |
| GridLegendWindow | 111 | SVGram | 235 |
| GShapeRec | 112 | TabulateAreaDialog | 243 |

# ◆ Table Document

| | | | |
|---|---|---|---|
| AttrRange | 10 | INFODir | 121 |
| AttrSingle | 11 | QueryWin | 194 |
| AttrUpdate | 12 | SQLCon | 230 |
| BitMap | 17 | SQLWin | 230 |
| DBICursor | 41 | SummaryDialog | 235 |
| Field | 69 | Table | 242 |
| FTab | 80 | VTab | 270 |

# ◆ User Interface

| | | | |
|---|---|---|---|
| AutoLabelDialog | 13 | MenuBar | 151 |
| Button | 19 | ModalDialog | 152 |
| ButtonBar | 20 | MovieWin | 154 |
| Choice | 27 | MsgBox | 154 |
| CodePage | 31 | Popup | 188 |
| Control | 37 | PopupSet | 188 |
| ControlSet | 38 | RematchDialog | 198 |
| Doc | 56 | SourceManager | 227 |
| DocGUI | 58 | Space | 228 |
| DocWin | 60 | TextWin | 248 |
| GeoCodeDialog | 89 | Tool | 256 |
| Help | 114 | ToolBar | 256 |
| ImageWin | 119 | ToolMenu | 257 |
| LabelButton | 125 | Window | 274 |
| Menu | 151 | | |

# ◆ *View Document*

| | | | |
|---|---|---|---|
| AreaOfInterestDialog | 10 | Labeler | 126 |
| BandStatistics | 14 | Layer | 128 |
| ChartSymbol | 26 | Legend | 130 |
| Classification | 29 | Librarian | 135 |
| ColumnChartSymbol | 34 | Library | 135 |
| Coverage | 39 | LinearLookup | 137 |
| DynName | 62 | MapDisplay | 143 |
| FTheme | 85 | MultiBandLegend | 156 |
| IdentifyWin | 117 | PieChartSymbol | 182 |
| IdentityLookup | 118 | SingleBandLegend | 226 |
| ImageLegend | 118 | SrcName | 231 |
| ImageLookup | 118 | Theme | 249 |
| ImgCat | 119 | ThemeOnThemeDialog | 253 |
| ImgSrc | 120 | Threshold | 253 |
| IntervalLookup | 123 | TOC | 255 |
| ISrc | 124 | View | 265 |
| ITheme | 125 | | |

# Topical Index of Avenue Scripts

## ◆ Applications

| | | | |
|---|---|---|---|
| CheckEnvironmentVariable | 281 | UserExtVariable | 283 |
| SetEnvironmentVariable | 282 | | |

## ◆ DocGUIs

| | | | |
|---|---|---|---|
| ButtonAdd | 283 | Menu | 295 |
| ButtonBar | 285 | MenuAdd | 297 |
| ButtonCopy | 286 | MenuCopy | 299 |
| ButtonDeleteEdit | 288 | MenuDelete | 300 |
| ChoiceAdd | 290 | ToolAdd | 302 |
| ChoiceDeleteEdit | 292 | ToolDeleteEdit | 304 |
| InstallaTool | 294 | | |

## ◆ Documentation

| | | | |
|---|---|---|---|
| ButtonBarDocumentation | 306 | MenuBarDocumentation | 326 |
| ButtonDocumentation | 308 | MenuDocumentation | 328 |
| ButtonReport | 309 | ProjectDocumentation | 330 |
| ChartDocumentation | 311 | SEdDocumentation | 338 |
| ChoiceDocumentation | 312 | TableDocumentation | 339 |
| ChoiceReport | 314 | TableReport | 341 |
| EmbeddedScriptDocumentation | 315 | TableReportFile | 344 |
| FThemeDocumentation | 316 | ToolBarDocumentation | 346 |
| GUIReport | 318 | ToolDocumentation | 348 |
| Indent | 323 | ToolReport | 350 |
| IThemeDocumentation | 323 | ViewDocumentation | 351 |
| LayoutDocumentation | 325 | ViewReport | 353 |

# ◆ Files

CurrentWorkingDirectoryInitialize    358
FileExists                           358
FileExistsInSearchPath               359

PLasXYEvent                          362
SetDirectories                       364
Tiger                                365

# ◆ Graphics

DrawSomething                        368
DrawText                             369

GraphicsButtons                      371
PrettyButtons                        375

# ◆ Message Boxes

MessageBoxFlowControl                383
MessageBoxInput                      384

MessageBoxReport                     385
MessageBoxSelection                  387

# ◆ Miscellaneous

Clone                                389
ColorDict                            392
Dictionary                           394
IconGet                              398

LabelButtons                         399
ReportCoordinates                    400
StackTest                            401

# ◆ Projects

ExtractScript                        402
ProjectInit                          405

ProjectSave                          409
ProjectSaveReset                     410

# ◆ SEds

AddComment                           412
AddScriptToEMail                     412
ConvertScript                        414
DeleteEmbeddedScripts                416
DeleteSEdsFromProject                417

ExecuteComment                       418
LineEditingDriver                    419
OpenScript                           423
ReadScriptsFromDiskette              426
ReplaceOldWithNew                    428

| | | | |
|---|---|---|---|
| ScriptExportAsEMail | 430 | SEd2TextFile | 436 |
| ScriptFileLoad | 432 | SEdSetName | 437 |
| ScriptManagerDriver | 433 | WriteScriptsToDisk | 438 |
| ScriptManagerSubset | 435 | | |

## ◆ Tables

| | | | |
|---|---|---|---|
| ApplySelection | 442 | GetSelectionToODB | 461 |
| BitmapClearSelection | 444 | GetVTabFields | 464 |
| BitmapCountSelected | 445 | SaveSelectionAddField | 465 |
| BitmapQuery | 446 | SaveSelectionCreateLookupTable | 468 |
| EditAttributes | 450 | SaveSelectionTableObjectTag | 470 |
| ExportSortedTable | 452 | SaveSelectionToODB | 472 |
| GetSelectionAddField | 455 | SelectFeatures | 474 |
| GetSelectionFromLookupTable | 457 | SelectFieldsFromVTAB | 477 |
| GetSelectionTableObjectTag | 459 | TableChange | 480 |

## ◆ Themes

| | | | |
|---|---|---|---|
| ImageMove | 482 | LegendSimple | 490 |
| InstallImage | 484 | LegendUnique | 492 |
| LegendMakeSimple | 487 | ScatterDiagram | 496 |

## ◆ Views

| | | | |
|---|---|---|---|
| ChooseViewTheme | 502 | HotlinkTableAdd | 510 |
| GetDoc | 504 | RotateWorld | 513 |
| GetViewTheme | 506 | ViewAdd | 515 |
| HotlinkClickSwitch | 508 | | |

# Class Hierarchy Quick Reference

# AddDBThemeDialog (Database Themes)

This class implements the dialog box to add database themes.

**Inherits from** ModalDialog

| Class Requests | |
|---|---|
| Show:aDBThemeList | Opens a dialog box to select and return database themes. |

*See also:* DBTheme

# AddressStyle

This class defines the address format and its required and optional fields.

**Inherits from** Obj

| Class Requests | |
|---|---|
| ClearStyles | Clears all defined address styles. |
| FindBestStyle (aThemeFTab) : anAddressStyle | Finds the best address style based on the fields of aTheme's feature table. |
| FindStyle (anAddressStyleName) : anAddressStyle | Finds an address style from the list of styles loaded by GetStyles. |
| GetDefStylesODB : aFileName | Gets the file name for the default address styles. |
| GetStyles (aFileName) : anAddressStyleList | Gets a list of address styles stored in a file. |
| SaveStyles (aFileName) | Saves all defined address styles to a file. |

| Instance Requests | |
|---|---|
| GetDefOffset : aNumber | Gets the default offset value. |
| GetDefSqueeze : aNumber | Gets the default squeeze value. |
| GetMatchFields : aMatchFieldList | Gets the list of match fields. |
| GetMaxScore : aNumber | Gets the address maximum match score. |
| GetMKeyZone : aString | Gets the match field name of the zone. |
| GetName : aString | Gets the name of the address style. |

3

| GetXDelimiter : aString | Gets the intersection delimiter. |
| GetXMaxScore : aNumber | Gets the intersection maximum match score. |
| HasIntersections : aBoolean | Returns true if the address style has an intersection component. |
| HasZone : aBoolean | Returns true if the address style has a zone component. |
| SetDefOffset (offsetValue) | Sets the default offset value. |
| SetDefSqueeze (squeezeValue) | Sets the default squeeze value. |

**See also:** MatchSource; MatchCand; MatchCase

# Albers

The Albers Equal Area Conic projection is implemented through this class.

### Inherits from Prj

| Class Requests | |
| --- | --- |
| CanDoSpheroid : aBoolean | Always returns true. |
| Make (aBox) : anAlbers | Creates a projection bounded by the specified box. |

| Instance Requests | |
| --- | --- |
| ProjectPt (aPoint) : aBoolean | Projects a point and returns true if successful. |
| Recalculate | Recalculates the derived constants. |
| ReturnCentralMeridian : aNumber | Returns the longitude degrees of the central meridian. |
| ReturnFalseEasting : aNumber | Returns the value of X at the central meridian. (Not available in 2.0.) |
| ReturnFalseNorthing : aNumber | Returns the value of Y at the central meridian. (Not available in 2.0.) |
| ReturnLowerStandardParallel : aNumber | Returns the lower latitude in degrees. |
| ReturnReferenceLatitude : aNumber | Returns the reference latitude in degrees. |
| ReturnUpperStandardParallel : aNumber | Returns the upper latitude in degrees. |
| SetCentralMeridian (aLongitude) | Sets the central meridian to the specified degrees. |
| SetFalseEasting (aNumber) | Sets the value of X at the central meridian. (Not available in 2.0.) |

| | |
|---|---|
| SetFalseNorthing (aNumber) | Sets the value of Y at the central meridian. (Not available in 2.0.) |
| SetLowerStandardParallel (lowerLatitude) | Sets the lower latitude to the specified degrees. |
| SetReferenceLatitude (aLatitude) | Sets the reference latitude to the specified degrees. |
| SetUpperStandardParallel (upperLatitude) | Sets the upper latitude to the specified degrees. |
| UnProjectPt (aPoint) : Boolean | Unprojects a point and returns true if successful. |

***See also:*** CoordSys; Spheroid; View

# *AnalysisEnvironment (Spatial Analyst)*

This class models the analysis environment for the spatial analyst.

**Inherits from** DocumentExtension

| *Instance Requests* | |
|---|---|
| Activate | Activates an analysis environment for the spatial analyst. |
| GetCellSize (aCellSizeNumber) : AnalysisEnvEnum | Indicates the type of cell size set for an analysis environment. |
| GetExtent (aRect) : AnalysisEnvEnum | Indicates the type of extent set for an analysis environment. |
| GetMask : GTheme | Gets the grid theme used as the mask in an analysis environment. |
| SetCellSize (AnalysisEnvEnum, aCellSizeNumber) | Sets the cell size for an analysis environment. |
| SetExtent (AnalysisEnvEnum, aRect) | Sets the extent for an analysis environment. |
| SetMask (aGTheme) | Sets the mask for an analysis environment. |
| Store : aBoolean | Returns true if the object is to be stored. |
| Enumerations AnalysisEnvEnum | |
| #ANALYSISENV_MINOF | Sets the cell size or extent to the minimum of all input Grids. |
| #ANALYSISENV_MAXOF | Sets the cell size or extent to the maximum of all input Grids. |
| #ANALYSISENV_VALUE | Sets the cell size or extent to the input value. |

***See also:*** GTheme; Grid

# AnalysisPropertiesDialog (Spatial Analyst)

This class implements the dialog box that sets the properties of an analysis environment.

**Inherits from** ModalDialog

| Class Requests | |
|---|---|
| Show (aView, forConversionBoolean, aTitleString) : anAnalysisEnvironment | Displays the dialog box for setting properties of an analysis environment within a view. If forConversionBoolean is true, a new analysis is created. |

***See also:*** AnalysisEnvironment

# Annotation

This class represents ARC/INFO annotation layers.

**Inherits from** Polygon

| Instance Requests | |
|---|---|
| = anotherObj : aBoolean | Returns true if an annotation is the same as anotherObject. |
| AsString : aString | Returns the string representation of an annotation. |
| GetBaseline : aPolyLine | Gets a polyline object that defines the baseline for the annotation's text. |
| GetHeight : aNumber | Gets annotation's text height in either map or page units. |
| GetLevel : aNumber | Gets an annotation level from its attribute table. |
| GetSymbol : aNumber | Gets the ARC/INFO symbol number used to define this annotation. |
| GetText : aString | Gets the text in an annotation. |

***See also:*** Labeler

# AppleEvent

This class implements client and server communication within the Macintosh environment. *AppleEvent* is used only in the Macintosh environment.

**Inherits from** Obj

| Class Requests | |
|---|---|
| GetErrorID : aNumber | Gets the identification code for the last error. |
| HasError : aBoolean | Returns true when the last Send request encounters an error. |
| PickProcess (promptString, titleString) : destinationString | Picks a process and returns a string in the form *Zone:Machine:Program* to be used as the destination for the Send request. |
| Send (destinationString, eventString, argumentList) : aList | Sends an Apple event and its argument list to an application. |

*See also:* AppleScript; System

# *AppleScript*

This class implements scripts in AppleScript language. *AppleScript* is used only in the Macintosh environment.

**Inherits from** Obj

| Class Requests | |
|---|---|
| Make (scriptString) : anAppleScript | Creates and compiles an Apple script object from the specified source. |

| Instance Requests | |
|---|---|
| DoIt (argumentList) : aString | Executes an Apple script. |
| GetErrorMsg : aString | Gets the description of the last error. |
| GetErrorNumber : aNumber | Gets the system's error number for the last error. |
| GetErrorPos : aNumber | Gets the character position of where the last error occurred. |
| GetErrorRange : aNumber | Gets the number of characters in the script that caused the last error. |
| HasError : aBoolean | Returns true when compilation or execution encounters an error. |

*See also:* Script

# Application

This class implements the ArcView application. Referred to as *av*, there is only one instance of this class.

## Inherits from Obj

| Instance Requests | |
|---|---|
| About | Displays the ArcView "about" dialog box. |
| ArrangeIcons | Arranges the icons of the minimized windows. |
| CascadeWindows | Cascades the open windows. |
| ClearGlobals | Clears global variables from memory. |
| ClearMsg | Clears the message area at the bottom of the ArcView window. |
| ClearStatus | Clears the status bar at the bottom of the ArcView window. |
| ClearWorkingStatus | Clears a working status bar. |
| DelayedRun (aScriptName, anObjOwner, aDelay) | Executes an Avenue script after specified seconds of delay under the ownership of an object. |
| FindDoc (aDocName) : aDoc | Searches the current project and loaded extensions and returns an ArcView document. |
| FindGUI (aGUIName) : aDocGUI | Returns a predefined document GUI object from the current or default projects. |
| FindGUIsFor (aClass) : aDocGUIList | Searches the project and loaded extensions and returns the DocGUIs associated with the input class name. |
| FindScript (aScriptName) : aScript | Returns the specified Avenue script. |
| GetActiveDoc : aDoc | Gets the active ArcView document. |
| GetActiveGUI : aDocGUI | Gets the GUI object of the active document. |
| GetActiveWin : aDocWin | Gets the window object of the active document. |
| GetCodepageConvert : aBoolean | Returns true if code page conversion is enabled. |
| GetExtensionWin : ExtensionWin | Gets the modal dialog box that allows loading or unloading extensions. |
| GetGUI : applicationGUIName | Gets the name of a DocGUI used when no project is open. |
| GetLicensedSite : aString | Gets the site name from the license file. |
| GetLicensedUser : aString | Gets the name of the licensed user. |
| GetName : aString | Gets the application name (ArcView). |
| GetProject : aProject | Gets the current project. |

| | |
|---|---|
| GetSerialNumber : aString | Gets the ArcView serial number. |
| GetSymbolWin : theSymbolWin | Gets the window object of symbol window. |
| GetSysDefault : aProject | Gets the system default project. |
| GetUserDefault : aProject | Gets the user default project. |
| GetVersion : aString | Returns ArcView's version number. (Not available in 2.0.) |
| Help : aHelp | Gets the system help object. |
| IsCustomizable : aBoolean | Returns true if ArcView can be customized. |
| IsHelping : aBoolean | Returns true if help mode is on. |
| Maximize | Maximizes the ArcView window. |
| Minimize | Minimizes the ArcView window. |
| Move (anX, aY) | Moves the ArcView window by the specified pixels. |
| MoveTo (anX, aY) | Moves the ArcView window to the specified pixel location. |
| PurgeObjects | Triggers a purge event that removes objects no longer in use. (Not available in 2.0.) |
| Quit | Ends ArcView. |
| ProcessAllInvals | Allows all invalidate requests to proceed. |
| Resize (aWidth, aHeight) | Resizes the ArcView window to the specified pixel dimensions. |
| Restore | Restores the ArcView window from its icon or maximize state. |
| ReturnExtent : aPoint | Returns the size of the ArcView window in pixels. |
| ReturnOrigin : aPoint | Returns the pixel location of the ArcView window's upper left corner. |
| Run (aScriptName, anObjOwner) : anOptionalObj | Executes an Avenue script. |
| SetCodepageConvert (aBoolean) | If aBoolean is true, code conversion is enabled. |
| SetCustomizable (canCustomize) | Sets the application's ability to customize to the specified Boolean object. |
| SetErrorReporting (isReporting) | Sets how run-time errors are reported based on the specified Boolean object. |
| SetHelping (isHelping) | Sets the help mode to the specified Boolean object. |
| SetName (aName) | Sets the name that is displayed in the application window's title bar. |
| SetStatus (aPercent) : aBoolean | Sets status bar progress and returns false if user has requested a cancel. |

| SetWorkingStatus : aBoolean | Returns false when the stop button of working status is pressed. |
|---|---|
| ShowMsg (aHelpString) | Displays a string in the message area. |
| ShowStopButton | Displays the Stop button next to the status bar. |
| TileWindows | Arranges open windows as tiles. |
| UseWaitCursor | Changes the cursor to the hourglass shape for the remainder of the script. |

*See also:* Doc; DocGUI; DocWin; Project

*Code example:* page 423

# AreaOfInterestDialog

This class implements the dialog box for defining an area of interest.

**Inherits from** ModalDialog

**NOTE:** This class is not available in 2.0.

| Class Requests ||
|---|---|
| Show (aView) : aRect | Displays the area of interest dialog box and returns the specified area of interest. |

*See also:* Prj; View

# AttrRange

This class defines the update procedure for a range of attributes in a split or union operation of a feature table.

**Inherits from** AttrUpdate

| Class Requests ||
|---|---|
| Make (anFTab, aField, aPairedField, AttrRangeTypeEnum) : anAttrRange | Creates an object of class AttrRange to define the union or split rules for two numeric fields of a feature table. |
| Instance Requests ||
| GetPairedField : aField | Gets the paired field object. |
| GetRangeType : AttrRangeTypeEnum | Returns the range type for an object of AttrRange. |
| SetPairedField (aField) | Sets the paired field object. |

| SetRangeType (AttrRangeTypeEnum) | Sets the range type for an object of AttrRange. |

## *Enumerations*

| **AttrRangeTypeEnum** | |
| --- | --- |
| #ATTR_RANGETYPE_ADDRESS | Treats the paired fields as address ranges and maintains the parity of field values. |
| #ATTR_RANGETYPE_CONTINUOUS | Treats the paired fields as continuous values. |

**See also:** AttrSingle

# *AttrSingle*

This class defines the update procedure for a single attribute in a split or union operation of a feature table.

**Inherits from** AttrUpdate

| **Class Requests** | |
| --- | --- |
| Make (anFTab, aField, AttrUnionRuleEnum, AttrSplitRuleEnum) : anAttrSingle | Creates an update procedure for a single field of a feature table. |
| **Instance Requests** | |
| GetSplitRule : AttrSplitRuleEnum | Returns the updating rule when splitting features. |
| GetUnionRule : AttrUnionRuleEnum | Returns the updating rule when merging features. |
| SetSplitRule (AttrSplitRuleEnum) | Sets the updating rule when splitting features. |
| SetUnionRule (AttrUnionRuleEnum) | Sets the updating rule when merging features. |

## *Enumerations*

| **AttrSplitRuleEnum** | |
| --- | --- |
| #ATTR_SPLITRULE_BLANK | Field value is set to blank. |
| #ATTR_SPLITRULE_COPY | The original value is copied. |
| #ATTR_SPLITRULE_PROPORTION | Attribute value of a numeric field is proportionally assigned using area or length. |
| #ATTR_SPLITRULE_SHAPEAREA | Attribute value of a numeric field is assigned based on the shape's area. |
| #ATTR_SPLITRULE_SHAPELENGTH | Attribute value of a numeric field is assigned based on the shape's length. |

| AttrUnionRuleEnum | |
|---|---|
| #ATTR_UNIONRULE_BLANK | Field value is set to blank. |
| #ATTR_UNIONRULE_COPY | The original value is copied. |
| #ATTR_UNIONRULE_PROPORTION | Attribute value of a numeric field is proportionally assigned using area or length. |
| #ATTR_UNIONRULE_ADD | Value of a numeric field is assigned by adding the original fields. |
| #ATTR_UNIONRULE_AVERAGE | Value of a numeric field is assigned by averaging the original fields. |
| #ATTR_UNIONRULE_SHAPEAREA | Attribute value of a numeric field is assigned based on the shape's area. |
| #ATTR_UNIONRULE_SHAPELENGTH | Attribute value of a numeric field is assigned based on the shape's length. |

*See also:* AttrRange

# AttrUpdate

This abstract class defines the rules for updating attributes when a split or union operation is performed on records of a feature table.

**Inherits from** Obj

**Inherited by** AttrRange, AttrSingle

| Class Requests | |
|---|---|
| MakeDefault (UpdateSubclass, aField, anFTab) : anAttrUpdate | Makes a default single or range update rule for aField and anFTab. |
| AddMissingRules (aRuleList, anFTab) | Applies the list of rule enumerations to the fields or paired fields of a feature table. |
| ReturnDefaultRules (anFTab) : aRuleList | Returns the default rule enumerations assigned to a feature table. |
| Instance Requests | |
| GetField : Field | Gets the field in AttrUpdate object. |
| HasField (aField) : aBoolean | Returns true if a field is assigned to an object of AttrUpdate. |

# AutoLabelDialog

This class implements the modal dialog box used for auto-labeling.

**Inherits from** ModalDialog

| Class Request | |
|---|---|
| Show (aTheme) | Shows the auto-labeling dialog box for the given theme. |

*See also:* Labeler

# Axis

This is the superclass to the X and Y axis of a chart part.

**Inherits from** ChartPart

**Inherited by** XAxis and YAxis

| Instance Requests | |
|---|---|
| Edit | Displays the axis editor dialog box. |
| GetBoundsMax : aNumber | Gets the maximum value of an axis. |
| GetBoundsMin : aNumber | Gets the minimum value of an axis. |
| GetCrossValue : aNumber | Gets the value at which this axis crosses another. |
| GetMajorGridSpacing : aNumber | Sets the major grid spacing. |
| GetMinorGridSpacing : aNumber | Sets the minor grid spacing. |
| IsAxisVisible : aBoolean | Returns true if axis is visible. |
| IsBoundsUsed : aBoolean | Returns true if minimum and maximum bounds are used in drawing the axis. |
| IsCrossValueUsed : aBoolean | Returns true if the cross value is used to draw the axis. |
| IsLabelVisible : aBoolean | Returns true when axis label is displayed. |
| IsLog : aBoolean | Returns true if the axis is logarithmic. |
| IsMajorGridVisible : aBoolean | Returns true when the major grid is visible. |
| IsMinorGridVisible : aBoolean | Returns true when the minor grid is visible. |
| IsTickLabelsVisible : aBoolean | Returns true when the labels for the axis tick mark are visible. |

| | |
|---|---|
| IsValueAxis : aBoolean | Returns false for a group axis. |
| SetAxisVisible (isVisible) | Sets the visibility of an axis with a specified Boolean value. |
| SetBoundsMax (aNumber) | Sets the maximum value of an axis. |
| SetBoundsMin (aNumber) | Sets the minimum value of an axis. |
| SetBoundsUsed (isUsed) | Sets how axis values are computed. |
| SetCrossValue (aNumber) | Sets the value at which the other axis crosses. |
| SetCrossValueUsed (isUsed) | Sets utilization of cross value in drawing the axis. |
| SetLabelVisible (isVisible) | Sets the visibility of axis label with a specified Boolean value. |
| SetLog (isLog) | Sets the axis to a logarithmic type with a true Boolean value. |
| SetMajorGridSpacing (aNumber) | Sets the spacing of the major grid to a specified number. |
| SetMajorGridVisible (isVisible) | Sets the visibility of the major grid with a specified Boolean value. |
| SetMinorGridSpacing (aNumber) | Sets the spacing of the minor grid. |
| SetMinorGridVisible (isVisible) | Sets the visibility of the minor grid with a specified Boolean value. |
| SetTickLabelsVisible (isVisible) | Sets the visibility of the labels for the axis tick marks. |
| SetVisible (isVisible) | Sets the visibility of an axis to a Boolean value. |

*See also:* ChartDisplay

# BandStatistics

A BandStatistics object stores image statistics.

**Inherits from** Obj

| Instance Requests | |
|---|---|
| GetMax : aNumber | Gets the maximum pixel value. |
| GetMean : aNumber | Gets the mean pixel value. |
| GetMin : aNumber | Gets the minimum pixel value. |
| GetStandardDeviation : aNumber | Gets the standard deviation of pixel values. |
| IsValid : aBoolean | Returns true if statistics have been calculated. |
| SetMax (aNumber) | Sets the maximum pixel value to the specified number. |

| SetMean (aNumber) | Sets the mean pixel value to the specified number. |
|---|---|
| SetMin (aNumber) | Sets the minimum pixel value to the specified number. |
| SetStandardDeviation (aNumber) | Sets the standard deviation to the specified number. |

***See also:*** Isrc; Itheme

# BasicMarker

A BasicMarker can be an icon, stipple, or character to mark a point on the screen.

**Inherits from** Marker

| Class Requests | |
|---|---|
| Make : aBasicMarker | Creates a default pattern style for the basic marker. |

| Instance Requests | |
|---|---|
| = anObj : aBoolean | Returns true if the basic marker and an object are the same. |
| CanSetSize : aBoolean | Returns true if the size of a basic marker can be set. |
| Copy (anotherBasicMarker) | Copies another basic marker into the basic marker object. |
| GetAngle : aNumber | Gets the angle of a basic marker. |
| GetBgColor : aColor | Gets the background color. |
| GetCharacter : aNumber | Gets the ASCII value of the character in pattern style. |
| GetFont : aFont | Gets the font in a pattern style. |
| GetIcon : aIcon | Gets the icon in an icon style. |
| GetSize : aNumber | Gets the size of the basic marker. |
| GetStipple : aStipple | Gets the stipple in a stipple style. |
| GetStyle : BasicMarkerStyleEnum | Gets the style of the basic marker. |
| SetAngle (aNumber) | Sets the angle of a basic marker to the given value. |
| SetBgColor (aColor) | Sets the background color to the specified color object. |
| SetCharacter (aNumber) | Sets the character in the pattern style to the specified ASCII value. |

| | |
|---|---|
| SetFont (aFont) | Sets the font in a pattern style. |
| SetIcon (anIcon) | Sets the icon in an icon style. |
| SetSize (aSizeinPoints) | Sets the size of the basic marker. |
| SetStipple (aStipple) | Sets the stipple in a stipple style. |
| SetStyle (aStyle) | Sets the style of the basic marker. |
| UnHook | Unhooks a basic marker from its scale so that it is not resized when the scale changes. |

## *Enumerations*

| *BasicMarkerStyleEnum* | |
|---|---|
| #BASICMARKER_STYLE_ICON | An icon is drawn as the marker. |
| #BASICMARKER_STYLE_OPAQUESTIPPLE | A small opaque rectangle with a pattern is drawn as the marker. |
| #BASICMARKER_STYLE_PATTERN | A character is drawn as the marker. |
| #BASICMARKER_STYLE_STIPPLE | A small rectangle with a pattern and transparent background is drawn as the marker. |

**See also:** CompositeMarker; Font; Icon; Stipple

# *BasicPen*

BasicPen implements the drawing of solid or dashed lines.

**Inherits from** Pen

| Class Requests | |
|---|---|
| Make : aBasicPen | Creates a default solid blank line as the basic pen. |

| Instance Requests | |
|---|---|
| = anObj : aBoolean | Returns true if the basic pen and another object are the same. |
| CanSetSize : aBoolean | Returns true if the size of a basic pen can be set. |
| Copy (anotherBasicPen) | Copies properties of another basic pen into the basic pen object. |
| GetCapStyle : BasicPenCapEnum | Gets the style for how ends are drawn. |
| GetJoinStyle : BasicPenJoinEnum | Gets the style for how segments are drawn. |

| | |
|---|---|
| GetSize : aNumber | Gets the size of basic pen in points (1/72 in.). |
| GetWidth : aNumber | Gets the width of basic pen in points. |
| ReturnPattern : aList | Returns a list of numbers describing the basic pen's pattern. |
| SetCapStyle (BasicPenCapEnum) | Sets the style for how ends are drawn. |
| SetJoinStyle (BasicPenJoinEnum) | Sets the style for how segment joins are drawn. |
| SetPattern (aDashPatternList) | Sets the pattern to the specified list of numbers. |
| SetSize (aNumber) | Sets the basic pen size to the specified number in points. |
| SetWidth (aNumber) | Sets the basic pen width to the specified number in points. Size and width are the same. |
| UnHook | Unhooks a basic pen from its scale so that it is not resized when the scale changes. |

## Enumerations

| BasicPenCapEnum | |
|---|---|
| #BASICPEN_CAP_BUTT | Square ends at the node. |
| #BASICPEN_CAP_ROUND | Round ends extended past the node. |
| #BASICPEN_CAP_SQUARE | Square ends extended past the node. |

| BasicPenJoinEnum | |
|---|---|
| #BASICPEN_JOIN_BEVEL | Beveled at the vertices. |
| #BASICPEN_JOIN_MITER | Mitered at the vertices. |
| #BASICPEN_JOIN_ROUND | Rounded at the vertices. |

**See also:** CompositePen; VectorPen

# BitMap

A BitMap is an ordered, fixed size list of Boolean objects.

**Inherits from** Collection

| Class Requests | |
|---|---|
| Make (aNumber) : aBitMap | Creates a bitmap of the specified size; all bits are set to false. |

| Instance Requests | |
|---|---|
| anotherBitMap : aBoolean | Returns true when two bitmaps have the same size and value. |
| And (anotherBitMap) | Performs a bit-wise And. |
| AsList : booleanList | Converts a bitmap to a list of Boolean values. (Not available in 2.0.) |
| Clear (aBitNumber) | Sets the bit at the specified position to false. |
| ClearAll | Sets all bits to false. |
| ClearRange (startOffset, numBits) | Sets a specified number of bits to false starting at the specified position. |
| Copy (sourceBitMap) | Copies the specified bitmap to a bitmap object. |
| Count : aNumber | Returns the number of bits that are set to true. |
| Get (aBitOffset) : aBoolean | Returns the value of a bit at the specified position. |
| GetNextSet (aBitOffset) : aNumber | Returns the position of the first bit set to true after the specified position. |
| GetPrevSet (aBitOffset) : aNumber | Returns the position of the last bit set to true before the specified position. |
| GetSize : aNumber | Returns the number of bits in a bitmap. |
| Not | Performs a Boolean Not operation on a bitmap. |
| NotRange (startOffset, numBits) | Performs a boolean Not operation on a range of bitmaps. |
| Or (anotherBitMap) | Performs a bit-wise Or. |
| Set (aBitOffset) | Sets the bit to true at the specified position. |
| SetAll | Sets all bits to true. |
| SetRange (startOffset, numBits) | Sets a specified number of bits to true starting at the specified position. |
| SetSize (newNumberOfBits) | Changes the bitmap to specified size. |
| XOr (anotherBitMap) | Performs a bit-wise XOr. |

*See also:* Table; VTab

*Code example:* page 445

# Boolean

Implements true and false objects.

**Inherits from** Value

| Instance Requests | |
|---|---|
| = anObj : aBoolean | Returns true if object has the same value as another object. |
| And anotherBoolean : aBoolean | Performs a logical And. |
| AsString : aString | Returns "true," "false," or "Boolean null." |
| IsNull : aBoolean | Returns true when a Boolean is null. |
| Not : aBoolean | Performs a Boolean Not operation. |
| Or anotherBoolean : aBoolean | Performs a logical Or. |
| XOr anotherBoolean : aBoolean | Performs an exclusive Or. |

*See also:* Number, String

# Button

Buttons are the user interface controls placed on a button bar.

**Inherits from** Control

| Class Requests | |
|---|---|
| Make : aButton | Creates a button. |

| Instance Requests | |
|---|---|
| Click | Triggers the click event of the button. |
| GetClick : aString | Gets the Avenue script associated with the click event. |
| GetIcon : anIcon | Gets the icon associated with the button. |
| HasScript (aScriptName) : aBoolean | Returns true if the given script is associated with a button. |
| SetClick (aScriptName) | Associates the click event to the specified script name. |
| SetIcon (anIcon) | Sets the icon to the specified icon. |

*See also:* ButtonBar; Choice; Icon; Space; Tool

*Code example:* page 283

# ButtonBar

A button bar, which manages buttons, is displayed as the middle row of
a document GUI.

**Inherits from** ControlSet

| Instance Requests | |
| --- | --- |
| Empty | Removes all buttons from a button bar. |
| Remove (aButton) | Removes the specified button from a button bar. |

*See also:* DocGUI; Button; Space; ToolBar; MenuBar

*Code example:* page 285

# Cassini

The Cassini projection is implemented through this class.

**Inherits from** Prj

| Class Requests | |
| --- | --- |
| CanDoSpheroid : aBoolean | Always returns true. |
| Make (aBox) : aCassini | Creates a projection bounded by the specified box. |

| Instance Requests | |
| --- | --- |
| ProjectPt (aPoint) : aBoolean | Projects a point and returns true if successful. |
| Recalculate | Recalculates the derived constants. |
| ReturnCentralMeridian : aNumber | Returns the longitude degrees of the central meridian. |
| ReturnReferenceLatitude : aNumber | Returns the reference latitude in degrees. |
| SetCentralMeridian (aLongitude) | Sets the central meridian to the specified degrees. |
| SetReferenceLatitude (aLatitude) | Sets the reference latitude to the specified degrees. |
| UnProjectPt (aPoint) : aBoolean | Unprojects a point and returns true if successful. |

*See also:* CoordSys; Spheroid; View

*Chart* **21**

# Chart

A chart is an ArcView document to display tabular data in the form of business graphics.

**Inherits from** Doc

| Class Requests | |
|---|---|
| Make (aVTab, aFieldList) : aChart | Creates a chart with a series from the specified table, and groups from the specified fields. |
| MakeUsingDialog (aVTab, aGUIName) : aChart | Displays the chart dialog box based on the specified table. The new chart uses the specified document GUI. |
| Chart.MakeWithGUI (aVTab, aFieldList, aGUIName) : aChart | Creates a chart with a series from the specified table, and groups from the specified fields. The new chart uses the given document GUI. |

| Instance Requests | |
|---|---|
| BlinkRecord (aRecordNumber) | Makes the series or groups related to the specified record number blink once. |
| CanUndoErase : aBoolean | Returns true if the UndoErase request can reselect rows. |
| Edit | Displays the chart's edit dialog box. |
| EditChartPart | Displays the edit dialog box for a chart part; used in an apply event. |
| Erase | Unselects a record in an apply event. |
| EraseWithPolygon | Unselects records in an apply event. |
| Find (aString) : aNumber | Returns the record number where a specified string is found. |
| GetChartDisplay : aChartDisplay | Gets the chart display object. |
| GetChartLegend : aChartLegend | Gets the chart legend object. |
| GetFields : aFieldList | Gets the list of fields shown on the chart. |
| GetFindString : aString | Gets the last string used by the Find request. |
| GetGroupLabel (aGroupNumber) : aString | Gets the label of a group based on the specified group number. |
| GetRecordLabelField : aField | Gets the field assigned as the record label. |
| GetSeriesLabel (aSeriesNumber) : aString | Gets the label of a series based on the specified series number. |
| GetTitle : aTitle | Gets the chart's title object. |

| | |
|---|---|
| GetUserGroup : aNumber | Gets the group number clicked on; used in an apply event. |
| GetUserPart : aChartPart | Gets a chart part that the user clicks on in an apply event. |
| GetUserRecord : aNumber | Gets the record number of data elements that the user clicks on in an apply event. |
| GetUserSeries : aNumber | Gets the series number clicked on in an apply event. |
| GetVTab : aVTab | Gets the source table for the chart. |
| GetXAxis : anXAxis | Gets the chart's X axis object. |
| GetYAxis : aYAxis | Gets the chart's Y axis object. |
| IsChartScatter : aBoolean | Returns true if the chart style is scatter. |
| IsSeriesFromRecords : aBoolean | Returns true if the chart is grouped by fields. |
| Print | Prints a chart. |
| SetGroupLabel (aGroupNumber, aString) | Sets the group label at the specified number to the specified string. |
| SetRecordLabelField (aField) | Sets the record label field to the specified field. |
| SetSeriesFromRecords (aBoolean) | Sets the grouping by field if the specified Boolean is true. |
| SetSeriesLabel (aSeriesNumber, aString) | Sets the series label at the specified number to the specified string. |
| ShowGallery (ChartDisplayEnum) | Displays the gallery dialog box for the specified style. |
| UndoErase | Reselects the last unselected records. |

## Enumerations

| ChartDisplayEnum | |
|---|---|
| #CHARTDISPLAY_AREA | Filled area chart. |
| #CHARTDISPLAY_BAR | Chart with horizontal rectangles. |
| #CHARTDISPLAY_COLUMN | Chart with vertical rectangle. |
| #CHARTDISPLAY_LINE | Line chart. |
| #CHARTDISPLAY_PIE | Pie chart. |
| #CHARTDISPLAY_XYSCATTER | Point chart. |

**See also:** ChartDisplay; ChartPart

**Code example:** page 311

# *ChartDisplay*

This class implements the visual representation of a chart.

**Inherits from** Obj

| Instance Requests | |
|---|---|
| GetChart : aChart | Gets the chart that has this chart display. |
| GetCrossGroup : aNumber | Gets the number where group and value axes cross. |
| GetGroupGap : aNumber | Gets the separation percentage for groups. |
| GetMark : ChartDisplayMarkEnum | Gets the marker type used for data points. |
| GetNumGroups : aNumber | Gets the number of groups in a chart. |
| GetNumSeries : aNumber | Gets the number of series in a chart. |
| GetSeriesColor (aSeriesNumber) : aColor | Gets the color of a series based on the specified series number. |
| GetSeriesOverlap : aNumber | Gets the overlap percentage between series. |
| GetStatus : aString | Gets chart error message. |
| GetStyle : ChartDisplayViewEnum | Gets the chart style. |
| GetType : ChartDisplayEnum | Gets the chart type. |
| IsInColor : aBoolean | Returns true when a color instead of a pattern is used to display a chart. |
| IsOK : aBoolean | Returns false when inconsistency errors exist in a chart. |
| IsPlotMarks : aBoolean | Returns true when marks are drawn at each data point in a line chart. |
| IsSeriesSeparated (aSeriesNumber) : aBoolean | Returns true if the pie chart is exploded at the specified series location. |
| SetCrossGroup (aCrossGroupNumber) | Sets the group where value and group axes cross. |
| SetGroupGap (aPercentage) | Sets the separation percentage in groups. |
| SetInColor (aBoolean) | Sets the chart to use color with a specified true value. |
| SetMark (ChartDisplayMarkEnum) | Sets the chart mark. |
| SetPlotMarks (aBoolean) | Sets the chart to plot marks when the specified Boolean is true. |
| SetSeriesColor (aSeriesNumber, aColor) | Sets the series identified by the specified number to a specified color. |
| SetSeriesOverlap (aPercentage) | Sets the series overlap to a percentage ranging from −100 to +100. |

| SetSeriesSeparated (aSeriesNumber, aBoolean) | Explodes the pie chart at the specified series when the specified Boolean is true. |
| --- | --- |
| SetStyle (ChartDisplayViewEnum) | Sets the chart style. |
| SetType (ChartDisplayEnum) | Sets the chart type. |

## Enumerations

| **ChartDisplayEnum** | |
| --- | --- |
| #CHARTDISPLAY_AREA | Filled area chart. |
| #CHARTDISPLAY_BAR | Chart with horizontal rectangles. |
| #CHARTDISPLAY_COLUMN | Chart with vertical rectangle. |
| #CHARTDISPLAY_LINE | Line chart. |
| #CHARTDISPLAY_PIE | Pie chart. |
| #CHARTDISPLAY_XYSCATTER | Point chart. |

| **ChartDisplayMarkEnum** | |
| --- | --- |
| #CHARTDISPLAY_MARK_BIGSQUARE | Mark with a big square. |
| #CHARTDISPLAY_MARK_CROSS | Mark with an X-shaped cross. |
| #CHARTDISPLAY_MARK_DOT | Mark with a dot. |
| #CHARTDISPLAY_MARK_MALTA | Mark with a Maltese cross. |
| #CHARTDISPLAY_MARK_PLUS | Mark with a plus sign. |
| #CHARTDISPLAY_MARK_SMALLPLUS | Mark with a small plus sign. |
| #CHARTDISPLAY_MARK_SMALLSQUARE | Mark with a small square. |
| #CHARTDISPLAY_MARK_SQUARE | Mark with a square. |

| **ChartDisplayViewEnum** | |
| --- | --- |
| #CHARTDISPLAY_VIEW_CUMULATIVE | Group values are added. |
| #CHARTDISPLAY_VIEW_RELATIVE | Values are represented by a percentage within a group. |
| #CHARTDISPLAY_VIEW_SIDEBYSIDE | Each value is represented independently. |

***See also:*** Chart; ChartPart

# ChartLegend

This class is part of the chart that displays the name and color of each series.

**Inherits from** ChartPart

| Instance Requests | |
|---|---|
| Edit | Displays the chart legend editor. |
| GetLocation : ChartDisplayLocEnum | Gets the location of a chart legend. |
| ReturnRelativeLocation : aPoint | Returns the relative location of a legend. |
| SetLocation (ChartDisplayLocEnum) | Sets the location of a chart legend. |
| SetRelativeLocation (aPoint) | Sets the relative position of a chart legend. |
| SetVisible (aBoolean) | Makes the legend visible when a True Boolean is specified. |

## *Enumerations*

| ChartDisplayLocEnum | |
|---|---|
| #CHARTDISPLAY_LOC_BOTTOM | Places the chart part at the bottom of the display area. |
| #CHARTDISPLAY_LOC_LEFT | Places the chart part at the left margin of the display area. |
| #CHARTDISPLAY_LOC_RELATIVE | Uses the relative position values to place the chart part. |
| #CHARTDISPLAY_LOC_RIGHT | Places the chart part at the right margin of the display area. |
| #CHARTDISPLAY_LOC_TOP | Places the chart part on top of the display area. |

***See also:*** ChartDisplay

# *ChartPart*

Subclasses of ChartPart implement the non-data parts of a chart.

**Inherits from** Obj

**Inherited by** Axis, ChartLegend, and Title

| Instance Requests | |
|---|---|
| Edit | Displays the editor for one of the chart part subclasses. |
| GetChartDisplay : aChartDisplay | Gets the display object of a chart. |
| GetColor : aColor | Gets the color of a chart part. |

| | |
|---|---|
| IsVisible : aBoolean | Returns true if a chart part is displayed. |
| SetColor (aColor) | Sets the color of a chart part. |
| SetVisible (aBoolean) | Makes the chart part visible when a True Boolean is specified. |

***See also:*** Chart; ChartDisplay

# ChartSymbol

This class models the symbology used in drawing a chart on the screen.

**Inherits from** Symbol

**Inherited by** ColumnChartSymbol, PieChartSymbol

| *Instance Requests* | |
|---|---|
| Copy (anotherSymbol) | Copies another symbol into a chart symbol. |
| GetBgColor : aColor | Gets the background color of a chart symbol. |
| GetBgSym : aSymbol | Gets the symbol used for the chart symbol background. |
| GetMaxSize : aNumber | Gets the maximum pie radius or column height in points. |
| GetMaxValue : aNumber | Gets the maximum value in a chart symbol. |
| GetMinSize : aNumber | Gets the minimum pie radius or column height in points. |
| GetMinValue : aNumber | Gets the minimum value in a chart symbol. |
| GetNumValues : aNumber | Gets the number of values in a chart symbol. |
| GetOlColor : aColor | Gets the outline color of a chart symbol. |
| GetSize : aNumber | Gets the size of a chart symbol in points. |
| GetType : SymbolEnum | Always returns #SYMBOL_CHART for a chart symbol object. |
| SetBgColor (aColor) | Sets the background color to the given color. |
| SetBgSym (aSymbol) | Sets the background symbol to the specified symbol. |
| SetMaxSize (aNumber) | Sets the maximum pie radius or column height in points. |
| SetMaxValue (aNumber) | Sets the maximum value in a chart symbol. |

| | |
|---|---|
| SetMinSize (aNumber) | Sets the minimum pie radius or column height in points. |
| SetMinValue (aNumber) | Sets the minimum value in a chart symbol. |
| SetNumValues (aNumber) | Sets the number of values in a chart symbol. |
| SetOlColor (aColor) | Sets the outline color to the given color. |
| SetSize (aNumber) | Sets the size of a chart symbol in points. |
| SetSymbols (aSymbolList) | Sets the fill symbol for each part of a chart symbol. |
| SetValues (aNumberList) | Sets the value of each part in a chart symbol. |

***See also:*** ColumnChartSymbol, PieChartSymbol

# Choice

A *choice* is a menu item on a pull-down menu.

**Inherits from** Control

| Class Requests | |
|---|---|
| Make : aChoice | Creates a menu item. |

| Instance Requests | |
|---|---|
| Click | Triggers the click event of a menu item. |
| GetClick : aScriptName | Gets the name of the script associated with the click event. |
| GetLabel : aString | Gets the label of the menu item. |
| HasScript (aScriptName) : aBoolean | Returns true when the given script is associated with a menu item. |
| SetClick (aScriptName) | Sets the script for the click event. |
| SetLabel (aLabelString) | Sets the menu item label. |

***See also:*** Button; Menu; MenuBar; Popup; Tool

***Code example:*** page 292

# Circle

This class implements a geometric circle.

## Inherits from Shape

| Class Requests | |
|---|---|
| Make (centerPoint, aRadius) : aCircle | Creates a circle with the specified center and radius. |
| MakeNull : aCircle | Creates a circle without location and size. |

| Instance Requests | |
|---|---|
| = anObj : aBoolean | Returns true if a circle and an object represent the same circle. |
| AsMultiPoint : aMultiPoint | Returns a multi-point object representing a circle. |
| AsPolygon : aPolygon | Returns a polygon object representing a circle. |
| AsString : aString | Returns the string "Circle.Make (centerPoint, radiusValue)" with actual values. |
| Contains (anotherShape) : aBoolean | Returns true if the specified shape is contained within the circle. |
| ExpandBy (aDistance) : aCircle | Returns a circle expanded by a distance value. |
| GetDimension : aNumber | Always returns 2 for the circle's two dimensions. |
| GetRadius : aNumber | Gets circle's radius. |
| InsetBy (aDistance) : aCircle | Returns a circle inset by a specified distance. |
| Intersects (aShape) : aBoolean | Returns true if a specified shape intersects with the circle. |
| IsNull : aBoolean | Returns true if the circle has no center point or radius. |
| Move (moveX, moveY) : aCircle | Returns a circle moved by specified X and Y values. |
| ReturnCenter : aPoint | Returns the center point of a circle. |
| ReturnClipped (aRect) : aShape | Clips a circle by the specified rectangle. (Not available in 2.0.) |
| ReturnDifference (aShape) : aPolygon | Returns a polygon by subtracted the given shape from a circle. |
| ReturnIntersection (aShape) : aPolygon | Returns a polygon by intersecting the given shape with a circle. |
| ReturnMerged (aShape) : aPolygon | Returns a polygon by merging the specified shape into a circle. |
| ReturnProjected (aPrj) : aShape | Returns the projected shape of a circle based on the specified projection. |

| | |
|---|---|
| ReturnUnion (aShape) : aPolygon | Returns a polygon by combining the given shape with a circle. |
| ReturnUnprojected (aPrj) : aShape | Returns the unprojected shape of a circle. |
| SetCenter (aPoint) | Sets the circle's center to the specified point. |
| SetRadius (aRadius) | Sets the circle's radius to a specified number. |
| Snap (aShape, snapDistance) : aBoolean | Returns true if a circle can be snapped to a specified shape within a specified snap distance. |

***See also:*** Line; Oval; Point; Polygon; Rect

# *Classification*

Classification represents a value or range of values for symbolization.

### **Inherits from** Obj

| Class Requests | |
|---|---|
| Make (aMinValue, aMaxValue) : aClassification | Creates a classification ranging from specified minimum and maximum numerical values. |
| MakeDisjoint (aClassificationList) : aClassification | Makes a disjoint classification from simple Classifications. |

| Instance Requests | |
|---|---|
| Contains (aValue) : aBoolean | Returns true if the specified value is within a classification range. |
| GetComponents : aClassificationList | Gets the simple classifications for a disjoint classification. |
| GetLabel : aString | Gets the classification label. |
| GetPrecision : aNumber | Gets a classification's precision. |
| IsDisjoint : aBoolean | Returns true when a classification is disjoint. |
| IsEmpty : aBoolean | Returns true if a classification is empty. |
| IsNoData : aBoolean | Returns true when a classification as no data. |
| IsText : aBoolean | Returns true if a classification is text. |
| ReturnMaximum : aNumber | Returns the maximum value in a classification. |
| ReturnMinimum : aNumber | Returns the minimum value in a classification. |
| ReturnRangeString : aString | Returns a string representation of a classification range. |

| SetLabel (aLabel) | Sets the classification label. |
|---|---|
| SetMaximum (aNumber) | Sets the maximum value to the specified number. |
| SetMinimum (aNumber) | Sets the minimum value to the specified number. |
| SetPrecision (aNumber) | Sets a classification's decimal-precision. |
| SetRangeString (aString) : aBoolean | Returns true if the specified string is a valid range representation. |

*See also:* Legend

*Code example:* page 492

# Clipboard

This class implements ArcView's clipboard. The ArcView clipboard is not the same as the system clipboard.

**Inherits from** List

| Class Requests | |
|---|---|
| The : Clipboard | Returns ArcView's clipboard. |

| Instance Requests | |
|---|---|
| Add (anObject) : theClipboard | Adds a clone of the specified object to the clipboard. |
| Empty | Clears the clipboard. |
| HasKindOf (aClass) : aBoolean | Returns true if the clipboard has an object of the specified class. |
| ReturnKindOf (aClass, aRect) : aList | Returns a list of cloned objects from the clipboard with the matching class. |
| Update : aBoolean | Adds the system clipboard to ArcView's clipboard; returns true if an object is added. |

*See also:* Display; Graphic; String

# ClosestFacilityWin (Network Analyst)

This class implements the dialog box for defining and solving closest facility problems of Network Analyst.

**Inherits from** NetworkWin

| Class Requests | |
| --- | --- |
| Make (aView, aNetwork, aResultFTheme, aSourceFTheme) : aClosestFacilityWin | Creates a dialog box object for closest facility problems. |
| MakeNewResultFTab (aNetwork, aFileName) : anFTab | Makes an FTab to hold the data for a closest facility problem. |

| Instance Requests | |
| --- | --- |
| AddPoint (anFTab, aLabelString, aPoint) | Adds the point to the closest facility window. |
| CanSolve : aBoolean | Returns true when a closest facility window has sufficient information to solve the problem. |
| Open | Displays a closest facility window. |
| Refresh | Refreshes the closest facility window to display added or modified information. |
| Solve : aBoolean | Solves and returns true if the closest facility problem was solved successfully. |

*See also:* Network

# *CodePage*

ArcView provides language support through instances of CodePage.

**Inherits from** Obj

| Class Requests | |
| --- | --- |
| EditProfile | Displays the Code Page Profile dialog box. |
| Load (aFullpathString) : aCodepage | Creates a code page from the given profile file. |
| Make : aCodepage | Makes a code page object. |

| Instance Requests: | |
| --- | --- |
| IsCodepageValid (aCodepageName) : aNumber | Checks for supported code pages. |

*See also:* Application

# Collection

Collection is an abstract superclass for container classes.

**Inherits from** Obj

**Inherited by** BitMap, Dictionary, Interval (not in version 2.0), List, Name-Dictionary, and Stack

| Class Requests | |
| --- | --- |
| Make : Collection | Makes a container object. |

| Instance Requests | |
| --- | --- |
| + anotherCollection : aCollection | Combines two containers. |
| AsList : aList | Returns a list of objects in the container. |
| Count : aNumber | Returns the number of objects in a container. |
| Empty | Clears the container of objects. |
| IsEmpty : aBoolean | Returns true when there are no objects in the container. |
| Merge (anotherCollection) : aCollection | Combines two containers. |

# Color

This class implements the colors.

**Inherits from** Obj

| Class Requests | |
| --- | --- |
| GetBlack : aColor | Gets the default black color. |
| GetBlue : aColor | Gets the default blue color. |
| GetCyan : aColor | Gets the default cyan color. |
| GetGray : aColor | Gets the default gray color. |
| GetGreen : aColor | Gets the default green color. |
| GetMagenta : aColor | Gets the default magenta color. |
| GetRed : aColor | Gets the default red color. |
| GetWhite : aColor | Gets the default white color. |

| GetYellow : aColor | Gets the default yellow color. |
|---|---|
| Make : aColor | Makes a black color. |

| **Instance Requests** | |
|---|---|
| = anObj : aBoolean | Returns true if a color is equal to a specified object. |
| Copy (anotherColor) | Copies one color into another. |
| GetRgbList : aList | Gets the three numbers representing red, green, and blue values. |
| IsTransparent : aBoolean | Returns true if a transparent color. |
| SetRgbList (aRGBList) | Sets the red, green, blue values. |
| SetTransparent (aBoolean) | Sets a color transparent when a true Boolean value is specified. |

***See also:*** ColorMap; Symbol; SymbolWin

***Code example:*** page 375

# ColorMap

A colormap is a list of colors in a single band image.

**Inherits from** Obj

| **Class Requests** | |
|---|---|
| MakeColorGray (numColors, numShades) : aColorMap | Makes a colormap with the specified number of colors and gray shades. |
| MakeColorWheel (aSize) : aColorMap | Creates a colormap based on a color wheel with the specified size. |
| MakeGrayScale (aSize) : aColorMap | Creates a gray shade colormap with the specified number of shades. |
| MakeNominal (aSize) : aColorMap | Makes a colormap of the specified size from discrete colors. |
| MakeRamp (aSize, fromColor, toColor) : aColorMap | Creates a color ramp based on the specified colors and size. |
| MakeRandom (aSize) : aColorMap | Creates a random colormap based on the specified size. |

| Instance Requests | |
|---|---|
| AdjustBlue (aPercent) | Adjusts a colormap's blue component by a specified percentage. |
| AdjustGreen (aPercent) | Adjusts a colormap's green component by a specified percentage. |
| AdjustRed (aPercent) | Adjusts a colormap's red component by a specified percentage. |
| AdjustSaturation (aPercent) | Adjusts every color's saturation by a specified percentage. |
| AdjustValue (aPercent) | Adjusts the color value by a specified percentage. |
| Count : aNumber | Gets the number of colors in a colormap. |
| Get (anIndex) : aColor | Gets the color object at a specified location in the colormap. |
| GetNull : aColor | Gets the color used for null data. |
| Match (aColor) : aNumber | Returns the location on a colormap that has a color matching the specified color. |
| Set (anIndex, aColor) | Sets a specified location on the colormap to a color. |
| SetColorGray (numColors, numShades) | Sets the colormap to the specified number of colors and number of gray shades. |
| SetColorWheel | Sets the colors to a color wheel. |
| SetGrayScale | Sets the colors to a gray scale. |
| SetNominal | Sets the colors to discrete colors. |
| SetNull (aColor) | Sets the color to be used with null data. |
| SetRamp (fromColor, toColor) | Sets the start and end colors in a color ramp. |
| SetRandom | Sets colors randomly. |

*See also:* Color; ISrc; ITheme; SingleBandLegend

*Code example:* page 371

# *ColumnChartSymbol*

This class models a graphic symbology to draw a column in a chart.

**Inherits from** ChartSymbol

| Class Requests | |
| --- | --- |
| Make : aColumnChartSymbol | Creates an object of ColumnChartSymbol class. |
| Instance Requests | |
| GetColumnWidth : aNumber | Returns the width of a column chart symbol in points. |
| SetColumnWidth (aNumber) | Sets the width of a column chart symbol in points. |

***See also:*** PieChartSymbol

# *CompositeFill*

This class is a specialization of the fill symbol with multiple layers.

**Inherits from** Fill

| Class Requests | |
| --- | --- |
| Make (aSymbolList) : aCompositeFill | Creates a composite based on the specified symbol list. |

| Instance Requests | |
| --- | --- |
| = anObj : aBoolean | Returns true if a composite fill is the same as another specified object. |
| GetSymbols : aSymbolList | Gets the list of symbols in the composite fill. |
| SetColor (aColor) | Sets the foreground color. |
| SetOlColor (aColor) | Sets the outline color. |
| SetOlWidth (aWidth) | Sets the outline width. |
| SetOutline (aBoolean) | Draws the composite fill with outline if the specified Boolean value is true. |

***See also:*** RasterFill; SymbolList; VectorFill

# *CompositeMarker*

This class is a specialization of the basic marker with multiple layers.

**Inherits from** Marker

| Class Requests | |
| --- | --- |
| Make (aSymbolList) : aCompositeMarker | Creates a composite marker from a specified symbol list. |

| Instance Requests | |
| --- | --- |
| = anObj : aBoolean | Returns true if the specified object is the same as a composite marker. |
| CanSetSize : aBoolean | Returns true if the size can be set. |
| GetBgColor : aColor | Gets the background color. |
| GetSize : aNumber | Gets a marker size in points. |
| GetSymbols : aSymbolList | Gets a list of markers. |
| SetBgColor (aColor) | Sets the background color. |
| SetColor (aColor) | Sets the foreground color. |
| SetSize (aNumber) | Sets a marker's size in points. |

**See also:** BasicMarker; SymbolList

# CompositePen

This class is a specialization of a pen class with multiple layers.

**Inherits from** Pen

| Class Requests | |
| --- | --- |
| Make (aSymbolList) : aCompositePen | Makes a composite pen from the specified list of pens. |

| Instance Requests | |
| --- | --- |
| = anObj : aBoolean | Returns true if the specified object is the same as a composite pen. |
| CanSetSize : aBoolean | Returns true if the width of a composite pen can be set. |
| GetSize : aNumber | Gets the size of a composite pen. |
| GetSymbols : aSymbolList | Returns a list of pen symbols. |
| SetColor (aColor) | Sets the pen color. |
| SetSize (aSize) | Sets the pen size in points. |

**See also:** BasicPen; SymbolList; VectorPen

# *Control*

This is the abstract class for ArcView's user interface controls.

**Inherits from** Obj

**Inherited by** Button, Choice, ControlSet, LabelButton, Space, and Tool

| Instance Requests | |
|---|---|
| GetControlSet : aControlSet | Gets the control set object that a control belongs to. |
| GetHelp : aString | Gets the help string of a control. |
| GetHelpTopic : aString | Gets the help topic of a control. |
| GetObjectTag : anObj | Gets the object tagged to a control. |
| GetTag : aString | Gets the string tag of a control. |
| GetUpdate : aScriptString | Gets the Avenue script that is executed for an update event. |
| HasScript (aScriptName) : aBoolean | Returns true when the given script is associated with a control. |
| IsEnabled : aBoolean | Returns true if a control is enabled. |
| IsModified : aBoolean | Returns true when a control property or position has been modified. |
| IsVisible : aBoolean | Returns true if a control is visible. |
| SetEnabled (aBoolean) | Enables a control if the specified Boolean value is true. |
| SetHelp (aString) | Sets the help string of a control. |
| SetHelpTopic (aString) | Sets the help topic string of a control. |
| SetModified (aBoolean) | Sets the modification flag for a control to the specified Boolean value. |
| SetObjectTag (anObj) | Tags an object to a control. |
| SetTag (aString) | Tags a string to a control. |
| SetUpdate (aScriptName) | Associates an update event to a specified script name. |
| SetVisible (aBoolean) | Makes a control visible if the specified Boolean value is true. |
| Update | Triggers the update event for a control. |

***See also:*** DocGUI; Script

# ControlSet

This is the abstract class for managing a group of controls.

**Inherits from** Control

**Inherited by** ButtonBar, Menu, MenuBar, PopupSet, and ToolBar

| Instance Requests | |
|---|---|
| Add (aControl, anIndex) | Adds a specified control at the specified position. |
| Empty | Removes all controls from a control set. |
| FindByName (aControlName) : aControl | Gets the control identified by the given name in a control set. |
| FindByScript (aScriptName) : aControl | Gets the control identified by the given associated-script in a control set. |
| GetControls : aList | Gets a list of controls associated with a set. |
| New (anIndex) : aControl | Creates a new control at the specified index. |
| Remove (aControl) | Removes the specified control from the set. |
| Shift (aControl, offsetNumber) | Moves a control by the specified offset. |
| Update | Triggers the update event for all controls in the set. |

*See also:* DocGUI; Icon

# CoordSys

This class implements a coordinate system.

**Inherits from** Obj

| Class Requests | |
|---|---|
| Make : aCoordSys | Makes a coordinate system without a projection. |

| Instance Requests | |
|---|---|
| GetDefault : aPrj | Gets the default projection. |
| GetProjections : aList | Gets a list of projections. |
| SetDefault (aPrj) | Sets the default projection. |

*See also:* Prj; Spheroid

# Coverage

This class implements ARC/INFO coverages.

**Inherits from** Obj

| Class Requests | |
|---|---|
| Exists (aCoveragePathString) : aBoolean | Returns true if the specified path leads to an ARC/INFO coverage. |
| IsdBASE (aCoveragePathString) : aBoolean | Returns true if the specified path leads to a PC ARC/INFO coverage. |
| IsDouble (aCoveragePathString) : aBoolean | Returns true if the specified path leads to a double precision ARC/INFO coverage. |
| IsINFO (aCoveragePathString) : Boolean | Returns true if the specified path leads to an INFO coverage. |
| ReturnSrcNames (aCoveragePathString) : aList | Returns a list of source names at the specified path. |
| SortSrcNames (aSrcNamesList) | Sorts by complexity the sources in a coverage according to the specified list. |

*See also:* Layer; SrcName

# Date

This class implements date and time information.

**Inherits from** Obj

| Class Requests | |
|---|---|
| GetDefFormat : aString | Returns default format, or null if using the system default. |
| Make (dateString, formatString) : aDate | Creates a date object from a date string and format. |
| MakeFromNumber (aNumber) : aDate | Creates a date from a number in yyyyMMdd format. |
| MakeNull : aDate | Creates a date object without a value. |
| Now : aDate | Returns the system date and time. |
| SetDefFormat (aFormatString) : aString | Sets and returns the default format. |

| **Instance Requests** | |
|---|---|
| = anotherDate : aBoolean | Returns true when the values of two dates are the same. |
| .. anotherDate : aDuration | Returns a duration between two dates. |
| + aDuration : aDate | Returns a new date. |
| < anotherDate : aBoolean | Returns true when a date is prior to another date. |
| <= anotherDate : aBoolean | Returns true when a date is the same or prior to another date. |
| > anotherDate : aBoolean | Returns true when a date is after another date. |
| >= anotherDate : aBoolean | Returns true when a date is the same or after another date. |
| - aDuration : aDate | Returns a new date. |
| - anotherDate : aDuration | Returns the duration between two dates. |
| AsSeconds : aNumber | Returns the number of seconds since January 1, 1970 to a date. |
| AsString : aString | Returns the date in a string format. |
| GetDay : aString | Gets the full day name. |
| GetDayOfYear : aNumber | Gets the number of days since January 1; if the date is January 1, returns 0. |
| GetFormat : aString | Gets the format string of a date. |
| GetHours : aNumber | Gets the number of whole hours since midnight. |
| GetMinutes : aNumber | Gets the number of minutes since the beginning of the hour. |
| GetMonth : aString | Gets the full month name. |
| GetMonthOfYear : aNumber | Gets a number from 1 to 12 representing the month of the year. |
| GetSeconds : aNumber | Gets the number of seconds since the beginning of a minute. |
| GetYear : aNumber | Gets the year as a number. |
| IsNull : aBoolean | Returns true when a date has no value. |
| SetFormat (aFormatString) : aDate | Sets a date's format; the default format is overwritten. |

| **Formats** | |
|---|---|
| M | Represents a month as an integer from 1 to 12. |
| MM | Same as M with a leading zero if a single digit (e.g., 02 for February). |

| MMM | Represents the short form of the month name (e.g., FEB for February). |
|---|---|
| MMMM | Represents the full month name. |
| d | Represents the day of the month as an integer from 1 to 31. |
| dd | Same as *d*, but with a leading zero for single digit days. |
| ddd | Represents the short form of the day name (e.g., SUN for Sunday). |
| dddd | Represents the long form of the day name. |
| y | Represents the year without the century (e.g., 97). |
| yy | Same as *y*, but with a leading zero for single digit year numbers. |
| yyy | Represents the year with the century (e.g., 1997). |
| yyyy | Same as *yyy*. |
| m | Represents the minute value as an integer from 0 to 59. |
| s | Represents the seconds value as an integer from 0 to 59. |
| h | Represents the hour as an integer from 1 to 12. |
| hh | Same as *h*, but with a leading zero. |
| hhh | Represents the hour from 0 to 23. |
| hhhh | Same as *hhh*, but with a leading zero. |
| AMPM | Represents the AM or PM string. |
| TZ | Represents the time zone string. |

***See also:*** Duration; String

# *DBICursor*

This class implements an SQL query through the ARC/INFO Database Integrator in the UNIX environment.

**Inherits from** Obj

| *Class Requests* | |
|---|---|
| Make (aSQLCon, queryString) : aDBICursor | Creates an SQL cursor through the database integrator. (Not available in 2.0.) |

| Instance Requests | |
|---|---|
| Close | Closes a DBI cursor. (Not available in 2.0.) |
| GetFields : aList | Gets a list of fields in a cursor. |
| GetNextRecord : aBoolean | Gets the next record in a query and returns true if successful. |
| GetValueNumber (aField) : aNumber | Gets the value of a numerical field. |
| GetValueString (aField) : aString | Gets the value of a character field. |
| HasError : aBoolean | Returns true if the last cursor operation caused an error. |
| ReturnValue (aField) : anObj | Returns the value of the field in an object of the appropriate class. |

*See also:* SQLCon; SQLWin

# DBTheme (Database Theme)

This is a specialized class of theme for SDE layers.

**Inherits from** Theme

| Class Requests | |
|---|---|
| Make (anSDELayer, aFeatureClass) : aDBTheme | Makes an object of class DBTheme from the specified SDELayer using the given feature class. |

| Instance Requests | |
|---|---|
| BlinkSDEFeature (aFeatureIDNumber) | Blinks the features identified by the given number in a DBTheme. |
| BuildQuery | Opens the SQL Query Builder dialog box. |
| CanEdit : aBoolean | Always returns false. |
| CanExportToFTab : aBoolean | Returns true when a DBTheme has selected feature. |
| CanHotLink : aBoolean | Always returns true. |
| CanLabel : aBoolean | Always returns true. |
| CanProject : aBoolean | Always returns true. |
| CanReturnClassCounts : aBoolean | Returns true when the number of features in each legend classification can be computed for a DBTheme. |
| CanSelect : aBoolean | Always returns true. |
| ClearSelection | Clears all selected features. |

| | |
|---|---|
| EditLegend | Displays the Legend Editor for aDBTheme. |
| ExportToFTab (aFileName) : anFTab | Converts the selected features of a DBTheme to a feature table stored in the given filename. |
| ExportToVTab (aFileName, aTableClass) : aVTab | Converts the selected features of a DBTheme to a table identifies by the given filename and class type. |
| FindByPoint (aPoint) : aNumberList | Returns a list of feature IDs containing a point in a DBTheme. |
| GetAttributesFirst : aBoolean | Returns false when SDE first finds all the features in a theme that fit in a view, and applies the Definition before ArcView draws the resulting features. |
| GetDefinition : aString | Returns a string representing the where clause of the SQL statement. |
| GetFixedSizeText : aBoolean | returns true when graphic text objects do not change size as the display scale of a DBTheme changes. |
| GetHotField : aField | Gets the field associated with the hot link property of a DBTheme. |
| GetHotScriptName : aString | Gets the script associated with the hot link property of a DBTheme. |
| GetLabelField : aField | Gets the label field of a DBTheme. |
| GetLabelTextSym : aSymbol | Gets the symbol used for labeling a DBTheme. |
| GetLastSelection : anSDELog | Gets an SDELog object representing the selection set saved using RememberSelection request. |
| GetSDELayer : anSDELayer | Gets the source of a DBTheme. |
| GetSelectedExtent : aRect | Gets the extent of the selected features in a DBTheme. |
| GetSelection : anSDELog | Gets an SDELog object representing the current selection set. |
| GetShapeClass : aClass | Gets the feature class in a DBTheme. |
| GetTextPositioner : aTextPositioner | Gets the text positioner object for the labels in a DBTheme. |
| HasAttributes : aBoolean | Always returns true. |
| Identify (aFeatureID, aLabelString) | Opens the identify window to display the specified feature ID. |
| RememberSelection | Saves the current selection to anSDELog so that it can be retrieved later using the GetLastSelection request. |

| | |
|---|---|
| ReturnClassCounts : aNumberList | Returns a list of numbers counting features in each legend classification of a DBTheme. |
| ReturnDefaultLegend : aLegend | Returns the default single-symbol legend for a DBTheme. |
| ReturnExtent : aRect | Returns the extent of a DBTheme. |
| ReturnLabel (aDisplay, aPoint) : aGraphicText | Returns a GraphicText label from the given display for a feature of a DBTheme found at the specified point. |
| ReturnSymbolSize (aFeatureID) : aNumber | Returns the size of the font in points used for labeling the feature of a DBTheme with the specified ID number. |
| SelectByAttributes (anSQLWhereClause, VTabSelTypeEnum) | Selects based on the given where clause. |
| SelectByPoint (aPoint, VTabSelTypeEnum) | Selects features of a DBTheme containing the specified point. |
| SelectByPolygon (aPolygon, VTabSelTypeEnum) | Selects features of a DBTheme that intersect the given polygon. |
| SelectByRect (aRect, VTabSelTypeEnum) | Selects features of a DBTheme that intersect the given rectangle. |
| SelectByShapes (Shapes, VTabSelTypeEnum) | Selects features of a DBTheme that intersect the given shape. |
| SelectByTheme (aTheme, FTabRelTypeEnum, aDistance,VTabSelTypeEnum) | Selects the features of a DBTheme that have the given relationship within the specified distance with the selected features of another theme. |
| SetAttributesFirst (aBoolean) | If the Boolean is false, first all the features that fit the view are found, then the Definition is applied before ArcView draws the resulting features. |
| SetDefaultLegend | Sets a DBTheme to a single symbol legend. |
| SetDefinition (anSQLWhereClause) | Sets the definition of a DBTheme. |
| SetFixedSizeText (aBoolean) | When set to true, graphic text does not change size when the scale changes. |
| SetHotField (aField) | Sets the field associated with the hot link property of a DBTheme. |
| SetHotScriptName (aScriptName) | Sets the script associated with the hot link property of a DBTheme. |
| SetLabelField (aField) | Sets the field used in labeling features of a DBTheme. |
| SetLabelTextSym (aSymbol) | Sets the text symbol used in labeling. |
| SetName (aName) | Sets the name of a DBTheme. |

| SetSelection (anSDELog) | Changes the selection set to those stored in the given SDELong object. |
|---|---|
| UpdateAttributes (aVTab) : aBoolean | Updates the specified VTab with the changes in a DBTheme. Returns true if successful. |
| UpdateLegend | Applies the legend changes. |

## Enumerations

| FTabRelTypeEnum | |
|---|---|
| #FTAB_RELTYPE_INTERSECTS | Intersects. |
| #FTAB_RELTYPE_COMPLETELYCONTAINS | Completely contains. |
| #FTAB_RELTYPE_CONTAINSTHECENTEROF | Contains the center of. |
| #FTAB_RELTYPE_ISCOMPLETELYWITHIN | Is completely within. |
| #FTAB_RELTYPE_HASCENTERWITHIN | Has center within. |
| #FTAB_RELTYPE_ISWITHINDISTANCEOF | Is within a distance of. |

| VTabSelTypeEnum | |
|---|---|
| #VTAB_SELTYPE_NEW | Replaces the existing selection with a new selection. |
| #VTAB_SELTYPE_AND | Selects from the current selection set. |
| #VTAB_SELTYPE_OR | Adds to the current selection set. |
| #VTAB_SELTYPE_XOR | Adds if not selected, and removes if already selected. |

***See also:*** SDELayer; SDELog; SDEFeature; SDEDataset

# DBTQueryWin (Database Theme)

This class models the dialog box that defines a query for DBThemes.

**Inherits from** Window

| Class Requests | |
|---|---|
| Make (aDBTheme) : aDBTQueryWin | Creates a query window associated with the given DBTHeme. |
| ReturnQuery (aDBTheme, aDefaultQueryString) : aQueryString | Displays a modal dialog box for defining and returning a query for the specified DBTheme. |

| Instance Requests | |
| --- | --- |
| Close | Closes a query window. |
| Open | Displays a query window. |

**See also:** QueryWin; DBTheme

# DDEClient

This class, used in the Windows environment only, implements the client application of Microsoft's Dynamic Data Exchange (DDE).

## Inherits from Obj

| Class Requests | |
| --- | --- |
| Make (applicationNameString, topicNameString) : aDDEClient | Creates a DDE client for communication with the specified application using the specified topic name. |

| Instance Requests | |
| --- | --- |
| Close | Closes the DDE communication. |
| Execute (aTaskString) | Implements a DDE execute transaction based on the specified task. |
| GetErrorMsg : aString | Gets the last DDE error description. |
| GetTimeout : aNumber | Gets the time-out value in seconds. |
| GetTopic : aString | Gets the topic string of the communication. |
| HasError : aBoolean | Returns true if the last DDE transaction contained an error. |
| IsTopic (topicNameString) : aBoolean | Returns true if the specified topic string matches a DDE topic string. |
| Poke (anItemString, dataValue) | Passes the specified number or string to the specified item of the server. |
| Request (anItemString) : aString | Returns the string equivalent of the data in the server's item. |
| SetTimeout (numberOfSeconds) | Sets the time-out to the specified seconds. |

**See also:** DDEServer; RPCClient; RPCServer

# DDEServer

This class, used in the Windows environment only, implements the server application of Microsoft's Dynamic Data Exchange (DDE).

**Inherits from** Obj

| Class Requests | |
| --- | --- |
| Start | Starts ArcView's DDE server. |
| StartNamed (applicationNameString) | Re-starts ArcView's DDE server with the specified application name. |
| Stop | Stops ArcView's DDE server and releases associated resources. |

*See also:* DDEClient; RPCClient; RPCServer

# DensitySurfaceDialog (Spatial Analyst)

This class implements the dialog box for defining surface density parameters in Spatial Analyst.

**Inherits from** ModalDialog

| Class Requests | |
| --- | --- |
| Show (aPointTheme) : anObjList | Displays the Density Surface dialog box for the given theme and returns the items specified on the dialog box. |

*See also:* Grid

# Dictionary

A dictionary implements an unordered data structure of associated keys and values.

**Inherits from** Collection

| Class Requests | |
| --- | --- |
| Make (hashSize) : aDictionary | Creates a dictionary object based on the specified hash size. |

| Instance Requests | |
|---|---|
| + anotherCollection : aCollection | Returns a new dictionary or list. |
| Add (aKey, anObj) : aBoolean | Returns true if the specified key and value object are successfully added. |
| AsList : aList | Returns a list of value objects in the dictionary. |
| Count : aNumber | Returns the number of key and value pairs in a dictionary. |
| Empty | Clears keys and values from a dictionary. |
| Get (aKey) : anObj | Gets the object associated with the specified key in a dictionary. |
| GetLoadAverage : aNumber | Returns the average number of associations chained from each hash table entry. |
| GetLoadBalance : aNumber | Returns the variance of the load average. |
| GetLoadMax : aNumber | Returns the size of the longest collision list for a dictionary. |
| GetLoadMin : aNumber | Returns the size of the shortest collision list for a dictionary. |
| GetSize : aNumber | Gets the hash size of a dictionary. |
| GetUsed : aNumber | Returns the percentage of used entries in the hash table of a dictionary. |
| Merge (anotherCollection) : aCollection | Merges another list or dictionary into a dictionary. |
| Remove (aKey) : aBoolean | Returns true if the specified key is found and successfully removed along with its value object. |
| ReturnKeys : aList | Returns a list of keys. |
| Set (aKey, anObj) | Adds or replaces the specified key and object(s) in a dictionary. |
| SetSize (newHashSize) | Changes the size of a dictionary. |

***See also:*** List; NameDictionary

***Code examples:*** page 368, page 394

## Digit (Digitizer)

NEW

This class supports the interface to a digitizer.

## Inherits from Obj

| Class Requests | |
|---|---|
| CanSetup : aBoolean | Returns true when a digitizer driver is available. |
| Make : aDigit | Creates the sole instance of digitizer interface. |
| The : aDigit | Returns the sole instance of digitizer interface. |

| Instance Requests | |
|---|---|
| GetDigitTics : aPointList | Gets the digitizer's tic points. |
| GetLastViewDoc : aView | Returns the last view document saved with the SetCurrentDocWin request. |
| GetMode : DigitTypeEnum | Returns the digitizer's input mode. |
| GetPoint (aPoint) : DigitBtnEnum | Retruns the puck button clicked at the given point. |
| GetPointsFromDig : aPointList | Gets digitizer tic points. |
| GetPuckClick : aPoint | Gets the last point clicked on the puck. |
| GetPuckLoc : aPoint | Gets the last location of the puck on the tablet. |
| GetRmsDigErr : aNumber | Returns root mean square error in digitizer units. |
| GetRmsMapErr : aNumber | Returns root mean square error in map units. |
| GetTicDigList : aPointList | Gets tic points on the digitizers. |
| GetTicMapList : aPointList | Gets tic points on the map. |
| GetUnits : DigitUnitsEnum | Returns the digitizer units. |
| IsAvailable : aBoolean | Returns true if the digitizer can be used with the current ArcView document. |
| IsBtnLeftClick : aBoolean | Returns true when the left button is single-clicked. |
| IsBtnLeftDblClick : aBoolean | Returns true when the left button is double-clicked. |
| IsBtnLeftDrag : aBoolean | Returns true when pointer is dragged with the left button pressed. |
| IsBtnMiddleClick : aBoolean | Returns true when the middle button is single-clicked. |
| IsBtnMiddleDblClick : aBoolean | Returns true when the middle button is double-clicked. |
| IsBtnMiddleDrag : aBoolean | Returns true when pointer is dragged with the middle button pressed. |
| IsBtnRightClick : aBoolean | Returns true when the right button is single-clicked. |
| IsBtnRightDblClick : aBoolean | Returns true when the right button is double-clicked. |
| IsBtnRightDrag : aBoolean | Returns true when pointer is dragged with the right button pressed. |

| | |
|---|---|
| IsButtonAssignedForAction (DigitBtnEnum) : aBoolean | Returns true if a button on the puck has been assigned to cause the specified action. |
| IsTracking : aBoolean | Returns true if the puck is tracking. |
| RegTics : aBoolean | Registers existing tic points and returns true if successful. |
| RegTicsFromList (mapTicList, digitizerTicList) : aBoolean | Registers tics using the given list of map and digitizer tic points. |
| ReturnBtnClick : DigitBtnEnum | Returns the last clicked button. |
| ReturnPhysicalBtnClick : aNumber | Returns the button number that was last clicked. |
| SetBeep (aBoolean) | If the given Boolean is true, clicking the puck generates a beep. |
| SetCurrentDocWin | Saves the current view document. |
| SetMode (DigitTypeEnum) | Sets the digitizer to relative or absolute mode. |
| SetSwitchButton (DigitBtnEnum) | Sets the button that switches between mouse and puck modes. |

## Enumerations

| DigitBtnEnum | |
|---|---|
| #DIGIT_BTN_NO_ACTION | No button action. |
| #DIGIT_BTN_LEFT_CLICK | Left mouse button click. |
| #DIGIT_BTN_LEFT_DBL_CLICK | Left mouse button double click. |
| #DIGIT_BTN_LEFT_DRAG | Left mouse button drag. |
| #DIGIT_BTN_RIGHT_CLICK | Right mouse button click. |
| #DIGIT_BTN_RIGHT_DBL_CLICK | Right mouse button double click. |
| #DIGIT_BTN_RIGHT_DRAG | Right mouse button drag. |
| #DIGIT_BTN_MIDDLE_CLICK | Middle mouse button click. |
| #DIGIT_BTN_MIDDLE_DBL_CLICK | Middle mouse button double click. |
| #DIGIT_BTN_MIDDLE_DRAG | Middle mouse button drag. |

| DigitTypeEnum | |
|---|---|
| #DIGIT_TYPE_MOUSE | Mouse-relative mode. |
| #DIGIT_TYPE_PUCK | Puck-absolute mode. |

| DigitUnitsEnum | |
|---|---|
| #DIGIT_UNITS_UNKNOWN | Digitizer returns location in unknown units. |

| #DIGIT_UNITS_INCHES | Digitizer returns location in Inches. |
| #DIGIT_UNITS_CMS | Digitizer returns location in Centimeters. |

**See also:** View

# Display

A display is the abstract superclass for the drawable areas of the screen.

**Inherits from** Obj

**Inherited by** ListDisplay, MapDisplay, and PageDisplay

| *Class Requests* | |
| --- | --- |
| CanRenderString : aBoolean | Returns true when string rendering is turned on. |
| SetCanRenderString (canRenderBoolean) | Turns on string rendering when the given Boolean parameter is true. |
| SetColorMapSize (numberOfColors, numberOfGrays) | Sets ArcView's colormap at the start of an application. |

| *Instance Requests* | |
| --- | --- |
| BeginClip : aBoolean | Makes a display "aware" of its boundaries. |
| CanUndoZoom : aBoolean | Returns true if a previous zoom or pan can be undone. |
| DeleteUserPoint | Used in an Apply script for a pop-up menu, it deletes the last point when entering a PolyLine or a Polygon. |
| DisableRedraw | Disables the processing that redraws a display. |
| DisableZoomUndo | Disables the zoom undo capability. |
| DrawCircle (aCircle, aSymbol) | Draws the specified circle using the specified symbol on a display. |
| DrawIcon (aRect, anIcon) | Draws an icon in the specified rectangle of a display. |
| DrawLabeledRect (aRect, aString, aSymbol) | Draws a rectangle with the specified symbol and places the specified string in its center. |
| DrawLine (aLine, aSymbol) | Draws a line on a display with the specified symbol. |
| DrawMultiPoint (aMultiPoint, aSymbol) | Draws a multi-point shape on a display with the specified symbol. |

| | |
|---|---|
| DrawOval (aOval, aSymbol) | Draws an oval on a display with the specified symbol. |
| DrawPoint (aPoint, aSymbol) | Draws a point on a display with the specified symbol. |
| DrawPolygon (aPolygon, aSymbol, showOutLineBoolean) | Draws a polygon on a display using the specified symbol. |
| DrawPolyLine (aPolyLine, aSymbol) | Draws a polyline on a display using the specified symbol. |
| DrawRect (aRect, aSymbol) | Draws a rectangle shape on a display with the specified symbol. |
| DrawShape (aShape, aSymbol) | Draws a shape on a display using the specified symbol. |
| DrawSymbolSample (aSymbol, aRect) | Draws a sample on a display of the specified symbol in the specified rectangle. |
| DrawText (aPointSize, aString, aSymbol) | Draws a string of the specified size and symbol. |
| DrawTextLine (aString, aSymbol, aPoint) | Draws text lines identified by the specified string starting at the specified point. |
| DrawTextPolyLine (aPolyLine, aString, aSymbol) | Draws text along a polyline on a display using the specified symbol. |
| EnableRedraw | Enables the redraw processing. |
| EnableZoomUndo | Enables the zoom undo capability. |
| EndClip | Informs the window manager that drawing on the current display is complete. |
| ExportEnd | Stops exporting graphics. |
| ExportStartFN (aFileName, aFormatString) : aDisplay | Exports graphics to a file in the specified format. |
| ExportStartParams (aFileName, aFormatString, exportOptionsList) : aDisplay | Starts exporting graphics drawn on a display to the given filename. |
| FlashShape (aShape, aSymbol, milliSeconds, numberOfTimes) | Flashes the specified shape on a display for the specified number of times and milliseconds. |
| Flush | Flushes all drawings to a display. |
| GetDistanceUnits : UnitsLinearEnum | Get a display's distance units. |
| GetFlushCount : aNumber | Gets the number of vertices buffered before drawing a segment on a display. |
| GetMouseLoc : aPoint | Gets the location of the mouse on a display. |
| GetReportUnits : UnitsLinearEnum | Gets the display's report measurement unit. |
| GetResolution : DisplayResEnum | Gets the resolution of a display. |

| | |
|---|---|
| GetUnits : UnitsLinearEnum | Gets the linear measurement unit of a display. |
| HookupSymbol (aSymbol) | Attaches the given symbol to a display's scale. |
| Invalidate (aBoolean) | Invalidates the entire display if a true Boolean value is specified. |
| InvalidateRect (aRect) | Invalidates the specified rectangular area of a display. |
| Pan | Allows the user to pan a display. |
| PanTo (aPoint) | Pans the display to make the specified point its center. |
| ReturnExportFormats : stringList | Returns a list of available graphic formats. |
| ReturnExtent : aRect | Returns the extent of a display. |
| ReturnMarginExtent : aRect | Returns the margin extent of a display. |
| ReturnPageExtent : aRect | Returns the page extent of a display. |
| ReturnPixMap (aRect) | Returns a pixmap contained in the specified rectangle. |
| ReturnUserCircle : aCircle | Returns the circle drawn by the user on a display. |
| ReturnUserLine : aLine | Returns the line drawn by the user on a display. |
| ReturnUserPoint : aPoint | Returns the point clicked on by the user on a display. |
| ReturnUserPolygon : aPolygon | Returns the polygon drawn by the user on a display. |
| ReturnUserPolyLine : aPolyLine | Returns the polyline drawn by the user on a display. |
| ReturnUserRect : aRect | Returns the rectangle drawn by the user on a display. |
| ReturnVisExtent : aRect | Returns the visible extent of a map display. |
| SetDistanceUnits (UnitsLinearEnum) | Sets a display's distance unit. |
| SetExtent (aRect) | Sets the extent of a display. |
| SetFlushCount (aNumber) | Sets the number of buffers before flushing to a display. |
| SetReportUnits (UnitsLinearEnum) | Sets a display's report unit. |
| SetResolution (DisplayResEnum) | Sets the resolution of a display. |
| SetUnits (UnitsLinearEnum) | Sets the linear measurement unit of a display. |
| UndoZoom | Undoes the last zoom or pan operation. |
| Validate | Validates a display. |
| ZoomIn (aPercentageNumber) | Zooms to the center of a display by the specified percentage number. |

| | |
|---|---|
| ZoomOut (aPercentageNumber) | Zooms out from the center of a display by the specified percentage number. |
| ZoomToRect (aRect) | Zooms a display to fit the specified rectangle. |
| ZoomToScale (aScaleNumber) | Zooms a display to the specified scale. |

## Enumerations

| DisplayResEnum | |
|---|---|
| #DISPLAY_RES_LOW | Use low display resolution. |
| #DISPLAY_RES_NORMAL | Use normal display resolution. |
| #DISPLAY_RES_HIGH | Use high display resolution. |

| UnitsLinearEnum | |
|---|---|
| #UNITS_LINEAR_UNKNOWN | Unknown. |
| #UNITS_LINEAR_INCHES | Use inches. |
| #UNITS_LINEAR_FEET | Use feet. |
| #UNITS_LINEAR_YARDS | Use yards. |
| #UNITS_LINEAR_MILES | Use miles. |
| #UNITS_LINEAR_MILLIMETERS | Use millimeters. |
| #UNITS_LINEAR_CENTIMETERS | Use centimeters. |
| #UNITS_LINEAR_METERS | Use meters. |
| #UNITS_LINEAR_KILOMETERS | Use kilometers. |
| #UNITS_LINEAR_NAUTICALMILES | Use nautical miles. |
| #UNITS_LINEAR_DEGREES | Use decimal degrees. |

**See also:** GraphicList; Layout; View

# DLL

This class represents the Dynamic Link Libraries. The *DLL* class is used only in the Windows environment.

**Inherits from** Obj

| Class Requests | |
| --- | --- |
| GetAVWindowHandle : aNumber | Returns ArcView's window handle. |
| IsRefresh : aBoolean | Returns true when ArcView's repaint mode is set. |
| Make (aFileName) : aDLL | Establishes a DLL object from the specified DLL file. |
| SetRefresh (isRefreshBoolean) | Sets the repaint mode for ArcView. |

| Instance Requests | |
| --- | --- |
| FindProc (functionName) : aDLLProc | Returns a DLL procedure with the specified name. |
| GetFileName : aFileName | Gets the disk file name of a DLL object. |
| GetProcs : aListOfDLLProcs | Gets a list of procedures within a DLL. |
| Is32BitDLL : aBoolean | Returns true if a DLL supports 32-bit functionality. |

**See also:** DLLProc

# *DLLProc*

This class implements individual functions with a DLL. The *DLLProc* class is used only in the Windows environment.

**Inherits from** Obj

| Class Requests | |
| --- | --- |
| Make (aDLL, functionName, returnDLLProcTypeEnum, argumentListOfDllProcTypeEnum) : aDLLProc | Establishes a DLL procedure object from the specified function in the specified DLL. Return and argument list data types are also defined |

| Instance Requests | |
| --- | --- |
| Call (argumentList) : anObj | Calls a DLL procedure. |

## *Enumerations*

| DLLProcTypeEnum | |
| --- | --- |
| #DLLPROC_TYPE_FLOAT | A decimal number. |

| | |
|---|---|
| #DLLPROC_TYPE_INT16 | A short integer. |
| #DLLPROC_TYPE_INT32 | An integer. |
| #DLLPROC_TYPE_PFLOAT | Pointer to a decimal number. |
| #DLLPROC_TYPE_PINT16 | Pointer to a short integer. |
| #DLLPROC_TYPE_PINT32 | Pointer to an integer. |
| #DLLPROC_TYPE_POINTER | A 4-byte pointer to a structure. |
| #DLLPROC_TYPE_STR | Pointer to a null terminated character string. |
| #DLLPROC_TYPE_VOID | No return value. |

*See also:* DLL

# Doc

This abstract class implements the common properties of ArcView documents: views, tables, charts, layouts, and scripts.

**Inherits from** Obj

**Inherited by** Chart, Layout, Project, SEd, Table, and View

| Class Requests | |
|---|---|
| Make : aDoc | Creates a new document. |
| MakeWithGUI (aGUIName) : aDoc | Creates a new document for the given GUI name. |
| RegisterExtension (anExtensionClass) | Registers the given extension class with the Document class. |
| UnregisterExtension (anExtensionClass) | Unregisters the given extension class from the Document class. |

| Instance Requests | |
|---|---|
| GetCloseScript : aScriptName | Gets the name of a script executed when a document is closed. |
| GetComments : aString | Gets a document's comments. |
| GetCreationDate : aString | Gets a document's creation date in string format. |
| GetCreator : aString | Gets the name of a document's creator. |
| GetExtension (anExtensionClass) : aDocumentExtension | Gets the document extension of the specified class. |

| | |
|---|---|
| GetGroupGUI : aString | Returns the name of DocGUI that a document is associated with. |
| GetGUI : aString | Gets the name of the document GUI for a document. |
| GetObjectTag : anObj | Gets the object tagged to a document. |
| GetOpenScript : aScriptName | Gets the name of the script executed when a document is opened. |
| GetUpdateScript : aScriptName | Gets the name of the script executed when the document GUI of a document is updated. |
| GetWin : aDocWin | Gets a document's window object. |
| IsActive : aBoolean | Returns true if a document is active. |
| IsSymWinClient : aBoolean | Returns true if a document is a client of a symbol window. |
| Print | Prints the document. |
| RemoveExtension (anExtensionClass) | Removes the extension of specified class from a document. |
| SetCloseScript (aScriptName) | Sets the name of the script to execute when a document is closed. |
| SetComments (aString) | Sets a document's comment to the specified string. |
| SetCreator (aString) | Sets the document creator to the specified string. |
| SetExtension (anExtension) | Sets a document's extension. |
| SetGroupGUI (aGUIName) | Sets the DocGUI that a document is associated with. |
| SetGUI (aString) | Sets the document GUI of a document to the specified document GUI name. |
| SetName (aString) | Sets the name of a document to the specified string. |
| SetObjectTag (anObj) | Tags a document with the specified object. |
| SetOpenScript (aScriptName) | Sets the name of the script to execute when a document is opened. |
| SetUpdateScript (aScriptName) | Sets the name of the script to execute when a document's GUI is updated. |

**See also:** Application; DocGUI; DocWin; ExtensionObject; Project

# DocFrame

This class implements the drawing of a table or chart on a layout document.

**Inherits from** PictureFrame

| Class Requests | |
|---|---|
| Make (aRect, aDocClass) : aDocFrame | Creates a document frame in the specified rectangle and for the specified document class. |

| Instance Requests | |
|---|---|
| CanUndo : aBoolean | Always returns false. |
| Draw | Draws the document frame. |
| Edit (aListofGraphics) : aBoolean | Displays the graphic editor. |
| GetFillObject : anObj | Gets a document frame's fill object. |
| GetFramedDoc : aDoc | Gets the document associated with the document frame. |
| GetFramedDocClass : Class | Gets the class of the document associated with the document frame. |
| IsFilled : aBoolean | Returns true if the document frame is filled. |
| IsFilledBy (aClass) : aBoolean | Returns true if a document frame is filled with objects of a specified class. |
| SetFillObject (anObj) | Sets the fill object of a document frame to the specified object. |
| SetFramedDoc (aDoc) | Sets the document of a document frame to the specified document. |
| SetFramedDocClass (aDocClass) | Sets the document class for a document frame. |

***See also:*** Chart; Display; GraphicList; Table; Template

# DocGUI

This class implements the GUI for documents.

**Inherits from** Obj

| Class Requests | |
|---|---|
| Make (moduleName) : aDocGUI | Creates a document GUI for a document class represented by the specified string. |

| **Instance Requests** | |
|---|---|
| Activate | Applies changes the GUI to a document GUI. |
| GetActionScript : aString | Gets a DocGUI's action-script name. |
| GetActionUpdateScript : aString | Gets a DocGUI's action-update script name. |
| GetButtonBar : aButtonBar | Gets the button bar object for a document GUI. |
| GetDocBaseName : aString | Gets the document base name. |
| GetIcon : anIcon | Gets the icon used for a DocGUI on the project window. |
| GetMenuBar : aMenuBar | Gets the menu bar object for a document GUI. |
| GetNewScript : aString | Gets a DocGUI's new-script name. |
| GetNewUpdateScript : aString | Gets a DocGUI's new-update script name. |
| GetOpenScript : aString | Gets a DocGUI's open-script name. |
| GetOpenUpdateScript : aString | Gets a DocGUI's open-update script name. |
| GetPopups : aPopupSet | Gets the pop-up set associated with aDocGUI. |
| GetTitle : aString | Gets a DocGUI's name used in the project window. |
| GetObjectTag | Gets the object tagged to a document GUI. (Not available in 2.0.) |
| GetToolBar : aToolBar | Gets the tool bar object for a document GUI. |
| GetType : moduleName | Returns a string representing the type of document GUI. |
| IsModified : aBoolean | Returns true if a document GUI has been modified. |
| IsVisible : aBoolean | Returns true when a DocGUI is visible on the project window. |
| SetActionScript (aScriptName) | Sets a DocGUI's action-script. |
| SetActionUpdateScript (aScriptName) | Sets a DocGUI's action-update script. |
| SetDocBaseName (aBaseName) | Sets a DocGUI's base name. |
| SetIcon (anIcon) | Sets a DocGUI's icon. |
| SetModified (aBoolean) | Sets a document GUI's modified flag to the specified Boolean value. |
| SetName (aName) | Sets a DocGUI's name. |
| SetNewScript (aScript) | Sets a DocGUI's new-script. |
| SetNewUpdateScript (aScript) | Sets a DocGUI's update-script. |
| SetObjectTag (anObj) | Tags the specified object to a document GUI object. |
| SetOpenScript (aScript) | Sets a DocGUI's open-script. |
| SetOpenUpdateScript (aScript) | Sets a DocGUI's open-update script. |

| | |
|---|---|
| SetTitle (aTitleString) | Sets a DocGUI's title shown in the project window. |
| SetVisible (isVisibleBoolean) | Makes a DocGUI visible in the project window if the Boolean value is true. |
| Update | Triggers the update event for all controls in a document GUI. |

***See also:*** Application; ButtonBar; ControlSet; Doc; MenuBar; Project; ToolBar

***Code example:*** page 286

# DocumentExtension

An abstract class for specialized documents added by ESRI's ArcView extension products.

## Inherits from ExtensionObject

| Instance Requests | |
|---|---|
| Activate | Makes a document-extension the active environment for an extension. |

# DocWin

This class implements the window through which a document can be viewed.

## Inherits from Obj

| Instance Requests | |
|---|---|
| Activate | Opens and activates the document associated with a document window. |
| Close | Closes the document associated with a document window. |
| GetNext : aDocWin | Gets the next document window in the project. |
| GetNextDoc : aDoc | Gets the next document with the same class as the document of a document window. |
| GetOwner : aDoc | Gets the associated document of a document window. |
| GetTitle : aString | Gets the title displayed in the title bar of a document window. |
| Invalidate : aBoolean | Forces a redraw event for a document window. |

| IsOpen : aBoolean | Returns true if the document associated with a document window is open. |
|---|---|
| Maximize | Maximizes a document window. |
| Minimize | Minimizes a document window. |
| Move (deltaX, deltaY) | Moves a document window by the specified X and Y values. |
| MoveTo (aX, aY) | Moves a document window to the specified X and Y values. |
| Open | Opens a document window. |
| Resize (aWidth, aHeight) | Resizes a document window to the specified values. |
| Restore | Restores a document window to its regular size. |
| ReturnExtent : aPoint | Returns the extent of a document window in pixels. |
| ReturnOrigin : aPoint | Returns the position of a document window in pixels. |
| SetTitle (aString) | Sets the title of a document window to the specified string. |

***See also:*** Application; Doc; DocGUI

# *Duration*

This class implements time span.

**Inherits from** Interval

| Instance Requests | |
|---|---|
| = anObj : aBoolean | Returns true when two durations have the same value. |
| by aPeriodofTime : aDuration | Sets the step value for looping through a duration. |
| AsSeconds : aNumber | Returns the number of seconds in a duration. |
| AsString : aString | Returns a string representing a duration. |
| SetIncrement (aPeriodOfTime) : aDuration | Sets the increment in a duration. (Not available in 2.0.) |

***See also:*** Date; Number

# DynName

This class implements the dynamic segmentation data source.

**Inherits from** SrcName

| Class Requests | |
| --- | --- |
| Make (anFTab, routeIdField) : aDynName | Creates a dynamic segmentation source from the specified theme table and field. |

| Instance Requests | |
| --- | --- |
| GetEventRouteId : aField | Gets the event table's route ID field. |
| GetEventTable : aVTab | Gets the event table. |
| GetFromField : aField | Gets the from field in a linear dynamic segmentation. |
| GetMeasureField : aField | Gets the measure field in a point dynamic segmentation. |
| GetOffsetField : aField | Gets the field used to offset an event. |
| GetRatRouteId : aField | Gets the route ID field from the route table. |
| GetRouteSys : anFTab | Gets the route table. |
| GetToField : aField | Gets the to field in a linear dynamic segmentation. |
| IsContinuous : aBoolean | Returns true for a continuous linear dynamic segmentation. |
| IsLinear : aBoolean | Returns true for a linear dynamic segmentation. |
| IsPoint : aBoolean | Returns true for a point dynamic segmentation. |
| SetLineEvent (anEventTable, RouteId, From, To) | Makes a dynamic segmentation linear. |
| SetOffsetField (aField) | Sets the field used to offset events. |
| SetPointEvent (anEventTable, RouteId, Measure) | Makes a dynamic segmentation a point event source. |

*See also:* Field; FTab; Theme

# EncryptedScript

This class implements scripts with an encrypted source code.

**Inherits from** Script

| Class Requests | |
| --- | --- |
| Make (anAvenueSourceCodeString) : anEncryptedScript | Makes an encrypted script from the specified source code string. |
| MakeFromScript (aScript) : anEncryptedScript | Makes a new encrypted script from the specified script object. |

| Instance Requests | |
| --- | --- |
| AsString : aString | Always returns an empty string. |

*See also:* Application; Project

# EnumerationElt

This class implements enumerations in Avenue.

**Inherits from** Value

| Class Requests | |
| --- | --- |
| ReturnElements (anEnumTypeStr) : aList | Returns the elements of specified enumeration. |

| Instance Requests | |
| --- | --- |
| = anObj : aBoolean | Returns true when the object is the same as an enumeration. |
| AsString : aString | Returns the string representation of an enumeration. |

# EqualAreaAzimuthal

This class implements the Lambert Equal-Area Azimuthal projection.

**Inherits from** Prj

| Class Requests | |
| --- | --- |
| Make (aRect) : anEqarazi | Creates a projection bounded by the specified rectangle. |

| Instance Requests | |
|---|---|
| CalcHorizon | Calculates the horizon polygon. |
| ProjectPt (aPoint) : aBoolean | Projects a point. |
| Recalculate | Recalculates the derived constants. |
| ReturnCentralMeridian : aNumber | Returns the central meridian in decimal degrees. |
| ReturnReferenceLatitude : aNumber | Returns the reference latitude in decimal degrees. |
| SetCentralMeridian (aLongitude) | Sets the central meridian to the specified degrees. |
| SetReferenceLatitude (aLatitude) | Sets the reference latitude to the specified degrees. |
| UnProjectPt (aPoint) : aBoolean | Unprojects a point. |

***See also:*** CoordSys; Spheroid

# *EqualAreaCylindrical*

This class implements the Equidistant Cylindrical projection.

## Inherits from Prj

| Class Requests | |
|---|---|
| Make (aRect) : anEqarcyl | Creates a projection bounded by the specified rectangle. |

| Instance Requests | |
|---|---|
| ProjectPt (aPoint) : aBoolean | Projects a point and returns true if successful. |
| Recalculate | Recalculates the derived constants. |
| ReturnCentralMeridian : aNumber | Returns the longitudinal degrees of the central meridian. |
| ReturnStandardParallel : aNumber | Returns the standard parallel in decimal degrees. |
| SetBehrmann | Sets the standard parallel to 30 degrees. |
| SetCentralMeridian (aLongitude) | Sets the central meridian to the specified degrees. |
| SetPeters | Sets the standard parallel to 45 degrees. |
| SetStandardParallel (aLatitude) | Sets the standard parallel to the specified degrees. |
| UnProjectPt (aPoint) : aBoolean | Unprojects a point and returns true if successful. |

***See also:*** CoordSys; Spheroid

# EquidistantAzimuthal

This class implements the Equidistant Azimuthal projection.

**Inherits from** Prj

| Class Requests | |
|---|---|
| Make (aRect) : anEquiazi | Creates a projection bounded by the specified rectangle. |

| Instance Requests | |
|---|---|
| CalcHorizon | Calculates the horizon polygon. |
| ProjectPt (aPoint) : aBoolean | Projects a point and returns true if successful. |
| Recalculate | Calculates the derived constants. |
| ReturnCentralMeridian : aNumber | Returns the central meridian in degrees. |
| ReturnReferenceLatitude : aNumber | Returns the reference latitude in decimal degrees. |
| SetCentralMeridian (aLongitude) | Sets the central meridian to the specified degrees. |
| SetReferenceLatitude (aLatitude) | Sets the reference latitude to the specified degrees. |
| UnProjectPt (aPoint) : aBoolean | Unprojects a point and returns true if successful. |

***See also:*** CoordSys; Spheroid

# EquidistantConic

This class implements the Equidistant Conic projection.

**Inherits from** Prj

| Class Requests | |
|---|---|
| CanDoSpheroid : aBoolean | Always returns true. |
| Make (aRect) : anEquicon | Creates a projection bounded by the specified rectangle. |

| Instance Requests | |
|---|---|
| ProjectPt (aPoint) : aBoolean | Projects a point and returns true if successful. |
| Recalculate | Calculates the derived constants. |
| ReturnCentralMeridian : aNumber | Returns the central meridian in decimal degrees. |

| | |
|---|---|
| ReturnLowerStandardParallel : aNumber | Returns the lower standard parallel in degrees. |
| ReturnReferenceLatitude : aNumber | Returns the reference latitude in decimal degrees. |
| ReturnUpperStandardParallel : aNumber | Returns the upper standard in degrees. |
| SetCentralMeridian (aLongitude) | Sets the central meridian to the specified decimal degrees. |
| SetLowerStandardParallel (aLatitude) | Sets the lower standard parallel to the specified degrees. |
| SetReferenceLatitude (aLatitude) | Sets the reference latitude to the specified degrees. |
| SetUpperStandardParallel (aLatitude) | Sets the upper standard parallel to the specified degrees. |
| UnProjectPt (aPoint) : aBoolean | Unprojects a point and returns true if successful. |

***See also:*** CoordSys; Spheroid

# EquidistantCylindrical

This class implements the Equidistant Cylindrical projection.

**Inherits from** Prj

| Class Requests | |
|---|---|
| Make (aRect) : aEquicyl | Creates a projection bounded by the specified rectangle. |

| Instance Requests | |
|---|---|
| ProjectPt (aPoint) : aBoolean | Projects a point and returns true if successful. |
| Recalculate | Calculates the derived constants. |
| ReturnCentralMeridian : aNumber | Returns the central meridian in decimal degrees. |
| ReturnReferenceLatitude : aNumber | Returns the reference latitude in decimal degrees. |
| SetCentralMeridian (aLongitude) | Sets the central meridian to the specified degrees. |
| SetReferenceLatitude (aLatitude) | Sets the reference latitude to the specified degrees. |
| UnProjectPt (aPoint) : aBoolean | Unprojects a point and returns true if successful. |

***See also:*** CoordSys; Spheroid

# EventDialog

This class implements the modal dialog box to create event themes.

**Inherits from** ModalDialog

| Class Requests | |
|---|---|
| Show (aView) : anFTab | Shows the event dialog box. |

# Extension

This class models ArcView extensions.

**Inherits from** ODB

| Class Requests | |
|---|---|
| Find (aNameOrFileName) : aExtension | Returns a loaded extension that is identified by the given string name or extension filename. |
| FindByName (aName) : aObj | Returns the object identified by the given name from the loaded extensions. |
| FindDoc (aDocName) : aDoc | Finds and returns the ArcView document identified by the given name in the loaded extensions. |
| FindGUI (aGUIName) : aDocGUI | Finds and returns the document GUI identified by the given name in the loaded extensions. |
| FindGUIsFor (aDocClass) : aDocGUIList | Returns a list of DocGUIs that are associated with a given document class in the loaded extensions. |
| GetDocs : aNameDictionary | Returns a name-dictionary of ArcView documents in the loaded extensions. |
| GetExtensions : aNameDictionary | Returns a name-dictionary of available extensions. |
| GetGUIs : aNameDictionary | Returns a name-dictionary of documents GUIs in the loaded extensions. |
| GetScripts : aNameDictionary | Returns a name-dictionary of scripts in the loaded extensions. |
| Make (aFileName, extensionName, installScript, uninstallScript, extensionDependenciesList) : anExtension | Creates an extension from the supplied information. |
| Open (aFileName) : anExtension | Loads the extension in the given filename. |

| Instance Requests | |
|---|---|
| CanUnload : aBoolean | Executes an extension's CanUnload script and returns true if the extension can be unloaded. |
| GetAbout : aString | Returns the about-string for an extension. |
| GetCanUnloadScript : aScript | Gets the CanUnload script for an extension. |
| GetExtVersion : aNumber | Returns the version number for the contents of an extension. |
| GetInstallScript : aScript | Gets the Install script of an extension. |
| GetLoadScript : aScript | Gets the Load script of an extension. |
| GetPreferences : aDictionary | Gets the user-preference dictionary for an extension. |
| GetProjectSaveScript : aScript | Gets the script that is executed each time the project is saved while an extension is loaded. |
| GetUninstallScript : aScript | Gets the Uninstall script of an extension. |
| GetUnloadScript : aScript | Gets the Unload script of an extension. |
| IsADependency : aBoolean | Returns true when another extension depends on this extension. |
| SetAbout (aboutString) | Sets the about string of an extension. |
| SetCanUnloadScript (aScript) | Sets the Unload script of an extension to the given script. |
| SetExtVersion (aVersionNumber) | Sets the version number of the content for an extension. |
| SetInstallScript (aScript) | Sets the install script of an extension. |
| SetLoadScript (aScript) | Sets the Load script of an extension to the specified script. |
| SetProjectSaveScript (aScript) | Sets the script that is executed each time the project is saved while an extension is loaded. |
| SetUninstallScript (aScript) | Sets the uninstall script of an extension. |
| SetUnloadScript (aScript) | Sets the Unload script of an extension. |
| Unload : aBoolean | Removes an extension from the ArcView project. |

***See also:*** Application; Project; ExtensionWin

# *ExtensionObject*

This abstract class allows ESRI's ArcView extensions to add properties to existing classes.

**Inherits from** Obj

**Inherited by** DocumentExtension, LegendExtension, SrcExtension, and ThemeExtension

***See also:*** Doc; Legend; Theme.

# *ExtensionWin*

This class models the modal dialog used in loading and unloading extensions.

**Inherits from** ModalDialog

| Instance Request | |
|---|---|
| Show : aBoolean | Opens the extension dialog box and returns true if the OK button is clicked. |

***See also:*** Extension; Project

# *Field*

This class implements an attribute in feature or virtual tables.

**Inherits from** Obj

| Class Requests | |
|---|---|
| Make (aName, FieldEnum, aWidthNumber, aPrecisionNumber) : aField | Creates a field object from the specified name, type, and size. |

| Instance Requests | |
|---|---|
| Clone : aField | Clones a field. |
| Copy (anotherField) | Copies the characteristics of one field into another. |
| GetAlias : aString | Gets the alias name of a field. |
| GetPixelWidth : aNumber | Gets the field width on the screen in pixels. |
| GetPrecision : aNumber | Gets the decimal precision of a numeric field. |
| GetStatus : FieldStatusEnum | Gets the status of a field. |
| GetType : FieldEnum | Gets a field's type. |
| GetWidth : aNumber | Gets the width of a field in number of characters. |

| | |
|---|---|
| IsEditable : aBoolean | Returns true when a field's value can be modified. |
| IsRedefined : aBoolean | Returns true for INFO redefined fields. |
| IsTypeNumber : aBoolean | Returns true if a field type is numeric. |
| IsTypeShape : aBoolean | Returns true if field values are shapes. |
| IsTypeString : aBoolean | Returns true if the field type is a string. |
| IsVisible : aBoolean | Returns true if the field is visible in a table. |
| SetAlias (aString) | Sets a field's alias to the specified string. |
| SetEditable (aBoolean) | Makes a field editable if true is specified as the Boolean value. |
| SetPixelWidth (aWidthNumber) | Sets the pixel width of a field to the specified values. |
| SetVisible (aBoolean) | Sets the visibility of the field to the specified Boolean value. |
| SetWidth (aWidth) | Sets the output width of a field to the specified amount. |

## Enumerations

| FieldEnum | |
|---|---|
| #FIELD_BYTE | One-byte integer. |
| #FIELD_CHAR | Fixed length string. |
| #FIELD_DATE | Date in string format of yyyymmdd. |
| #FIELD_DECIMAL | Real number in binary coded decimal format. |
| #FIELD_DOUBLE | Double precision number. |
| #FIELD_FLOAT | Single precision number. |
| #FIELD_ISODATE | Date is string format of yyyy-mm-dd. |
| #FIELD_ISODATTETIME | Date and time in string format of yyyy-mm-dd-hh:mm:ss. |
| #FIELD_ISOTIME | Time in string format of hh:mm:ss. |
| #FIELD_LOGICAL | One-byte Boolean. |
| #FIELD_LONG | Long integer. |
| #FIELD_MONEY | Real number as string with two decimal digits. |
| #FIELD_SHAPELINE | Polyline. |
| #FIELD_SHAPEMULTIPOINT | Multi-point. |
| #FIELD_SHAPEPOINT | Point. |

| #FIELD_SHAPEPOLY | Polygon. |
|---|---|
| #FIELD_SHORT | Short integer. |
| #FIELD_UNSUPPORTED | Unsupported data type. |
| #FIELD_VCHAR | Variable length string. |

| *FieldStatusEnum* | |
|---|---|
| #FIELD_STATUS_DELETED | Field deleted by ArcView. |
| #FIELD_STATUS_MISSING | Field deleted outside ArcView. |
| #FIELD_STATUS_NEW | New field. |
| #FIELD_STATUS_OLD | Field unchanged since last save. |

*See also:* FTab; VTab

*Code example:* page 455

# File

File is an abstract class for implementing access to and management of disk files.

**Inherits from** Obj

**Inherited by** LineFile and TextFile

| *Class Requests* | |
|---|---|
| CanDelete (aFileName) : aBoolean | Returns true if the given filename exists and can be deleted. |
| Copy (fromFileName, toFileName) | Copies the contents of one file to another. |
| Delete (aFileName) | Deletes a disk file identified by the specified file name object. |
| Exists (aFileName) : aBoolean | Returns true if a file exists. |
| GetFileDescriptorLimit : aNumber | Gets the maximum number of operating system descriptors that a file will use. |
| IsWritable (aFileName) : aBoolean | Returns true if the file identified by the specified file name object can be written to. |
| SetFileDescriptorLimit (aNumber) | Sets the maximum number of operating system descriptors that a file will use. |

| Instance Requests | |
|---|---|
| Close | Flushes the memory and closes the file. |
| Flush | Writes the memory cache to the disk file. |
| GetEolType : FileEolTypeEnum | Gets the type of line termination. |
| GetFileName : aFileName | Gets the file name object of a file. |
| GetName : aString | Gets the full path of a file. |
| GetPos : aNumber | Gets the position of the file pointer in a file. |
| GetSize : aNumber | Gets the number of elements in a file. |
| GotoBeg | Sets the file pointer to the beginning of a file. |
| GotoEnd | Sets the file pointer to the end of a file. |
| IsAtEnd : aBoolean | Returns true when file pointer has passed the last element in a file. |
| IsScratch : aBoolean | Returns true if a file is marked as scratch. |
| ReadElt : anObj | Reads an element from a file. |
| SetEolType (FileEolTypeEnum) | Sets the end of line termination type. |
| SetPos (aNumber) | Sets the pointer to an element number starting from element zero. |
| SetScratch (aBoolean) | Identifies a file as scratch so that the file is deleted after it is closed. |
| WriteElt (anObj) | Writes an element to a file. |

## Enumerations

| FileEolTypeEnum | |
|---|---|
| #FILE_MSW_EOL | Lines are terminated with a carriage return and a newline character. |
| #FILE_MAC_EOL | Lines are terminated with a carriage return character. |
| #FILE_UNIX_EOL | Lines are terminated with a newline character. |

| FilePermEnum | |
|---|---|
| #FILE_PERM_APPEND | Opens an existing or new file to append elements. |
| #FILE_PERM_CLEARMODIFY | Overwrites an existing file, or creates a new file for reading and writing. |
| #FILE_PERM_MODIFY | Opens an existing or new file for reading and writing. |

| #FILE_PERM_READ | Opens an existing file for reading. |
|---|---|
| #FILE_PERM_WRITE | Overwrites an existing file or creates a new file for writing. |

***See also:*** FileName; ODB

***Code example:*** page 358

# FileDialog

This class implements the open file dialog box.

**Inherits from** ModalDialog

| Class Requests | |
|---|---|
| Put (defaultFileName, aPatternString, titleString) : aFileName | Displays file dialog box for creating new files. |
| ReturnFiles (patternsStringList, labelsStringList, aTitleString, defaultPatternIndexNumber) : filenamesList | Displays a file dialog and returns a list of selected file names. |
| Show (aPatternString, aPatternLabelString, titleString) : aFileName | Displays the file dialog for selecting an existing file. |

***See also:*** FileName

# FileName

This class implements the file name object used in accessing disk files or directories.

**Inherits from** Obj

| Class Requests | |
|---|---|
| ExistsInPaths (aFileNameString) : aBoolean | Returns true if the specified file name is found in the search path. |
| FileName.FindInSystemSearchPath (aNameString) | Creates a filename object if the given name is found in the system's search path. |
| GetCWD : aFileName | Gets the current working directory. |
| GetSaveAsUNIX : aBoolean | Returns true when file names are being converted to the UNIX style. |

| | |
|---|---|
| GetSearchPaths : aList | Gets a list of directories that make up the search path. |
| GetTmpDir : aFileName | Gets the system directory for temporary files. |
| Make (aString) : aFileName | Creates a file name based on the specified string. |
| MakeExisting (aString) : aFileName | Creates a file name based on the specified string if it exists in the search path. |
| Merge (aPathString, aFileString) : aFileName | Creates a file by merging the specified strings for the path and file name. |
| SetSaveAsUNIX (aBoolean) | Converts file names to the UNIX style if the specified Boolean value is true. |
| SetSearchPaths (aFileNameList) | Sets the search path to a list of directories specified by file name objects. |
| UniqueInPaths (aFileNameString) : aBoolean | Returns true if only one instance of a file exists in the search path. |

### Instance Requests

| | |
|---|---|
| = anotherFileName : aBoolean | Returns true when two file name objects reference the same file. |
| AsString : aString | Returns the complete path name. |
| Copy (fromFileName) | Copies the name from the specified file name object to a file name. |
| GetBaseName : aString | Gets the file name without its path. |
| GetExtension : aString | Gets a file name extension. |
| GetFullName : aString | Gets a file name with its complete path, and expands all environment variables in the path. |
| GetName : aString | Gets a file name with a path relative to the current directory. |
| IsDir : aBoolean | Returns true if a file name object references an existing directory. |
| IsFile : aBoolean | Returns true when a file name object references an existing file. |
| IsRoot : aBoolean | Returns true when a file name references the root directory. |
| LowerCase | Converts the file name and path to lower-case characters. |
| MakeTmp (aPrefixString, anExtensionString) : aFileName | Returns a unique temporary file name in a directory. |

| | |
|---|---|
| MergeFile (aFileString) | Adds the specified string to the name of a file name. |
| MergePath (aPathString) | Adds the specified string to a path. |
| Read (filterString) : filenamesList | Returns a list of file names matching the specified filter and a list of directories from a directory. |
| ReadFiles (filterString) : filenamesList | Returns a list of files matching the filter string from a directory. |
| ReturnDir : aFileName | Returns the directory portion of a file name. |
| SetCWD | Sets the current working directory to a directory. |
| SetExtension (anExtensionString) | Sets the extension of a file name to the specified string. |
| SetName (aName) | Sets the full name in a file name object to the specified name string. |
| StripFile | Removes the base name of a file name resulting in a file name object that references a directory. |

**See also:** File; ODB; String

**Code example:** page 359

# *Fill*

This class implements symbolization of polygons.

**Inherits from** Symbol

**Inherited by** CompositeFill, RasterFill, and VectorFill

| Instance Requests | |
|---|---|
| GetOlColor : aColor | Gets the outline color of a fill. |
| GetOlWidth : aNumber | Gets the outline width of a fill in points (1/72 in.). |
| GetType : SymbolEnum | Always returns #SYMBOL_FILL. |
| IsOutlined : aBoolean | Returns true if an outline for the fill exists. |
| SetOlColor (aColor) | Sets the outline color. |
| SetOlWidth (aNumber) | Sets the outline width in points. |
| SetOutlined (aBoolean) | Adds outline to a fill if the specified Boolean value is true. |

# FocalStatisticsDialog (Spatial Analyst)

This class represents the dialog for specifying parameters of neighborhood statistics for a grid.

**Inherits from** ModalDialog

| Class Requests | |
|---|---|
| Show (aTheme, GridStaTypeEnum, aCellSize) : a List | Displays the neighborhood statistics dialog box and returns any input parameters. |

## Enumerations

| GridStaTypeEnum | |
|---|---|
| #GRID_STATYPE_MAJORITY | Majority. |
| #GRID_STATYPE_MAX | Maximum. |
| #GRID_STATYPE_MEAN | Mean. |
| #GRID_STATYPE_MEDIAN | Median. |
| #GRID_STATYPE_MIN | Minimum. |
| #GRID_STATYPE_MINORITY | Minority. |
| #GRID_STATYPE_RANGE | Range. |
| #GRID_STATYPE_STD | Standard deviation. |
| #GRID_STATYPE_SUM | Sum. |
| #GRID_STATYPE_VARIETY | Variety or number of unique occurrences. |

*See also:* Grid; NbrHood

# Font

This class implements the typeface in text symbols.

**Inherits from** Obj

| Class Requests | |
|---|---|
| GetDefault : aFont | Gets the default font object. |
| Make (aFamilyName, aStyleName) : aFont | Creates a font based on the specified family and style names. |

| MakeDefault : aFont | Creates a standard style Helvetica font. |
| MakeStandard (FontEnum) : aFont | Creates a standard font based on the specified enumeration. |
| SetDefault (aFamilyName, aStyleName) : aFont | Sets and returns the default font object. |

| **Instance Requests** | |
| --- | --- |
| = anotherFont : aBoolean | Returns true if a font is the same as another. |
| GetFamily : aString | Gets the family of a font. |
| GetStyle : aString | Gets the style of a font. |
| SetStyle (aStyleName) | Sets a font's style. |

## Enumerations

| **FontEnum** | |
| --- | --- |
| #FONT_COUR | Courier Roman. |
| #FONT_COURB | Bold Courier Roman. |
| #FONT_COURBI | Bold, Italic Courier Roman. |
| #FONT_COURI | Italic Courier Roman. |
| #FONT_GOTHM | Gothic Medium |
| #FONT_GOTHMH | Gothic Medium Hankaku |
| #FONT_HELV | Helvetica Roman. |
| #FONT_HELVB | Bold Helvetica Roman. |
| #FONT_HELVBI | Bold, italic Helvetica Roman. |
| #FONT_HELVI | Italic Helvetica Roman. |
| #FONT_TIME | Times Roman. |
| #FONT_TIMEB | Bold Times Roman. |
| #FONT_TIMEBI | Bold, italic Times Roman. |
| #FONT_TIMEI | Italic Times Roman. |

***See also:*** FontManager; TextSymbol

***Code example:*** page 369

# FontManager

The single instance of this class manages available system fonts.

**Inherits from** Obj

| Class Requests | |
|---|---|
| The : FontManager | Gets the single instance of the font manager. |

| Instance Requests | |
|---|---|
| GetWeight (aFamilyName, aStyleName) : FontWgtEnum | Gets the weight in the specified font family and style. |
| GetWideness (aFamilyName, aStyleName) : FontWidenessEnum | Gets the width in the specified font family and style. |
| HasStyle (aFamilyName, aStyleName) : aBoolean | Returns true if the specified family name has the specified style name. |
| IsBold (aFamilyName, aStyleName) : aBoolean | Returns true when a family and style names are bold. |
| IsItalic (aFamilyName, aStyleName) : aBoolean | Returns true when a family and style names are italic. |
| IsMonoSpaced (aFamilyName, aStyleName) : aBoolean | Returns true when a family and style names are monospaced. |
| IsSerif (aFamilyName, aStyleName) : aBoolean | Returns true when a family and style names are serif fonts. |
| ReturnCommonFamilyName (aFamilyName) : aString | Returns the common name of the specified family name. |
| ReturnFamilies : aList | Returns a list of font families available on the system. |
| ReturnFontName (aFamilyName, aStyleName) : aString | Returns the font name from the specified family and style names. |
| ReturnStyles (aFamilyName) : aList | Returns a list of style names for the specified family name. |

## Enumerations

| FontWgtEnum | |
|---|---|
| #FONT_WGT_NORMAL | Font weight is normal. |
| #FONT_WGT_THICK | Font weight is thick. |
| #FONT_WGT_THIN | Font weight is thin. |

| **FontWidenessEnum** | |
|---|---|
| #FONT_WIDENESS_CONDENSED | Font width is condensed. |
| #FONT_WIDENESS_NORMAL | Font width is normal. |
| #FONT_WIDENESS_WIDE | Font is wide. |

*See also:* Font

# Frame

Objects of this class contain elements in a layout document.

**Inherits from** Graphic

**Inherited by** LegendFrame, PictureFrame, and ViewFrame

| **Class Requests** | |
|---|---|
| Make (aRect) : aFrame | Creates a frame bounded by the specified rectangle. |

| **Instance Requests** | |
|---|---|
| Draw | Draws a frame. |
| Edit (framesList) : aBoolean | Opens the frame editor. |
| GetQuality : FrameQualityEnum | Gets the drawing quality of a frame. |
| GetRefresh : FrameRefreshEnum | Gets the refresh style of a frame. |
| Offset (aPoint) | Moves a frame by the specified amount. |
| SetBounds (aRect) | Sets the frame boundary to the specified rectangle. |
| SetQuality (FrameQualityEnum) | Sets the drawing quality of a frame. |
| SetRefresh (FrameRefreshEnum) | Sets the refresh style of a frame. |

## Enumerations

| **FrameQualityEnum** | |
|---|---|
| #FRAME_QUALITY_DRAFT | Data or picture is not drawn in the frame. |
| #FRAME_QUALITY_PRESENTATION | Frame's contents are drawn on the display. |

| **FrameRefreshEnum** | |
|---|---|
| #FRAME_REFRESH_ALWAYS | Frame is redrawn as necessary. |
| #FRAME_REFRESH_WHENACTIVE | Frame is redrawn only when the layout document is active. |

**See also:** Display; GraphicList; Layout; Template

# FTab

Objects of a feature table store attribute and shape information for feature themes.

## Inherits from VTab

| **Class Requests** | |
|---|---|
| CanManageDataSet : aBoolean | Always returns true. |
| GetAnnoAddFieldsPreference : aBoolean | Returns true if annotation fields are automatically added. |
| GetDataSetName : aString | Returns the data set name of FTab Class. |
| Make (aSrcName) : anFTab | Creates a feature table from an existing feature source. |
| MakeNew (aFilename, aClass) : anFTab | Creates a new feature table stored in the specified file name; the specified class determines the feature type. |
| SetAnnoAddFieldsPreference (aBoolean) | If set to true annotation fields are automatically added. |

| **Instance Requests** | |
|---|---|
| AddFields (fieldsList) | Adds the specified fields to a feature table. |
| AddAnnoFields | Creates an annotation style file and joins it to a feature table. |
| AddRecord : aNumber | Adds a null record to a feature table and returns the record number. |
| AsGrid (aPrj, aField, gridSize) : aGrid | Creates a grid by rasterizing a feature table. |
| BatchMatch | Performs batch mode address matching on a geocodable feature theme. |
| BeginTransaction | Starts a transaction for changing a feature table. |
| CalculateDensity (aPrj, aField, densitySpecList, gridSize) : aGrid | Creates a grid from a point or multipoint feature table. |

| | |
|---|---|
| CanAddFields : aBoolean | Returns true if fields can be added to a feature table. |
| CanAddRecord : aBoolean | Returns true if new records can be added to a feature table. |
| CanEdit : aBoolean | Returns true if a feature table can be edited. |
| CanModifyIndex (aField) : aBoolean | Returns true if the index based on the specified field can be modified. |
| CanRedo : aBoolean | Returns true when there are undo operations that can be redone. |
| CanRemoveFields : aBoolean | Returns true if fields of a feature table can be removed. |
| CanRemoveRecords : aBoolean | Returns true if records can be removed from a feature table. |
| CanUndo : aBoolean | Returns true when there are changes to a feature table that can be undone. |
| CreateIndex (aField) | Creates an index for the specified field. |
| EditMatch | Displays the match editor for a geocoded feature table. |
| EndTransaction | Indicates the end of a transaction. |
| Export (aFileName, aClass, selectedRecordsOnlyBoolean) : aVTab | Creates a file based on the specified class, and exports the selected records or all records to that file. |
| ExportClean (aFileName, SelectedRecordsOnlyBoolean) : anFTab | Exports a feature table to a shapefile identified by the given filename of type Shape. |
| ExportProjected (aFileName, aPrj, SelectedRecordsOnlyBoolean) : anFTab | Exports to a projected shapefile. |
| ExportUnprojected (aFileName, aPrj, SelectedRecordsOnlyBoolean) : anFtab | Unprojects and exports a feature table to a shapefile. |
| Flush | Synchronizes the contents of the disk file with the memory. |
| FocalStats (aPrj , aField , statsSpecList , gridSize) : aGrid | Creates a grid by calculating neighborhood statistics for the points in a point or multipoint feature table. |
| GetExtension (anExtensionClass) : aSrcExtension | Gets the extension object of the specified class. |
| GetLabelPoint (recordNumber) : aPoint | Gets the label point of a specified record number in a feature table. |
| GetMatchPref : aMatchPref | Returns the matching preferences of a geocoded feature table. |

| | |
|---|---|
| GetShapeClass : aClass | Gets the shape class of a feature table. |
| GetSrcName : aSrcName | Gets the source name of a feature table. |
| HasError : aBoolean | Returns true when a feature table is missing data components. |
| Interpolate (aPrj , zField , anInterp , gridSize) : aGrid | Creates a grid from a point or multipoint feature table based on the specified interpolation method. |
| IsAnno : aBoolean | Returns true if a feature table was created from an annotation coverage. |
| IsBeingEditedWithRecovery : aBoolean | Returns true during editing sessions. |
| IsBlockDisplayEnabled : aBoolean | Returns true if displaying blocks in a CAD drawing is enabled. (Not available in 2.0.) |
| IsDrawing : aBoolean | Returns true if a feature table is a CAD drawing. (Not available in 2.0.) |
| IsEditable : aBoolean | Returns true if a feature table can be edited. |
| IsFieldIndexed (aField) : aBoolean | Returns true if an index exists for the specified field. |
| IsGeocoded : aBoolean | Returns true if a feature table is geocoded. |
| IsWorldFileEnabled : aBoolean | Returns true if drawing to world transformation is enabled. (Not available in 2.0.) |
| Join (aToField, aFromVTab, aFromField) | Joins another table to a feature table based on the specified fields. |
| QueryShape (aRecordNumber, aPrj, anObj) : aBoolean | Returns true if it can project the shape in the specified record number, while using the specified projection into the specified existing object. |
| QueryValue (aField, aRecordNumber, anObj) : aBoolean | Returns true if it can place the value of a specified field at the specified record number into the specified existing object. (Available in 2.0 only.) |
| Redo | Carries out the redo operation. |
| RemoveExtension (anExtensionClass) | Removes the source extension specified by the extension class. |
| Refresh | Synchronizes the contents of the memory with the disk file. |
| RemoveFields (fieldsList) | Removes the specified list from a feature table. |
| RemoveIndex (aField) | Removes the index of a field from a feature table. |
| RemoveRecord (aRecordNumber) | Removes the record at the specified record number from a feature table. |
| RemoveRecords (aBitmap) | Removes the records set at the specified bitmap. |
| ReturnActiveLayers : layersList | Returns a list of active layers. |
| ReturnLayers : layersList | Returns a list of active layers. |

| | |
|---|---|
| ReturnLocation (aPoint, aSearchToleranceNumber) : aList | Returns a list of route numbers found at the specified point using the specified tolerance. |
| ReturnRouteMeasure (aRouteNumber, aMeasureNumber) : aList | Returns a list of points where the specified route number and measure can be found. |
| ReturnSelectedLayers : layersList | Returns a list of selected layers. |
| ReturnValue (aField, aRecordNumber) : anObj | Returns the value of a field at the specified record number. |
| ReturnValueString (aField, aRecordNumber) : aString | Returns the string representation of a field value at the specified record number. |
| ReturnWorldFile : aFileName | Returns the name of the file with the drawing to world transformation parameters. |
| SaveEditsAs (aFileName) : aVTab | Saves a feature table's changes to a file instead of applying it to the table. |
| SelectByFTab (anotherFTab, FTabRelTypeEnum, aDistanceNumber, VTabSelTypeEnum) | Selects records of a feature table based on another feature table. |
| SelectByLine (aLine, VTabSelTypeEnum) | Selects a feature table's records that touch the specified line. |
| SelectByPoint (aPoint, aToleranceNumber, VTabSelTypeEnum) | Selects a feature table's records that touch the specified point or its tolerance area. |
| SelectByPolygon (aPolygon, VTabSelTypeEnum) | Selects a feature table's records that touch the specified polygon. |
| SelectByPolyLine (aPolyLine, VTabSelTypeEnum) | Selects a feature table's records that touch the specified polyline. |
| SelectByRect (aRect, VTabSelTypeEnum) | Selects a feature table's records that touch the specified rectangle. |
| SelectByShapes (shapesList, VTabSelTypeEnum) | Selects a feature table's records that touch the shapes in the specified list. |
| SetBlockDisplayEnable (aBoolean) | Enables the display of blocks in a CAD drawing if the specified Boolean value is true. (Not available in 2.0.) |
| SetEditable (aBoolean) | Makes the feature table editable if the specified Boolean value is true. |
| SetExtension (aSrcExtension) | Sets a source extension for a feature table. |
| StartEditingWithRecovery : aBoolean | Starts editing session with recovery features. |
| StopEditingWithRecovery (saveEditsBoolean) : aBoolean | Ends an editing session, returns true if changes are saved. |
| SetMatchPref (aMatchPref) | Sets the matching preferences of a geocoded feature table to the specified match preference object. |

| SetSelectedLayers (layersList) | Selects the layers specified in the list. (Not available in 2.0.) |
| SetValue (aField, aRecordNumber, anObj) | Sets the value of the specified field in the specified record number of an editable feature theme to the value of the specified object. |
| SetValueString (aField, aRecordNumber, aString) | Sets the value of the specified field in the specified record number of an editable feature theme to the specified string. |
| SetWorldFile (aFileName) | Sets the file with the drawing to world transformation parameters. (Not available in 2.0.) |
| SetWorldFileEnable (aBoolean) | Enables the drawing to world transformation when the specified Boolean value is true. (Not available in 2.0.) |
| Summarize (aFileName, aClass, aField, fieldsList, VTabSummaryEnumList) : VTab | Returns a new table created with the specified file name after grouping records by the specified field and performing the list of summary operations on the specified list of fields. |
| SupportsAOI : aBoolean | Returns true if a feature table can have an area of interest property. (Not available in 2.0.) |
| Undo | Carries out the undo operation. |
| UnjoinAll | Removes all joins from a feature table. |

## Enumerations

| **FTabRelTypeEnum** | |
| --- | --- |
| #FTAB_RELTYPE_INTERSECTS | Intersects. |
| #FTAB_RELTYPE_COMPLETELYCONTAINS | Completely contains. |
| #FTAB_RELTYPE_CONTAINSTHECENTEROF | Contains the center of. |
| #FTAB_RELTYPE_ISCOMPLETELYWITHIN | Is completely within. |
| #FTAB_RELTYPE_HASCENTERWITHIN | Has center within. |
| #FTAB_RELTYPE_ISWITHINDISTANCEOF | Is within a distance of. |

| **VTabSelTypeEnum** | |
| --- | --- |
| #VTAB_SELTYPE_NEW | Replaces existing selections with new selections. |
| #VTAB_SELTYPE_AND | Selects only from the existing selections. |
| #VTAB_SELTYPE_OR | Adds the selection to the existing selections. |
| #VTAB_SELTYPE_XOR | Adds if not selected, and removes if already selected. |

| VTabSummaryEnum | |
| --- | --- |
| #VTAB_SUMMARY_AVG | Average of a group. |
| #VTAB_SUMMARY_COUNT | Count of non-null values in a group. |
| #VTAB_SUMMARY_FIRST | First value in a group. |
| #VTAB_SUMMARY_LAST | Last value in a group. |
| #VTAB_SUMMARY_MAX | Maximum value in a group. |
| #VTAB_SUMMARY_MIN | Minimum value in a group. |
| #VTAB_SUMMARY_STDEV | Standard deviation of a group. |
| #VTAB_SUMMARY_SUM | Summation value of a group. |
| #VTAB_SUMMARY_VAR | Variance in a group. |

***See also:*** BitMap; Field; FTheme; SrcName; ThemeOnThemeDialog

# FTheme

The feature theme class implements the visual representation of a coverage, map library layer, or shape file.

**Inherits from** Theme

| Class Requests | |
| --- | --- |
| Make (anFTab) : anFTheme | Creates a feature theme from the specified feature table. |

| Instance Requests | |
| --- | --- |
| AutoComplete (aLine) | Adds new feature shapes to an editable polygon feature using the given line. |
| BlinkRecord (aRecordNumber) | Blinks the feature with the specified record number in the feature table. |
| BuildQuery | Displays the query builder dialog box. |
| CanDeleteFromView : aBoolean | Returns true if a theme can be deleted from its view. |
| CanExportToFTab : aBoolean | Always returns true. |
| CanEdit : aBoolean | Returns true if a feature theme can be edited. |
| CanHotLink : aBoolean | Returns true whe a feature theme can have hot link. |
| CanLabel : aBoolean | Always returns true. |
| CanProject : aBoolean | Returns true if a feature theme can be projected. |

| | |
|---|---|
| CanReturnClassCounts : aBoolean | Returns true when the number of features in each class in the legend can be computed. |
| CancelEditing | Stops an editing session and discards the changes. |
| CanSelect : aBoolean | Returns true when features in a feature theme can be selected. |
| ClearSelected | Unselects the features in a feature theme. |
| ClearSelection | Unselects the selected features. |
| ClipSelected (aShape) | Clips the selected features of an editable feature theme using the specified shape. |
| Clone : anFTheme | Returns a cloned feature theme. |
| CopySelected | Copies the selected features of a feature theme to the application's clipboard. |
| CutSelected | Cuts the selected features of a feature theme to the application's clipboard. |
| EditLegend | Displays the legend editor. |
| EditTable | Opens the feature table associated with a feature theme. |
| ExportToFTab (aFileName) : anFTab | Exports the feature theme to a shapefile. |
| FindByPoint (aPoint) : aList | Returns a list of features that contain the specified point. |
| GetAttrUpdateRules : aList | Returns a list of AttrUpdate objects. |
| GetFeatureSymbol (aRecordNumber) : aSymbol | Gets the feature symbol from the Legend. |
| GetFixedSizeText : aBoolean | Returns the size of fixed text for a feature theme. |
| GetFTab : anFTab | Gets the feature table of a feature theme. |
| GetHotField : aField | Gets the hot link field of a feature theme. |
| GetHotScriptName : aScriptName | Gets the script name associated with the hot link in a feature theme. |
| GetInteractiveSnapTolerance : aNumber | Gets the tolerance for interactive snap. |
| GetLabelField : aField | Gets the label field of a feature theme. |
| GetLabelTextSym : aSymbol | Gets the symbol for label text. |
| GetMatchSource : aMatchSource | Gets the match source object of a feature theme. |
| GetSelectedExtent : aRect | Gets the extent of the selected features. |
| GetSnapTolerance : aNumber | Gets the snap tolerance of a feature theme. |
| GetTextPositioner : TextPositioner | Gets the text positioner object of a feature theme. |
| HasTable : aBoolean | Returns true if a table document is associated with the feature theme. |

| | |
|---|---|
| Identify (aRecordNumber, titleString) | Displays the specified record in an identify window. |
| IntersectSelected | Converts the intersected features of an editable theme into a single feature. |
| IsInteractiveSnapping : aBoolean | Returns true when interactive snapping is enabled. |
| IsMatchable : aBoolean | Returns true if a feature theme is matchable. |
| IsSnapping : aBoolean | Returns true if snapping for a feature theme is on. |
| MergeSelected | Merges the selected feature into a single feature. |
| Paste | Pastes the contents of the application clipboard into a feature theme. |
| ReturnClassCounts : aNumberList | Returns the count of features in each class of the legend. |
| ReturnDefaultLegend : aLegend | Returns the single symbol legend for a feature theme. |
| ReturnExtent : aRect | Returns the extent of a feature theme. |
| ReturnLabel (aDisplay, aPoint) : aGraphicText | Returns a graphic text object on the specified display at the specified point. |
| ReturnSymbolSize (aRecordNumber) : aNumber | Returns the font size for the label at the specified record number. |
| SaveEditsAs (aFileName) : anFTheme | Saves changes to a file without applying them to a feature theme. |
| Select | Allows the selection of features in a feature table. |
| SelectByLine (aLine, VTabSelTypeEnum) | Selects features that intersect the specified line. |
| SelectByPoint (aPoint, VTabSelTypeEnum) | Selects features that touch the specified point. |
| SelectByPolygon (aPolygon, VTabSelTypeEnum) | Selects features that touch the specified polygon. |
| SelectByPolyLine (aPolyLine, VTabSelTypeEnum) | Selects features that touch the specified polyline. |
| SelectByRect (aRect, VTabSelTypeEnum) | Selects features that touch the specified rectangle. |
| SelectByShapes (shapesList, VTabSelTypeEnum) | Selects features that touch the shapes in the specified list. |
| SelectByTheme (anotherTheme, FTabRelTypeEnum, aDistanceNumber, VTabSelTypeEnum) | Selects features of a theme based on another theme. |
| SelectToEdit | Selects a feature through the apply event to reshape. |
| SelectToEditByPoint (aPoint) | Selects the feature at the given point to reshape. |
| SetAOI : aBoolean | Sets an area of interest for a feature theme. |

| | |
|---|---|
| SetAttrUpdateRules (attrUpdateList) | Sets the rules for union and split operations. |
| SetDefaultLegend | Sets a feature theme to its single symbol legend. |
| SetFixedSizeText (isScaledBoolean) | The fixed size text is scaled when set to true. |
| SetHotField (aField) | Sets the hot link field of a feature theme to the specified field. |
| SetHotScriptName (aScriptName) | Sets the hot link script name of a feature theme. |
| SetInteractiveSnapTolerance (aNumber) | Sets the tolerance for interactive snapping. |
| SetInteractiveSnapping (aBoolean) | Enables and disables interactive snapping. |
| SetLabelField (aField) | Sets the label field of a feature theme. |
| SetLabelTextSym (aSymbol) | Sets the symbol for the label text. |
| SetMatchSource (aMatchSource) | Sets the match source object of a feature theme. |
| SetName (aString) | Sets the name of a feature theme. |
| SetSnapping (allowSnappingBoolean) | Activates or deactivates snapping. |
| SetSnapTolerance (aNumber) | Sets the snap tolerance. |
| Split (aLine) | Splits features intersecting the specified line. |
| StopEditing (saveBoolean) : aBoolean | Stop editing and save or discard the edits. |
| SubtractShape (aShape) | Subtracts the specified shape from the selected features. |
| SupportsAOI : aBoolean | Returns true if an area of interest can be set for a feature theme. (Not available in 2.0.) |
| UnionSelected | Joins the selected features. |
| UpdateLegend | Updates the legend of a feature theme. |
| UserSnapPoint (PointSnapEnum, aPoint) | Snaps the specified point to a feature following the given rule. |

## *Enumerations*

| *FTabRelTypeEnum* | |
|---|---|
| #FTAB_RELTYPE_INTERSECTS | Intersects. |
| #FTAB_RELTYPE_COMPLETELYCONTAINS | Completely contains. |
| #FTAB_RELTYPE_CONTAINSTHECENTEROF | Contains the center of. |
| #FTAB_RELTYPE_ISCOMPLETELYWITHIN | Is completely within. |
| #FTAB_RELTYPE_HASCENTERWITHIN | Has center within. |
| #FTAB_RELTYPE_ISWITHINDISTANCEOF | Is within a distance of. |

| PointSnapEnum | |
|---|---|
| #POINT_SNAP_NONE | No point snapping. |
| #POINT_SNAP_VERTEX | Snaps point to closest vertex of a shape. |
| #POINT_SNAP_BOUNDARY | Snaps point to closest boundary line of a shape. |
| #POINT_SNAP_ENDPOINT | Snaps point to closest endpoint of a PolyLine. |
| #POINT_SNAP_INTERSECTION | Snaps point to nearest intersection of shapes. |

| VTabSelTypeEnum | |
|---|---|
| #VTAB_SELTYPE_NEW | Replaces existing selections with new selections. |
| #VTAB_SELTYPE_AND | Selects only from the existing selections. |
| #VTAB_SELTYPE_OR | Adds the selection to the existing selection. |
| #VTAB_SELTYPE_XOR | Adds if not selected, and removes if already selected. |

***See also:*** FTab; Legend; SrcName; View

***Code example:*** page 474

# GEdgeRec

This class models a proxy record for parts of shapes corresponding to a shared edge along two or more shapes stored in a feature table.

**Inherits from** GraphicShape

| Instance Requests | |
|---|---|
| GetSymbol : aSymbol | Gets the symbol used for rendering a GEdgeRec. |

# GeocodeDialog

This class represents the modal dialog box used to create geocoded themes.

**Inherits from** ModalDialog

| Class Requests | |
|---|---|
| Show (aView) : anFtab | Opens the geocoding dialog box for the specified view. |

***See also:*** GeoName

# GeoName

Through this class a match source object is associated to an event table.

**Inherits from** SrcName

| Class Requests | |
| --- | --- |
| Make (aMatchSource, anEventVTab, anEventField, aZoneField) : aGeoName | Creates a geoname from the specified parameters. |

| Instance Requests | |
| --- | --- |
| GetAliasTable : aVTab | Gets the alias table of a geoname. |
| GetDisplayField : aField | Gets a GeoName's display field. |
| GetEventField : aField | Gets the event field of a geoname. |
| GetEventTable : aVTab | Gets the event table of a geoname. |
| GetJoinField : aField | Gets the join field of a geoname. |
| GetMatchPref : aMatchPref | Gets the match preferences object of a geoname. |
| GetMatchSource : aMatchSource | Gets the match source object of a geoname. |
| GetOffset : aNumber | Gets the offset value of a geoname. |
| GetOutFileName : aFileName | Gets the output file name of a geoname. |
| GetSqueeze : aNumber | Gets the squeeze value of a geoname. |
| GetZoneField : aField | Gets the zone field of a geoname. |
| SetAliasTable (aVTab) | Associates place names in an alias table to the addresses of a geoname. |
| SetDisplayField (aField) | Sets the field displayed in geocoding editor. |
| SetJoinField (aJoinField) | Sets the field to be copied from the matchable theme to the geocoded theme for the purpose of joining the two tables. |
| SetMatchPref (aMatchPref) | Sets the address matching preferences. |
| SetOffset (anOffsetNumber) | Sets the marker distance from the matched location. |
| SetOutFileName (anOutputFileName) | Sets the name of the shape file that will store matched marks. |
| SetSqueeze (aSqueezeNumber) | Sets the marker distance for intersection addresses. |

***See also:*** Field; MatchSource; VTab

# GNodeRec

This class models a proxy record for parts of shapes correspond to a shared node along two or more shapes stored in a feature table.

**Inherits from** GraphicShape

| Instance Requests | |
|---|---|
| GetSymbol : aSymbol | Gets the symbol used in rendering a GNodeRec. |

# Gnomonic

This class implements the Gnomonic projection.

**Inherits from** Perspective

| Class Requests | |
|---|---|
| CanDoCustom : aBoolean | Always returns true because it can be customized. |
| Make (aRect) : aGnomonic | Create a projection within the specified rectangular boundary. |

*See also:* CoordSys; Spheroid

# Graphic

This abstract class implements the objects that can be seen on the display with the exception of features and images.

**Inherits from** Obj

**Inherited by** Frame, GraphicGroup, GraphicShape, GraphicText, and ScaleBarFrame

| Instance Requests | |
|---|---|
| CanEditText : aBoolean | Returns true when graphic object has editable text. |
| CanProject : aBoolean | Returns true when a graphic object can be projected. |
| CanSimplify : aBoolean | Returns true when a graphic object can be simplified. |

| | |
|---|---|
| CanUndo : aBoolean | Returns true if changes to a graphic object can be undone. |
| ConstrainBounds (aPoint) | Sets the aspect ratio of bounds in a graphic object to the specified values. |
| Draw | Draws a graphic object. |
| DrawHandles | Displays the selection handles of a graphic object. |
| Edit (documentGraphicsList) : aBoolean | Displays the graphic editor. |
| EditName : aBoolean | Displays the name editor. |
| EditSizeAndPos : aBoolean | Displays the size and position editor. |
| EditText : aBoolean | Displays the text editor. |
| GetBounds : aRect | Gets the bounds of a graphic object. |
| GetDisplay : aDisplay | Gets the display object associated with a graphic object. |
| GetDocument : aDoc | Gets the document that owns a graphic display object. |
| GetExtent : aPoint | Gets the extent of a graphic object. |
| GetFillObject : anObj | Gets the fill object of a graphic. |
| GetObjectTag : anObj | Gets the object tag of a graphic object. |
| GetOrigin : aPoint | Gets the origin of a graphic object. |
| GetShape : aShape | Gets the shape object of a graphic. |
| GetSiblingList : aList | Gets a GraphicLabel's siblings. |
| GetSymbol : aSymbol | Gets the symbol object of a graphic. |
| HasSiblings : aBoolean | Returns true if GraphicLabel object has siblings. Other graphic objects cannot have siblings. |
| Invalidate | Invalidates the display area covered by a graphic object. |
| IsFilled : aBoolean | Returns true when a graphic is filled. |
| IsFilledBy (aClass) : aBoolean | Returns true when a graphic is filled by the objects of a specified class. |
| IsGroup : aBoolean | Returns true if the group is a graphic group. |
| IsHit (aRect) : aBoolean | Returns true if the specified rectangle intersects a graphic object. |
| IsMoveable : aBoolean | Returns true when a graphic object can be moved. |
| IsNowVisible : aBoolean | Returns true if a graphic needs to be drawn. |
| IsOwnerActive : aBoolean | Returns true when the document that holds a graphic is active. |
| IsSelected : aBoolean | Returns true if a graphic is selected. |

| | |
|---|---|
| IsTouchedBy (aRect) : aBoolean | Returns true if a graphic touches the specified rectangle. |
| IsVisible : aBoolean | Returns true when a graphic is visible. |
| Offset (aPoint) | Moves a graphic by the specified point. |
| Project (aPrj) | Projects a graphic object using the specified projection. |
| ReturnSymbols : symbolsList | Returns a list of symbols used in a graphic object. |
| ReturnUniqueName : aString | Generate a unique name for a graphic. |
| Select | Selects a graphic object. |
| SelectToEdit | Selects a graphic object to edit. |
| SetBounds (aRect) | Sets the graphic bounds to the specified rectangle. |
| SetDisplay (aDisplay) | Associates a graphic to the specified display. |
| SetExtent (aPoint) | Sets the extent of a graphic object. |
| SetFillObject (anObj) | Fills a graphic with the specified object. |
| SetObjectTag (anObj) | Sets the object tag of a graphic. |
| SetOrigin (aPoint) | Sets the origin point of a graphic object. |
| SetSelected (aBoolean) | Sets the selected state of a graphic object to the specified Boolean value. |
| SetSymbol (aSymbol) | Sets the symbol of a graphic object. |
| SetSymbols (aSymbolList) | Sets the symbols of a graphic object. |
| SetVisible (aBoolean) | Sets the visibility of a graphic object to the specified Boolean value. |
| UnProject (aPrj) | Unprojects a graphic object. |
| Unselect | Unselects a graphic object. |

**See also:** Display; GraphicList; Template

# *GraphicFlag (Network Analyst)*

This class models a location on a network.

**Inherits from** GraphicShape

| Instance Requests | |
|---|---|
| GetInvalidSymbol : aSymbol | Gets the symbol used for identifying invalid GraphicFlags. |
| GetValidSymbol : aSymbol | Gets the symbol used for identifying valid GraphicFlags. |

| | |
|---|---|
| SetInvalidSymbol (aSymbol) | Sets the symbol used for identifying invalid GraphicFlags. |
| SetShape (aPoint) | Sets the shape of a GraphicFlag to the given point. |
| SetSymbol (aSymbol) | Sets the symbol of a GraphicFlag to the given symbol. |
| SetSymbols (aSymbolList) | Sets the symbols of a GraphicFlag to the list of specified symbols. |
| SetValidSymbol (aSymbol) | Sets the symbol used for identifying valid GraphicFlags. |

**See also:** GraphicList, Network

# GraphicGroup

This class allows grouping graphic objects into a single graphic object.

**Inherits from** Graphic

**Inherited by** NorthArrow

| Class Requests | |
|---|---|
| Make : aGraphicGroup | Creates a graphic group object. |

| Instance Requests | |
|---|---|
| Add (aGraphic) | Adds the specified graphic object to a group. |
| CanProject : aBoolean | Returns true when all graphic objects in a group can be projected. |
| CanTransform : aBoolean | Returns true when a graphic group can be transformed. |
| Draw | Draws a graphic group. |
| Edit (documentGraphicsList) : aBoolean | Displays the graphic editor. |
| EditName : aBoolean | Displays the name editor. |
| Empty | Clears a graphic group. |
| GetGraphics : graphicsList | Gets the list of graphics in a group. |
| Insert (aGraphic) | Adds a graphic to the beginning of a group. |
| Invalidate | Invalidates the area of display covered by a graphic group. |
| IsGroup : aBoolean | Returns true if the object is a graphic group. |

| IsHit (aRect) : aBoolean | Returns true if the specified rectangle intersects a graphic group. |
|---|---|
| Offset (aPoint) | Moves the graphic group by the specified amount. |
| Project (aPrj) | Projects a graphic group using the specified projection. |
| ReturnSymbols : symbolsList | Returns a list of symbols used in a graphic group. |
| SetBounds (aRect) | Sets the bounds of a graphic group. |
| SetDisplay (aDisplay) | Associates a graphic group with the specified display. |
| SetSelected (aBoolean) | Sets the selection state of a group to the specified Boolean value. |
| SetSymbols (aSymbolList) | Sets the symbol list for a graphic group. |
| UnProject (aPrj) | Unprojects a graphic group. |

*See also:* Display; GraphicList; Template

# GraphicLabel

This class provides the ability to draw symbolized string objects.

**Inherits from** GraphicText

| Class Requests | |
|---|---|
| MakeWithSym (aString, aPoint, aSymbol, aGraphicText) : aGraphicLabel | Makes a graphic label at the specified point with the given string, symbol, and graphic text. |

| Instance Requests | |
|---|---|
| ChangeSets (anotherGraphicLabel) | Moves a graphic label to the set that given graphic label is in. |
| Clone : aGraphicLabel | Clones a graphic label. |
| GetSiblings : aList | Gets a list of other graphic labels that share the same text symbol. |
| HasSiblings : aBoolean | Returns true when there are other graphic labels that share the same text symbol. |
| SetBounds (aRect) | Sets the bounds of a graphic label. |

*See also:* GraphicList, GraphicText, TextPositioner

# GraphicList

Objects of this class contain graphic objects associated with a single display object.

**Inherits from** List

| Class Requests | |
| --- | --- |
| Make : GraphicList | Creates a graphic list. |

| Instance Requests | |
| --- | --- |
| Add (aGraphic) : aGraphicList | Adds the specified graphic object to the list and draws it. |
| AddBatch (aGraphic) | Adds the specified graphic to the list without drawing it. |
| AddName (aGraphic) | Adds the specified graphic to the list, draws it, and gives it a unique name. |
| AlignSelected | Aligns the selected graphics in a list. |
| AsList : aList | Returns a graphic list as a simple list. |
| BasicPaste (aClass) | Pastes objects of the specified class from the application clipboard to a graphic list. |
| CanRestoreGraphic : aBoolean | Returns true if there are changes that can be restored. |
| ClearSelected | Deletes the selected graphic objects of a graphic list. |
| ClearUndoStack | Clears the undo stack of a graphic list. |
| CopySelected | Copies the selected graphics to the application clipboard. |
| CutSelected | Cuts the selected graphic objects to the application clipboard. |
| Draw | Draws the graphic list. |
| Edit : aBoolean | Displays the graphic editor. |
| EditName : aBoolean | Displays the name editor. |
| EditSizeAndPos | Displays the size and position editor. |
| EditText : aBoolean | Displays the text editor. |
| Empty | Clears the graphic list. |
| EndBatch | Ends adding graphic objects in batch. |
| FillFrames (objectsList) | Fill the frame and graphic text objects with the objects of the specified list. |

| | |
|---|---|
| FindAllByClass (aClass) : aList | Returns a list of graphics from the specified class in a graphics list. |
| FindAllByLocation (aPoint) : aList | Returns a list of graphics at the specified location. |
| FindAllByName (aString) : aList | Returns a list of graphics with the specified name. |
| FindAllByObjectTag (anObj) : aList | Returns a list of graphics that have the specified object as respective object tags. |
| FindByClass (aClass) : aGraphic | Finds the first graphic object of the specified class. |
| FindByLocation (aPoint) : aGraphic | Finds the first graphic object at the specified point. |
| FindByName (aString) : aGraphic | Finds the first graphic object with the specified name. |
| FindByObjectTag (anObj) : aGraphic | Finds the first graphic object with the specified object tag. |
| GetDisplay : aDisplay | Gets the display object associated with a graphic list. |
| GetFrameObjs : aList | Gets a list of objects which fill the frame graphics. |
| GetMaxUndoTransactionSize : aNumber | Gets the maximum number of graphics that can be undone as part of one transaction. |
| GetSelected : aList | Gets the list of selected graphic objects. |
| GetUndoStackSize : aNumber | Gets the depth of undo stack. |
| GroupSelected | Groups the selected graphic objects. |
| HasSelected : aBoolean | Returns true when selected graphic objects exist. |
| Invalidate | Invalidates the display covered by a graphic list. |
| MoveSelectedToBack | Moves the selected graphic objects to the back. |
| MoveSelectedToFront | Moves the selected graphic objects to the front. |
| OffsetSelected (aPoint) | Moves the selected graphic objects by the specified value. |
| Paste | Pastes the graphic objects from the application clipboard to a graphic list. |
| Project (anOldPrj, aNewPrj) | Changes the projection of a graphic list. |
| RegisterSelected | Registers the symbols of the selected graphics with the symbol window. |
| RemoveGraphic (aGraphic) | Removes the graphic object from a graphic list. |
| RestoreGraphic | Performs the undo operation. |
| ReturnExtent : aRect | Returns the extent of a graphic list. |
| ReturnSelectedExtent : aRect | Returns the extent of the selected graphics in a list. |
| SaveGraphic (aGraphic, keepGraphicGroupBoolean) | Saves a graphic object in its group form or by its components. |
| SaveSelectedGraphics | Saves the changes to the selected graphic objects. |

| SelectAll | Selects all graphics in a graphic list. |
|---|---|
| SelectRect (aRect, addToSelectionBoolean) : aBoolean | Returns true if any graphic objects completely within the rectangle are selected. |
| SetDisplay (aDisplay) | Associates the list with a display object. |
| SetMaxUndoTransactionSize (aNumber) | Sets the maximum number of graphics that can be undone in one transaction. |
| SetRedrawDeferred (aBoolean) | Sets the defer redraw flag. |
| SetSelectedExtent (aRect) | Sets the extent of the selected graphics. |
| SetUndoStackSize (aNumber) | Sets the size of undo stack. |
| SimplifySelection | Reduces selected frames to graphic objects. |
| UngroupSelected | Ungroups the selected graphic objects. |
| UnselectAll | Unselects all graphic objects in a graphic list. |
| UnselectAllExcept (aGraphic) | Unselects all except the specified graphic object. |

*See also:* GraphicSet; GraphicShape; Palette

# GraphicSet

This class associates graphic elements to a theme.

**Inherits from** List

| Instance Requests | |
|---|---|
| CanConvertOverlappingLabels (anOverlapColor) : aBoolean | Returns true if a graphic set contains overlapping graphic labels. |
| ConvertOverlappingLabels (anOverlapColor) | Converts the overlapping labels to non-overlapping. |
| HasLabels : aBoolean | Returns true when a graphic set contains graphic labels or text. |
| HasOverlappingLabels (anOverlapColor) : aBoolean | Returns true when a graphic set contains overlapping labels. |
| Invalidate | Invalidates the display area covered by the graphics in a graphic set. |
| SelectLabels | Selects all graphic labels or text. |
| SelectOverlappingLabels (anOverlapColor) | Selects the overlapping labels. |
| SetSelected (aBoolean) | Sets the selection flag of the graphics in a set to the specified Boolean value. |
| SetVisible (aBoolean) | Sets the visibility of the graphics in a set to the specified Boolean value. |

*See also:* GraphicList; Theme

# *GraphicShape*

Objects of this class can draw shapes following a mathematical statement.

**Inherits from** Graphic

**Inherited by** GEdgeRec, GNodeRec, GraphicFlag, and GShapeRec

| Class Requests | |
|---|---|
| Make (aShape) : aGraphicShape | Creates a graphic shape from the specified shape object. |

| Instance Requests | |
|---|---|
| CanProject : aBoolean | Returns true if a graphic shape can be transformed. |
| CanTransform : aBoolean | Returns true if a graphic shape can be transformed. |
| Draw | Draws the shape in a graphic shape. |
| DrawHandles | Draws the selection handles of a graphic shape. |
| EditSizeAndPos : aBoolean | Displays the size and position editor. |
| GetOrigin : aPoint | Gets the lower left coordinates of a graphic shape. |
| GetShape : aShape | Gets the shape contained in a graphic shape. |
| IsHit (aRect) : aBoolean | Returns true if the specified rectangle intersects a graphic shape. |
| Offset (aPoint) | Moves a graphic shape by the specified amount. |
| Project (aPrj) | Projects a graphic shape. |
| SelectEdit | Selects a graphic shape for editing. |
| SetBounds (aRect) | Sets the bounds of a graphic shape to the specified rectangle. |
| SetDisplay (aDisplay) | Sets the display of a graphic shape. |
| SetOrigin (aPoint) | Sets the location of a graphic shape. |
| SetShape (aShape) | Sets the shape contained in a graphic shape. |
| UnProject (aPrj) | Unprojects a graphic shape. |

*See also:* Display; GraphicList; Shape

*Code example:* page 368

# GraphicText

Objects of this class can draw string objects.

**Inherits from** Graphic

**Inherited by** GraphicLabel

| Class Requests | |
| --- | --- |
| Make (aString, aPoint) : aGraphicText | Creates a graphic text from the specified string at the specified location. |

| Instance Requests | |
| --- | --- |
| CanEditText : aBoolean | Returns true if the text in a graphic text can be edited. |
| CanProject : aBoolean | Returns true if a graphic text can be projected. |
| ClearTransforms | Clears the transformation of a graphic text. |
| Draw | Draws a graphic text on its associated display. |
| Edit (aList) : aBoolean | Displays the graphic editor, and ignores the list. |
| EditSizeAndPos : aBoolean | Displays the size and position editor. |
| EditText : aBoolean | Displays the text editor. |
| GetAlignment : TextComposerJustEnum | Gets the text justification. |
| GetAngle : aNumber | Gets the rotation angle of a graphic text. |
| GetFillObject : anObj | Gets the fill object of a graphic text. |
| GetSpacing : aNumber | Gets the line spacing in points. |
| GetText : aString | Gets the string contained in a graphic text. |
| Invalidate | Invalidates the display area covered by a graphic text. |
| IsFilled : aBoolean | Returns true if an object has filled a graphic text. |
| IsFilledBy (aClass) : aBoolean | Returns true when objects of the specified class can fill a graphic text. |
| Project (aPrj) | Projects a graphic text. |
| ReturnSymbols : aList | Returns a list of symbols in a graphic text. |
| SetAlignment (TextComposerJustEnum) | Sets the text justification in a graphic text. |
| SetAngle (aNumber) | Sets the rotation angle to the specified number. |
| SetBounds (aRect) | Sets the bounds to the specified rectangle. |
| SetDisplay (aDisplay) | Sets the display of a graphic text. |

| | |
|---|---|
| SetFillObject (anObj) | Sets the fill object of a graphic text to the specified object. |
| SetSpacing (aNumber) | Sets the line spacing to the specified number of points. |
| SetSymbols (aSymbolList) | Sets the symbols of a graphic list. |
| SetText (aString) | Sets the string contained in a graphic text. |
| UnProject (aPrj) | Unprojects a graphic text. |

## Enumerations

| *TextComposerJustEnum* | |
|---|---|
| #TEXTCOMPOSER_JUST_CENTER | Center justification for graphic text. |
| #TEXTCOMPOSER_JUST_LEFT | Left justification for graphic text. |
| #TEXTCOMPOSER_JUST_RIGHT | Right justification for graphic text. |

**See also:** Display; GraphicList; Layout

**Code example:** page 369

# Grid (Spatial Analyst)

This class models the tiled raster ARC/INFO GRID data.

**Inherits from** Obj

| *Class Requests* | |
|---|---|
| CanManageDataSet : aBoolean | Returns true if the class supports data set management. |
| CopyDataSet (sourceGridFileName, destinationGridFileName) : aBoolean | Copies the source grid data set to the specified destination and returns true if successful. |
| DeleteDataSet (aGridFileName) : aBoolean | Deletes the specified grid data set and returns true if successful. |
| GetAnalysisCellSize (aCellSize) : GridEnvTypeEnum | Gets the current analysis environment cell size for Spatial Analyst. |
| GetAnalysisExtent (aRect) : GridEnvTypeEnum | Gets the current analysis environment extent for Spatial Analyst. |
| GetAnalysisMask : aGrid | Gets the current analysis environment mask for Spatial Analyst. |

| | |
|---|---|
| GetDataSetName : aString | Returns the data set name of Grid class. |
| GetNoDataTransparency : aNumber | Gets the percentage value for the number of no data cells in a grid. |
| GetVerify : GridVerifyEnum | Returns the type of verification used when an existing grid data set is overwritten. |
| IsValidDataSetFileName (aFileName) : aBoolean | Returns true if the given filename is a valid grid data set. |
| Make (aSrcName) : aGrid | Creates a grid from the specified source name. |
| MakeByInterpolation (pointFTab, aPrj, aField, anInterp, gridSize) : aGrid | Makes a grid by interpolating points using the specified interpolation method. |
| MakeDensitySurface (pointFTab, aPrj, aField, densitySpecList, gridSize) : aGrid | Makes a grid from the FTab points using the density function. |
| MakeFromASCII (aFileName, useFloatBoolean) : aGrid | Creates a floating or integer grid from the given filename. |
| MakeFromDEM (aFileName) : aGrid | Makes a grid from a USGS DEM file. |
| MakeFromDTED (aFileName) : aGrid | Makes a grid from a DTED file. |
| MakeFromFloat (aFileName) : aGrid | Creates a Grid from a binary floating point formatted raster file. |
| MakeFromFTab (anFTab, aPrj, aField, gridSize) : aGrid | Makes a grid by rasterizing selected or all features in anFTab. |
| MakeFromImage (anImageFileName, aBandNumber) : aGridOrGridList | Converts the specified band or all bands of an image file to a grids. |
| MakeFromNumb (aNumber) : aGrid | Creates a grid of uniform values. |
| MakeFromPointStats (pointFTab, aPrj, aField, statsSpecList, gridSize) : aGrid | Makes a Grid by calculating neighborhood statistics on the points in the given FTab. |
| MakeNormal : aGrid | Makes a Grid from normally distributed random numbers. |
| MakeRandom : aGrid | Makes a Grid from uniformly distributed random numbers between 0 and 1. |
| MakeSrcName (aString) : aSrcName | Creates a grid source name from the specified string. |
| RenameDataSet (sourceGridFileName, destinationGridFileName) : aBoolean | Renames a grid data set and returns true if successful. |
| Reset | Resets the Spatial Analyst extension's analysis environment to its default setting. |
| SetAnalysisCellSize (GridEnvTypeEnum, aCellSize) | Sets the Spatial Analyst's cell size. |
| SetAnalysisExtent (GridEnvTypeEnum, aRect) | Sets the Spatial Analyst's extent. |

| | |
|---|---|
| SetAnalysisMask (maskGrid) | Sets the Spatial Analyst's mask. |
| SetNoDataTransparency (maxPercentNoDataNumber) | Sets the percentage of no-data cells needed in a Grid to display the null symbol in the default legends of grid themes transparently. |
| SetVerify (GridVerifyEnum) | Sets the type of verification to use when an existing grid is overwritten. |

## Instance Requests

| | |
|---|---|
| ! anotherGrid : aGrid | Returns a new grid from the bitwise XOR operation between two grids. |
| + anotherGrid : aGrid | Adds cells of two grids. |
| * anotherGrid : aGrid | Multiplies cells of two grids. |
| ^ anotherGrid : aGrid | Raises cells of a grid to the power of cells from another grid. |
| - anotherGrid : aGrid | Subtracts cells of two grids. |
| / anotherGrid : aGrid | Divides cells of two grids. |
| < anotherGrid : aGrid | Returns a new grid by comparing two grids. |
| << anotherGrid : aGrid | Performs bitwise left-shift on cells of a grid by the amount specified in another grid. |
| <= anotherGrid : aGrid | Returns a new grid by comparing two grids. |
| <> anotherGrid : aGrid | Returns a new grid by comparing two grids. |
| = anotherGrid : aGrid | Returns a new grid by comparing two grids. |
| > anotherGrid : aGrid | Returns a new grid by comparing two grids. |
| >= anotherGrid : aGrid | Returns a new grid by comparing two grids. |
| >> anotherGrid : aGrid | Performs bitwise right-shift on cells of a grid by the amount specified in another grid. |
| % anotherGrid : aGrid | Returns a grid based on the integer remainder produced by dividing cells of a grid by cells of another grid. |
| & anotherGrid : aGrid | Returns a new grid from the bitwise AND operation between cells of two grids. |
| \| anotherGrid : aGrid | Returns a new grid from the bitwise OR operation between cells of two grids. |
| ~ : aGrid | Returns a new grid that is bitwise complement of cells in a grid. |
| Abs : aGrid | Returns a new grid that is the absolute value of cells in a grid. |

| | |
|---|---|
| ACos : aGrid | Returns a new grid by calculating the inverse cosine of cells in a grid. |
| ACosH : aGrid | Returns a new grid by calculating the inverse hyperbolic cosine of cells in a grid. |
| Aggregate (aCellFactorNumber, GridStaTypeEnum, noExpandBoolean, noDataBoolean) : aGrid | Generates a reduced-resolution copy of a grid by multiplying the cell size by the given cell factor. |
| And anotherGrid : aGrid | Returns a new grid from the bitwise AND operation between cells of two grids. |
| ASin : aGrid | Returns a new grid by calculating the inverse sine of cells in a grid. |
| ASinH : aGrid | Returns a new grid by calculating the inverse hyperbolic sine of cells in a grid. |
| Aspect : aGrid | Calculates surface aspect for cells of a grid. |
| AsPolygonFTab (aFileName, performWeedingBoolean, aPrj) : anFTab | Converts a grid in the given projection to a polygon feature table. |
| ATan : aGrid | Returns a new grid by calculating the inverse tangent of cells in a grid. |
| ATan2 (anotherGrid) : aGrid | Calculates the inverse tangent based on division of cells in a grid by the given grid. |
| ATanH : aGrid | Returns a new grid by calculating the inverse hyperbolic tangent of cells in a grid. |
| BlockStats (GridStaTypeEnum, aNbrHood, noDataBoolean) : aGid | Calculates the requested operation for the given neighborhood. |
| BoundaryClean (GridSortTypeEnum, onceOnlyBoolean) : aGrid | Smoothes the boundary between zones in a grid by expanding and shrinking. |
| BuildSTA : aBoolean | Builds grid statistics and returns true if successful. |
| BuildVAT : aBoolean | Builds value attribute table for an integer grid and returns true if successful. |
| Cell : aGrid | Converts the cell values of a grid to their next highest whole number. |
| CellValue (aPoint, aPrj) : aNumber | Returns the value of a cell in a grid that contains the specified point. |
| CellValueList (aPointList, aPrj) : aNumberList | Returns the list of values for cells in a grid that contain the specified points. |
| Combine (aGridList) : aGrid | Returns a grid of unique values by combining the given grids with a grid. |
| Con (yesGrid, noGrid) : aGrid | Returns the value in yesGrid if a grid cell is non-zero otherwise it returns the value found in noGrid. |

| | |
|---|---|
| Contour (aFileName, anIntervalNumber, baseContourNumber, aPrj) : anFTab | Generates a feature table contour based on the given parameters, and saves the feature table in the specified filename. |
| ContourList (aFileNname, aNumberList, aPrj) : anFTab | Generates a feature table contour based on the list of specified intervals, and saves the feature table in the specified filename. |
| Cos : aGrid | Returns a new grid by calculating the cosine of cells in a grid. |
| CosH : aGrid | Returns a new grid by calculating the hyperbolic cosine of cells in a grid. |
| CostDistance (costGrid, directionFileName, allocationFileName, maxDistance) : aGrid | Calculates the least-accumulative-cost distance over a cost surface for each cell of a grid. |
| CostPath (distanceGrid, directionGrid, byZoneBoolean) : aGrid | Calculates the shortest cost paths for a grid. |
| Curvature (profileCurvFileName, planCurvFileName, slopeFileName, aspectFileName) : aGrid | Calculates the curvature for each cell of a grid. |
| EqualTo (aGridList) : aGrid | Returns the number of cells in the specified list of grids that are equal to cells of a grid. |
| EucAllocation (distanceFileName, directionFileName, maxDistance) : aGrid | Calculates each cell's Euclidean allocation to the closest cell in a grid, and optionally produces distance and direction grid data sets. |
| EucDistance (directionFileName, allocationFileName, maxDistance) : aGrid | Calculates each cell's Euclidean distance to the closest cell in a grid, and optionally produces distance and direction grid data sets. |
| Exp : aGrid | Returns the base e exponential for cells of aGrid. |
| Exp10 : aGrid | Returns the base 10 exponential for cells of aGrid. |
| Exp2 : aGrid | Returns the base 2 exponential for cells of aGrid. |
| Expand (numberOfCells, aZoneValueList) : aGrid | Expands the specified zones of a grid by the given number of cells. |
| ExtractByAttributes (aVTabQueryString) : aGrid | Extracts cells from a grid using a Boolean VTab query string, a cell retains its value if the query returns true the cell is given the value of no data. |
| ExtractByCircle (aCircle, aPrj, selectOutsideBoolean) : aGrid | Extracts cells from a grid using the area inside or outside of a circle. |
| ExtractByMask (maskGrid) : aGrid | Extracts cells from a grid using a mask grid. |
| ExtractByPoints (aMultiPoint, aPrj, selectNotContainsBoolean) : aGrid | Extracts cells that contain or do not contain points of the specified multi-point object. |
| ExtractByPolygon (aPolygon, aPrj, selectOutsideBoolean) : aGrid | Extracts cells from a grid using the area inside or outside of a polygon. |

| | |
|---|---|
| ExtractByRect (aRect, aPrj, selectOutsideBoolean) : aGrid | Extracts cells from a grid using the area inside or outside of a rectangle. |
| ExtractSelection : aGrid | Extracts selected cells of a grid, non-selected cells are give no-data value. |
| Flip : aGrid | Flips a grid in the direction of Y axis. |
| Float : aGrid | Converts a grid to floating values. |
| Floor : aGrid | Converts the cell values to the greatest integer less than or equal to the cell. |
| FlowAccumulation (weightGrid) : aGrid | Calculates the flow accumulation for each cell of a grid. |
| FlowDirection (forceEdgeBoolean) : aGrid | Calculates the flow direction of each cell in a grid. |
| FlowLength (weightGrid, upStreamBoolean) : aGrid | Calculates the accumulated flow length to each cell up stream to the top of the drainage divide or down stream to a sink or outlet. |
| FMod (anotherGrid) : aGrid | Calculates each cell's floating point remainder after dividing a grid by another. |
| FocalFlow (aThresholdNumber) : aGrid | Returns the value of each cell to flow to its immediate neighborhood. |
| FocalStats (GridStaTypeEnum, aNbrHood, noDataBoolean) : aGrid | Calculates the requested statistics for cells' neighbor hood. |
| GetCellSize : aNumber | Gets the cell size of a grid in map units. |
| GetExtent : aRect | Gets the extent of a grid in map units. |
| GetNumRowsAndCols : aNumberList | Returns the number of rows and columns in a grid. |
| GetSrcName : aSrcName | Gets a grid's source name. |
| GetStatistics : aNumberList | Returns a set of pre-defined statistics for a grid. |
| GetVTab : aVTab | Gets a grid's VTab. |
| GridsGreaterThan (aGridList) : aGrid | Returns the number of cells in the specified list of grids that are greater than cells of a grid. |
| GridsLessThan (aGridList) : aGrid | Returns the number of cells in the specified list of grids that are less than cells of a grid. |
| HasError : aBoolean | Returns true when grid objects are not created properly. |
| HasTempDataSet : aBoolean | Returns true if a grid's data set is temporary. |
| HillShade (lightSourceAzimuth, anAltitude, zFactor) : aGrid | Generates a hill shaded Grid. |
| Int : aGrid | Truncates cell values. |
| IsInteger : aBoolean | Returns true for integer grids. |
| IsNull : aGrid | Returns a grid of ones and zeros, cells with no data are set to one. |

| | |
|---|---|
| LocalStats (GridStaTypeEnum, aGridList) : aGrid | Calculates the requested statistic for a grid. |
| Log : aGrid | Calculates the natural logarithm for each cell of a grid. |
| Log10 : aGrid | Calculates the base 10 logarithm for each cell of a grid. |
| Log2 : aGrid | Calculates the base 2 logarithm for each cell of a grid. |
| Lookup (aFieldName) : aGrid | Returns a new Grid based on a new value from the given field. |
| MajorityFilter (allNeighborsBoolean, halfOkBoolean) : aGrid | Returns a grid with cell values based on the majority value of its neighborhood. |
| Merge (aGridList) : aGrid | Merges grids. |
| Mirror : aGrid | Mirrors a grid in the direction of its X axis. |
| Mosaic (aGridList) : aGrid | Creates a smooth transition over the overlapping areas. |
| Negate : aGrid | Negates the value of each cell in a grid. |
| Nibble (maskGrid, dataOnlyTrue) : aGrid | Sets the cells of a grid that are no-data in the maskGrid with the values of their nearest neighbors. |
| Not : aGrid | Returns the logical complement of cells in a grid. |
| Or anotherGrid : aGrid | Returns the logical OR between two grids. |
| Pick (aGridList) : aGrid | Returns the cell value from the given grids whose position is equal to the value of a grid. |
| PointValue (aPoint, aPrj) : aNumber | Returns an interpolated value at the given point on the surface of a grid. |
| PointValueList (aPointList, aPrj) : aNumberList | Returns an interpolated value for each given point on the surface of a grid. |
| Pow (anotherGrid) : aGrid | Raises cells of a grid to the power of cells from another grid. |
| QueryContour (aPoint, zValue, anArc, aPrj) : aBoolean | Returns true if a contour exists at the given point. |
| Reclass (aVTab, fromField, toField, outField, noDataBoolean) : aGrid | Reclassifies the value of each cell that falls between values of from and to fields to the value of out field. |
| ReclassByClassList (aFieldName, aClassificationList, noDataBoolean) : aGrid | Reclassifies the values of cells in aGrid using the specified classifications. |
| RegionGroup (allNeighboorsBoolean, crossClassBoolean, excludedValue) : aGrid | Groups each cell of a grid into a connected region. |
| Rename (aNewFileNmae) | Renames a grid's data set. |

| | |
|---|---|
| Resample (aCellSize, GridResTypeEnum) : aGrid | Resamples a grid using the given cell size and technique. |
| ReturnContour (aPoint, zValue, aPrj) : aPolyLine | Returns the contour crossing at the specified point. |
| ReturnCostPath (directionGrid, aPoint) : aPolyLine | Returns the cost path from the given point to the nearest source cell. |
| Rotate (anAngle, GridResTypeEnum) : aGrid | Rotates and resamples a grid by the specified angle in a clockwise direction around the lower-left corner. |
| Sample (aGridList) : aVTab | Samples a list of grids using only the cells with values other than no-data in a grid. |
| SaveAsASCII (aFileName) | Saves a grid as a raster text file. |
| SaveAsFloat (aFileName) | Saves a grid as a raster binary file. |
| SaveDataSet (aNewFileName) : aBoolean | Saves a grid's data set to a new file. |
| SetNull (anotherGrid) : aGrid | Sets a cell to no-date value if a grid is non-zero, otherwise it sets the value found in another grid. |
| Shift (newOriginPoint, aCellSize) : aGrid | Shifts a grid to a new origin with a new cell size. |
| Shrink (numberCells, aZoneValueList) : aGrid | Shrinks the specified zones of a grid by specified number of cells. |
| Sin : aGrid | Calculates the sine for each cell value as radians. |
| SinH : aGrid | Calculates the hyperbolic sine for each cell value as radians. |
| Sink : aGrid | Calculates all areas of internal drainage. |
| Slice (GridSliceTypeEnum, numberOfZones, baseZoneNumber) : aGrid | Slices the values of a grid into number of equal zones starting from the base zone. |
| Slope (zFactor, percentRise) : aGrid | Calculates the slope for each cell of a grid. |
| SnapPourPoint (weightGrid, snapDistance) : aGrid | Snaps pour points to new locations. |
| Sqr : aGrid | Calculates the square of each cell. |
| Sqrt : aGrid | Calculates the square root of each cell. |
| StreamLink (flowDirectionGrid) : aGrid | Assigns unique values to sections of a raster linear network grid. |
| StreamOrder (flowDirectionGrid, useShreveMethodBoolean) : aGrid | Assigns a numeric order to segments of a grid. |
| TabulateArea (columnField, rowObj, rowPrj, rowField, switchColRowBoolean) : aVTab | Computes cross tabulated areas between two data sets. |
| Tan : aGrid | Calculates tangent for each cell value as radians. |
| TanH : aGrid | Calculates hyperbolic tangent for each cell value as radians. |

| | |
|---|---|
| Test (aVTabQueryString) : aGrid | Returns a new grid based on the query string. |
| Thin (notBinaryBoolean, doFilterBoolean, sharpenCornersBoolean, maxThickness) : aGrid | Thins raster linear features in a grid. |
| Visibility (anFTab, aPrj, cellObservedBoolean) : aGrid | Determines the visibility of each cell. |
| Warp (linkLineList, anOrderNumber, GridResTypeEnum, aCellSize) : aGrid | Applies a polynomial transformation of given order to a grid using specified links. |
| Watershed (sourceGrid) : aGrid | Determines the contributing area above each set of cells. |
| XOr anotherGrid : aGrid | Retruns the logical XOR of two grids. |
| ZonalFill (weightGrid) : aGrid | Fills cells with the same value in a Grid with the minimum value along their boundary found in the given weightGrid. |
| ZonalGeometry (GridGeomDescEnum) : aGrid | Calculates the requested geometric descriptor for the cells having the same value. |
| ZonalGeometryTable (outFileName) : aVTab | Creates a Vtab from the requested geometric descriptor for the cells having the same value. |
| ZonalStats (GridStaTypeEnum, zoneGridOrFTab, zonePrj, zoneField, noDataBoolean) : aGrid | Calculates the requested statistics. |
| ZonalStatsTable (zoneGridOrFTab, zonePrj, zonefield, noDataBoolean, outFileName) : aVTab | Creates a Vtab based on the requested statistics. |

## Enumerations

| GridDensTypeEnum | |
|---|---|
| #GRID_DENSTYPE_SIMPLE | Simple density. |
| #GRID_DENSTYPE_KERNEL | Kernel density. |

| GridEnvTypeEnum | |
|---|---|
| #GRID_ENVTYPE_MINOF | Sets to the minimum of all input grids. |
| #GRID_ENVTYPE_MAXOF | Sets to the maximum of all input Grids. |
| #GRID_ENVTYPE_VALUE | Sets to a given value. |

| GridGeomDescEnum | |
|---|---|
| #GRID_GEOMDESC_AREA | Area of a zone. |
| #GRID_GEOMDESC_PERIMETER | Perimeter of a zone. |
| #GRID_GEOMDESC_THICKNESS | Thickness of a zone. |
| #GRID_GEOMDESC_CENTROID | Characteristics of an ellipse that is fit to a zone. |

| GridResTypeEnum | |
|---|---|
| #GRID_RESTYPE_NEAREST | Nearest neighbor. |
| #GRID_RESTYPE_BILINEAR | Bilinear interpolation. |
| #GRID_RESTYPE_CUBIC | Cubic convolution. |

| GridSliceTypeEnum | |
|---|---|
| #GRID_SLICETYPE_EQINTERVAL | Equal interval ranges. |
| #GRID_SLICETYPE_EQAREA | Break into ranges of equal area. |

| GridSortTypeEnum | |
|---|---|
| #GRID_SORTTYPE_NOSORT | No sorting by size. |
| #GRID_SORTTYPE_DESCEND | Sorts zones in descending order by size. |
| #GRID_SORTTYPE_ASCEND | Sort zones in ascending order by size. |

| GridStaTypeEnum | |
|---|---|
| #GRID_STATYPE_MAJORITY | Majority. |
| #GRID_STATYPE_MAX | Maximum. |
| #GRID_STATYPE_MEAN | Mean. |
| #GRID_STATYPE_MEDIAN | Median. |
| #GRID_STATYPE_MIN | Minimum. |
| #GRID_STATYPE_MINORITY | Minority. |
| #GRID_STATYPE_RANGE | Range. |
| #GRID_STATYPE_STD | Standard deviation. |
| #GRID_STATYPE_SUM | Sum. |
| #GRID_STATYPE_VARIETY | Variety or number of unique occurrences. |

| GridVerifyEnum | |
|---|---|
| #GRID_VERIFY_ON | Asks for verification before overwritting the current data set. |

| #GRID_VERIFY_OFF | Always overwrite the existing grid data set. |
| --- | --- |
| #GRID_VERIFY_ERROR | Gives an error message and does not overwrite the existing grid data set. |

***See also:*** Gtheme; NbrHood; SVGram

# GridLegendExtension (Spatial Analyst)

This class models the storage of options available for grid theme legends in the legend editor.

**Inherits from** LegendExtension

| *Instance Requests* | |
| --- | --- |
| GetBrightnessGrid : aGrid | Gets the assigned brightness grid. |
| GetBrightnessThemeName : aString | Gets the theme name of the brightness grid. |
| GetBrightnessThemeView : aView | Gets the view with the brightness theme. |
| GetMaxBrightness : aNumber | Gets the maximum brightness value. |
| GetMinBrightness : aNumber | Gets the minimum brightness value. |
| OverrideAdvancedDialog : aBoolean | Always returns true. |
| QueryBrightnessRange (minBrightnessNumber, maxBrightnessNumber) | Queries the given range of brightness values. |
| SetBrightnessRange (minBrightnessNumber, maxBrightnessNumber) | Sets the range of brightness to the specified values. |
| SetBrightnessThemeName (aGThemeName) | Sets the name of the brightness theme. |
| SetBrightnessThemeView (aView) | Places the brightness theme in the given view. |
| ShowAdvancedDialog (theOwningGridTheme) | Displays the grid theme's legend editor. |

***See also:*** Grid; Gtheme; Legend

# GridLegendWindow (Spatial Analyst)

This class represents the dialog box for defining the advanced options of grid theme legends.

**Inherits from** ModalDialog

| Class Requests | |
|---|---|
| Show (aGridLegendExtension, aGTheme) | Displays the modal dialog box for editing legend options of a grid theme. |

*See also:* Grid; Legend

# GShapeRec

This class represents a proxy record for a shape stored in a feature table.

**Inherits from** GraphicShape

# GTheme (Spatial Analyst)

This class models the presentation of cell based raster data from an ARC/INFO grid data source.

**Inherits from** Theme

| Class Requests | |
|---|---|
| GetQueryColor : aColor | Gets the color that is used to display the result of MakeQueryMap request. |
| Make (aGrid) : aGTheme | Creates a grid theme from a grid object. |
| MakeCalculationMap (aView, aName) | Makes a grid theme from a user defined expression based on the grid map layers in the specified view. |
| MakeQueryMap (aView, aName) | Makes a grid theme from a user defined query based on the grid map layers in the specified view. |
| SetQueryColor (aColor) | Sets the color that is used to display the result of MakeQueryMap request. |

| Instance Requests | |
|---|---|
| BuildQuery | Opens the query builder dialog box for a grid theme. |
| CanEdit : aBoolean | Returns true if a grid theme can be edited. |

| | |
|---|---|
| CanExportToFTab : aBoolean | Returns true if a grid theme can be exported to a polygon feature table. |
| CanLabel : aBoolean | Always returns false. |
| CanSelect : aBoolean | Always returns false. |
| ClearSelection | Clears the cells selected through query. |
| Clone : aGTheme | Clones a grid theme. |
| EditExpression | Displays the map query or map calculator dialog box. |
| EditLegend | Displays the legend editor for a grid theme. |
| EditTable | Opens the VTab of a grid theme. |
| ExportToFTab (aFileName) : anFTab | Exports a grid to a polygon feature table. |
| FindByPoint (aPoint) : aList | Returns the key to the cells containing the given point. |
| GetGrid : aGrid | Gets the grid object of a grid theme. |
| GetVTab : aVTab | Gets the VTab object of a grid theme. |
| HasExpression : aBoolean | Returns true if a grid theme is created from a query or calculation. |
| HasTable : aBoolean | Returns true if a grid theme has a VTab. |
| Histogram (useLegendBoolean, aField, useGraphicsBoolean) : aChart | Creates a histogram chart from the given field of a grid theme. |
| Identify (aCellKeyValue, aTitleString) | Displays the identify window for the specified cell in a grid theme. |
| QueryValStr (aFieldName, aCellKeyValue, aString) : aBoolean | Queries the specified field of a grid theme. |
| ReturnCellValue (aPoint) : aNumber | Returns the value of a cell containing the specified point. |
| ReturnExtent : aRect | Returns the extent of a grid theme. |
| SetName (aName) | Sets the name of a grid theme. |
| UpdateLegend | Applies legend changes to a grid theme's legend. |
| ZoneHistogram (useLegendBoolean, aField, zoneTheme, zoneField) : aChart | Creates histograms for each zone value of a grid theme. |

**See also:** Grid; Legend; SrcName; View; VTab

# Hammer

This class implements the Hammer-Aitoff projection.

**Inherits from** Prj

| Class Requests | |
|---|---|
| Make (aRect) : aHammer | Makes a Creates a Hammer-Aitoff projection bounded by the specified rectangle. |
| Instance Requests | |
| ProjectPt (aPoint) : aBoolean | Projects the given point and returns true if the projected point is visible. |
| Recalculate | Applies the changes to the projection parameters. |
| ReturnCentralMeridian : aNumber | Returns the central meridian as a longitude in degrees. |
| SetCentralMeridian (aNumber) | Sets the central meridian using the given longitude in degrees. |
| UnProjectPt (aPoint) : aBoolean | Unprojects the given point and returns true if successful. |

***See also:*** CoordSys; Spheroid

# Help

This class implements ArcView's on-line help system.

**Inherits from** Obj

| Class Requests | |
|---|---|
| Make : aHelp | Makes a help object. |

| Instance Requests | |
|---|---|
| Close | Closes the help window. |
| Context | Displays context sensitive help based on the active window. |
| ExecuteMacro (macroString) | Executes the given WinHelp macro. |
| GetAVFile : aFileName | Gets ArcView's default help file. |
| GetClassesFile : aFileName | Gets ArcView's Avenue classes help file. |

| | |
|---|---|
| GetCurrentTopic : aString | Gets the topic name currently displayed. |
| GetFile : aFileName | Gets the associated help file. |
| GetTitle : aString | Gets the title of a help object. |
| IsOpen : aBoolean | Returns true when help file window is opened. |
| Maximize | Maximizes the help window. |
| Minimize | Minimizes the help window. |
| Move (deltaXNumber, delatYNumber) | Moves a help window by the specified amount. |
| MoveTo (XNumber, YNumber) | Moves the help window to a new specified location. |
| PrintTopic (aTopicName) | Prints the specified topic name from a help object. |
| RegisterExtension (helpModuleName) | Registers the extension help resource module for context-sensitive help with ArcView's help system. |
| Resize (Width, Height) | Resizes a help window to the specified pixel values. |
| Restore | Restores a help system. |
| ReturnExtent : aPoint | Returns a help window's extent in pixels. |
| ReturnOrigin : aPoint | Returns the lower left corner of the help window. |
| ReturnText (aTopicName) : aString | Returns the text of the specified topic name. |
| Search | Displays the search window. |
| SearchUsing (aSearchString) | Displays the search window with the specified string as the search default. |
| SetAVFile | Sets a help object as the default ArcView help file. |
| SetClassesFile | Sets a help object as the default help file for Avenue classes. |
| SetFile (FileName) | Sets the current help object to the specified file name. |
| SetHelping (stateBoolean) | Sets the state of a help file. |
| SetTitle (aTitle) | Sets the title of a help window. |
| Show | Displays the help object. |
| ShowTopic (aTopicName) | Displays the help text associated with the specified topic. |
| UnregisterExtension (helpModuleName) | Unregisters the specified module from ArcView's help system. |
| WriteAll (aFileName, aTopicName) | Writes the topic name of a help file to the specified disk file. |

***See also:*** Application

# Icon

Icons are the graphics on the faces of buttons (20 by 17 pixels) and tools (18 by 15 pixels).

**Inherits from** Obj

| Class Requests | |
|---|---|
| Empty : anIcon | Makes a blank icon. |
| IsFileValid (aFileName) : aBoolean | Returns true if the specified file contains valid graphics for an icon. |
| Make (aFileName) : anIcon | Creates an icon from the specified graphics file. |
| MakeFromMacClipboard : anIcon | Creates an icon from the icon family resource stored on the Macintosh clipboard. |
| MakeFromMacResFile (aFileName, resourceIDNumber) : anIcon | Creates an icon from the specified icon Macintosh family resource. |
| MakeFromResName (aResName) : anIcon | Creates an icon from the specified Open Interface icon resource. |

| Instance Requests | |
|---|---|
| = (anotherIcon) : aBoolean | Returns true when two icons are the same. |
| GetDepth : aNumber | Gets the number of bits per pixel. |
| GetHeight : aNumber | Gets the height of an icon in pixels. |
| GetWidth : aNumber | Gets the width of an icon in pixels. |
| IsMacIcon : aBoolean | Returns true for Macintosh icons. |
| IsOK : aBoolean | Returns true if an icon was created with no problems. |
| SetSize (widthNumber, heightNumber) | Sets the pixel height and width of an icon. |

*See also:* IconMgr

*Code example:* page 398

# IconMgr

This class manages the list of icons available in ArcView.

**Inherits from** ModalDialog

| Class Requests | |
| --- | --- |
| AddIcons (anIconList) | Adds the specified icons to the list of icons in ArcView. |
| GetIcons : aList | Gets the list of icons available in ArcView. |
| LoadIcon : anIcon | Displays a file window to select a file and create an icon. |
| Show (maxSize) : anIcon | Displays the icon manager dialog with icons that have a height or width under the maximum size. |
| ShowWithDefault (maxSize, defIcon) : anIcon | Displays the icon manager with the default icon selected. If the maximum size is zero, all icons are displayed. |

*See also:* Icon

*Code example:* page 294

# IdentifyWin

This class displays records from a table.

**Inherits from** Window

| Class Requests | |
| --- | --- |
| Make : anIdentifyWin | Creates an identity window. |
| The : anIdentifyWin | Gets ArcView's identity window. |

| Instance Requests | |
| --- | --- |
| Add (aVTabOrATheme, aRecordKey, aName) | Adds the name to the identify window and associates it to the specified record of the specified table or theme. |
| Clear | Clears the contents of an identify window. |
| Close | Closes an identify window. |
| Open | Opens an identify window. |

*See also:* VTab

# IdentityLookup

This class implements image look-up tables.

**Inherits from** ImageLookup

| Class Requests | |
|---|---|
| Make : anIdentityLookup | Creates an identity look-up table. |

| Instance Requests | |
|---|---|
| Clone : anIdentityLookup | Clones an identity look-up table. |
| Lookup (aValue) : aNumber | Returns the output associated with the specified value in an image look-up table. |

*See also:* ImageLegend; ISrc

# ImageLegend

This class implements legends for image themes.

**Inherits from** Obj

**Inherited by** MultiBandLegend and SingleBandLegend

*See also:* ITheme

# ImageLookup

An image look-up maps pixel values to display colors.

**Inherits from** Obj

**Inherited by** IdentityLookup, IntervalLookup, and LinearLookup

| Instance Requests | |
|---|---|
| Lookup (aValue) : aNumber | Returns the output associated with the specified value in an image look-up table. |

*See also:* ImageLegend; ITheme

# ImageWin

An image window displays the contents of an image file.

**Inherits from** Window

| Class Requests | |
| --- | --- |
| Make (aFileName, aTitleString) : anImageWin | Displays the image contained in the specified file name. |

| Instance Requests | |
| --- | --- |
| GetFileName : aFileName | Gets the name of the file associated with an image window. |
| GetScaled : aBoolean | Returns true when an image window is scaled to fit the image. |
| Open | Shows an image window. |
| SetFileName (aFileName) | Sets the image file for an image window. |
| SetScaled (aBoolean) | If the specified Boolean value is true, then the window is scaled to fit the image. |

*See also:* MsgBox

# ImgCat

This class represents an image data source based on an image catalog.

**Inherits from** ImgSrc

| Class Requests | |
| --- | --- |
| IsValidFileName (aFileName) : aBoolean | Returns true if the specified filename contains image source names. |
| Make (aSrcName) : aImgCat | Creates an image catalog from the specified source name. |
| SetChecking (checkingIsOnBoolean) | Turns checking of source names on or off. |

| Instance Requests | |
| --- | --- |
| ContainType (aTypeNumber) : aBoolean | Returns true if an image catalog contains the given type. |

| | |
|---|---|
| GetNumBands : aNumber | Gets the number of bands in the lead image source of an image catalog. |
| GetNumImages : aNumber | Returns the number of images in an image catalog. |
| GetSrcName : aSrcName | Gets the source name for an image catalog. |
| HasError : aBoolean | Returns true if an image catalog has errors. |
| ReturnClipExtent : aRect | Returns the clip extent of an image catalog. |
| ReturnExtent : aRect | Returns the extent of an image catalog. |
| ReturnImageNames : aSrcNameList | Returns the source names of an image catalog. |
| SetClipExtent (aRect) | Sets the clip extent in an image catalog. |
| ReturnMapExtent : aRect | Returns the map extent of an image catalog. |

***See also:*** BandStatistics; ImageLegend; ImageLookup; ImgSrc; Isrc; ITheme

# ImgSrc

This is an abstract class for image data sources that can be drawn as an image theme.

**Inherits from** Obj

**Inherited by** ImgCat, ISrc

| *Class Requests* | |
|---|---|
| IsValidFileName (aFileName) : aBoolean | Returns true if the specified file can be an image source. |
| Make (aSrcName) : anImgSrc | Creates an image source from the given source name. |

| *Instance Requests* | |
|---|---|
| GetNumBands : aNumber | Returns the number of bands in an image source. |
| GetSrcName : aSrcName | Gets the source name of an image source. |
| HasError : aBoolean | Returns true if an image source has errors. |
| ReturnClipExtent : aRect | Returns the clip extent of an image source. |
| ReturnExtent : aRect | Returns the full extent of an image source. |
| ReturnMapExtent : aRect | Returns the map extent of an image source. |
| SetClipExtent (aRect) | Sets the clip extent of an image source to the specified rectangle. |

***See also:*** BandStatistics; ColorMap; Display; ImageLegend; ImageLookup; ImgCat; Isrc; ITheme

# INFODir

This class implements INFO directories.

**Inherits from** Obj

| Class Requests | |
|---|---|
| FileExists (aFileName) : aBoolean | Returns true for INFO files. |
| IsValid (aDirectory) : aBoolean | Returns true for INFO directories. |
| Make (aDirectory) : anINFODir | Creates an INFO directory. |

| Instance Requests | |
|---|---|
| ContainsFile (aString) : aBoolean | Returns true if a file with the specified name is found in an INFO directory. |
| GetFiles : aFileNameList | Returns a list of files in an INFO directory. |
| Is8Dot3 : aBoolean | Returns true for INFO version 8.3. |

*See also:* VTab

# Interp (Spatial Analyst)

The interpolator object models how a grid should be interpolated from a point feature table.

**Inherits from** Obj

| Class Requests | |
|---|---|
| Make : anInterp | Makes a default interpolator which is Inverse Distance Weighted with a power of 2, no barriers, and a variable Radius of 12 points with no maximum distance. |
| MakeFromVariogram (aSVGram, aRadius, aBarrierFTab, outVarGridFileName) : anInterp | Makes a Kriging interpolator from a semi-variogram. |
| MakeIDW (powerNumber,aRadius, aBarrierFTab) : anInterp | Makes an Inverse Distance Weighted interpolator. |
| MakeKriging (KrigingEnum,aRadius, aBarrierFTab, outVarGridFileName) : anInterp | Makes a Kriging interpolator. |
| MakeSpline (SplineEnum, weightNumber, numberOfPoints) : anInterp | Makes a Spline interpolator. |

| | |
|---|---|
| MakeTrend (anOrder, useLogisticBoolean) : anInterp | Makes a Trend Surface interpolator. |

## *Enumerations*

| KrigingEnum | |
|---|---|
| #KRIGING_SPHERICAL | Spherical. |
| #KRIGING_CIRCULAR | Circular. |
| #KRIGING_EXPONENTIAL | Exponential. |
| #KRIGING_GAUSSIAN | Gaussian. |
| #KRIGING_LINEAR | Linear with sill. |
| #KRIGING_UNIVERSAL1 | Universal Kriging with linear drift. |
| #KRIGING_UNIVERSAL2 | Universal Kriging with quadratic drift. |

| SplineEnum | |
|---|---|
| #SPLINE_TENSION | Spline with Tension. |
| #SPLINE_REGULARIZED | Regularized Spline. |

***See also:*** Grid; Radius

# InterpolationDialog (Spatial Analyst)

This class represents the interpolation dialog box.

**Inherits from** ModalDialog

| Class Requests | |
|---|---|
| Show (aPointFTheme, aCellSize) : aList | Displays the interpolation dialog box and returns the specified field and defined interpolator. |

***See also:*** Grid; Interp; Radius

# Interval

This class implements a finite numeric set with a finite number of subdivisions.

**Inherits from** List

**Inherited by** Duration

| Instance Requests | |
|---|---|
| = anObj : aBoolean | Returns true when the specified object has the same value as an interval. |
| AsString : aString | Returns a description of the interval. |
| by aNumber : anInterval | Sets the increment size to the specified number. |
| Count : aNumber | Returns the number of subdivisions in an interval. |
| Find (aNumber) : aBoolean | Returns true when the specified number is equal to a number inside an interval. |
| Get (subdivisionNumber) : aNumber | Gets the number associated with the specified subdivision number of an interval. |
| GetIncrement : aNumber | Gets the increment value of an interval. |
| GetLower : aNumber | Gets the lower bound of an interval. |
| GetUpper : aNumber | Gets the upper bound of an interval. |
| SetIncrement (aNumber) : anInterval | Sets the increment size to the specified number. |
| SetLower (aNumber) : anInterval | Sets the lower bound of an interval. |
| SetUpper (aNumber) : anInterval | Sets the upper bound of an interval. |

*See also:* Number

# *IntervalLookup*

This class maps pixel values to an interval.

**Inherits from** ImageLookup

| Class Requests | |
|---|---|
| Make (minNumber, maxNumber, numberOfIntervals) : anIntervalLookup | Creates an interval look-up table based on the specified values. |

| Instance Requests | |
|---|---|
| Clone : anIntervalLookup | Clones an interval look-up table. |
| GetNumIntervals : aNumber | Gets the number of subdivisions in an interval look-up table. |
| Lookup (aValue) : aNumber | Returns the output number associated with the specified value. |

*See also:* ImageLegend; ISrc

# ISrc

This class implements image data sources that can be drawn in an image theme.

**Inherits from** ImgSrc

| Class Requests | |
| --- | --- |
| IsValidFileName (aFileName) : aBoolean | Returns true when the specified file can be a source for an image theme. |
| Make (aSrcName) : anISrc | Creates an image data source. |
| ReturnExtensions : stringList | Returns a list of acceptable image file extensions. |
| ReturnTypeFromFileExt (anExtensionString) : aNumber | Returns a number for the type of graphic file. |
| SetExtensionChecking (aBoolean) | Checks the extension of the image source file when the Boolean value is set to true. |
| SetExtensions (stringList) | Sets the list of valid image source file extensions. |

| Instance Requests | |
| --- | --- |
| BuildBandStatistics | Builds band statistics for an image data source. |
| GetBandStatistics : aList | Gets the band statistics for an image data source. |
| GetNumBands : aNumber | Gets the number of bands in an image data source. |
| GetNumColumns : aNumber | Gets the number of columns in an image data source. |
| GetNumRows : aNumber | Gets the number of rows in an image data source. |
| GetSrcName : aSrcName | Gets the source name for an image data source. |
| HasError : aBoolean | Returns true when an image data source contains an error. |
| MatchPoints (fromPoint, toPoint) | Changes the map extent of an image data source. |
| MatchRectangles (fromRect, toRect) | Scales the map extent of an image data source. |
| ReturnClipExtent : aRect | Returns the clip extent of an image source. |
| ReturnExtent : aRect | Returns the extent of an image source. |
| ReturnMapExtent : aRect | Returns the extent of an image data source. |
| ReturnType : aNumber | Returns the type of an image source. |
| SetClipExtent (aRect) | Sets the clip extent of an image source to the specified rectangle. |

| | |
|---|---|
| SetMapExtent (aRect) | Sets the extent of an image data source to the specified rectangle. |

***See also:*** BandStatistics; ColorMap; Display; ImageLegend; ImageLookup

***Code example:*** page 484

# *ITheme*

This class implements image themes.

**Inherits from** Theme

| Class Requests | |
|---|---|
| Make (anImgISrc) : anITheme | Creates an image theme based on the specified image data source. |
| MakeFromSrc (anImgSrc) : anITheme | Creates an image theme from the specified image source. |

| Instance Requests | |
|---|---|
| EditLegend | Displays the legend editor. |
| GetImgSrc : anImgSrc | Gets the image data source of an image theme. |
| InvalidateLegend | Invalidates the legend of an image theme. |
| ReturnExtent : aRect | Returns the extent of an image theme. |
| SetLegend (anImageLegend) | Sets the legend of an image theme. |
| UpdateLegend | Applies the legend changes to an image theme. |

***See also:*** ImageLegend; ImgSrc

***Code example:*** page 323

# *LabelButton*

This class implements the push-button controls on the project window.

**Inherits from** Control

| Class Requests | |
|---|---|
| Make : LabelButton | Makes a blank label button. |

| Instance Requests | |
|---|---|
| Click | Executes the click event of a label button. |
| GetClick : aScriptName | Gets the script name associated with the click event. |
| GetLabel : aString | Gets the string displayed on a button. |
| HasScript (aScriptName) : aBoolean | Returns true when the specified script is associated with a label button. |
| SetClick (aScriptName) | Associates the given script name with the click event of a label button. |
| SetLabel (aLabel) | Sets the label of a label button. |

*See also:* Button; Choice; Project

*Code example:* page 399

# Labeler

The objects of Labeler class label the features of a theme based on their parameters.

**Inherits from** Obj

| Class Requests | |
|---|---|
| Make (aRect) : aLabeler | Create a labeler to label the area bounded by the given rectangle. |

| Instance Requests | |
|---|---|
| GetFeatureWeight : LabelWeightEnum | Gets the feature weight in labeling. |
| GetLabelWeight : LabelWeightEnum | Gets the label weight in labeling. |
| GetLinePlacement : aNumber | Gets the line placement mask. |
| Load (aTheme) : aBoolean | Loads the specified theme into a labeler and returns true if successful. |
| RemoveDuplicates (aBoolean) | If the Boolean value is set to true, it removes duplicate labels. |
| RemovingDuplicates : aBoolean | Returns true if a labeler is set to remove duplicates. |
| SetFeatureWeight (LabelWeightEnum) | Sets the feature weight. |
| SetLabelWeight (LabelWeightEnum) | Sets the label weight. |
| SetLinePlacement (aNumber) | Sets the line placement mask. |

## *Enumerations*

| LabelWeightEnum | |
|---|---|
| #LABEL_WEIGHT_NO | No weight. |
| #LABEL_WEIGHT_LOW | Low weight. |
| #LABEL_WEIGHT_MEDIUM | Medium weight. |
| #LABEL_WEIGHT_HIGH | High weight. |

**See also:** Ftheme; GraphicList; View

# Lambert

This class implements the Lambert Conformal Conic projection.

**Inherits from** Prj

| Class Requests | |
|---|---|
| CanDoSpheroid : aBoolean | Always returns true. |
| Make (aRect) : aLambert | Creates a projection bounded by the specified rectangle. |

| Instance Requests | |
|---|---|
| ProjectPt (aPoint) : aBoolean | Projects a point and returns true if successful. |
| Recalculate | Recalculates the derived constants. |
| ReturnCentralMeridian : aNumber | Returns the longitude degrees of the central meridian. |
| ReturnFalseEasting : aNumber | Returns the X value at the central meridian. |
| ReturnFalseNorthing : aNumber | Returns the Y value at the reference latitude. |
| ReturnLowerStandardParallel : aNumber | Returns lower latitude in degrees. |
| ReturnReferenceLatitude : aNumber | Returns degrees of latitude where Y is equal to the false northing. |
| ReturnUpperStandardParallel : aNumber | Returns upper latitude in degrees. |
| SetCentralMeridian (aLongitude) | Sets the central meridian to the specified degrees. |
| SetFalseEasting (aNumber) | Sets the X value at central meridian. |
| SetFalseNorthing (aNumber) | Sets the Y value at reference latitude. |
| SetLowerStandardParallel (aLatitude) | Sets the lower parallel to the specified degrees. |
| SetReferenceLatitude (aLatitude) | Sets the reference latitude at the specified degrees. |

| | |
|---|---|
| SetUpperStandardParallel (aLatitude) | Sets the upper parallel to the specified degrees. |
| UnProjectPt (aPoint) : aBoolean | Unprojects a point and returns true if successful. |

*See also:* CoordSys; Spheroid

# LandMark (Network Analyst)

Objects of this class are locations shown in directions.

**Inherits from** Obj

*See also:* Network

# Layer

This class implements an ARC/INFO or ArcStorm library layer.

**Inherits from** Obj

| Class Requests | |
|---|---|
| ReturnSrcNames (aLayerName) : aList | Returns a list of source name objects. |

| Instance Requests | |
|---|---|
| GetFullName : aLayerName | Gets the full name of a layer. |
| GetName : aString | Gets the short name of a layer. |

*See also:* Librarian; Library; SrcName

# Layout

Layout is an ArcView document for composing a map sheet.

**Inherits from** Doc

| Class Requests | |
|---|---|
| Make : aLayout | Creates a layout document. |
| MakeWithGUI (aGUIName) : aLayout | Creates a layout document based on the specified GUI. |

| Instance Requests | |
|---|---|
| AsTemplate : aTemplate | Creates a template object from a layout. |
| Draw (aDisplay) | Draws a layout on the specified display. |
| EditProperties : aBoolean | Displays the property editor for a layout. |
| Export : aFileName | Exports the layout to a graphic file after displaying the file dialog box. |
| ExportToFile (aFileName, formatString, parameterList): aFileName | Exports a layout to the specified graphic file. |
| GetDisplay : aDisplay | Gets the display object associated with a layout. |
| GetGraphics : aGraphicList | Gets a layout's graphics list. |
| GetSelectMode : GraphicsSelectModeEnum | Returns the graphics selection mode for a layout. |
| Invalidate | Invalidates the layout display. |
| IsSymWinClient : aBoolean | Returns true if a layout is a client of the symbol window. |
| Print | Prints the layout document. |
| ReturnUserCircle : aCircle | Allows the user to draw a circle. |
| ReturnUserLine : aLine | Allows the user to draw a line. |
| ReturnUserPolygon : aPolygon | Allows the user to draw a polygon. |
| ReturnUserPolyLine : aPolyLine | Allows the user to draw a polyline. |
| ReturnUserRect : aRect | Allows the user to draw a rectangle. |
| Select | Selects a graphic element when the user clicks on the mouse. |
| SelectToEdit | Selects graphics for editing. |
| SetSelectMode (GraphicsSelectModeEnum) | Sets the mode for selecting graphics. |
| UseTemplate (aTemplate) | Uses the specified template as the basis for a layout. |

## Enumerations

| GraphicsSelectModeEnum | |
|---|---|
| #GRAPHICS_SELECT_NORMAL | Solid handles on a graphic's bounds can move, resize, or delete. |
| #GRAPHICS_SELECT_VERTEX | Hollow handles at graphic's vertices can reshape by adding, deleting, or moving vertices. |

***See also:*** Frame; GraphicList; PageSetupDialog; Shape; Template; Theme; View

***Code example:*** page 325

# Legend

This class implements the symbolization of a theme.

**Inherits from** Obj

| Class Requests | |
|---|---|
| GetMaxClassifications : aNumber | Gets the maximum number of classifications that a legend may have. |
| Make (SymbolEnum) : aLegend | Creates a legend for the specified symbol type. |
| SetMaxClassifications (aNumber) | Sets the maximum number of classifications that a legend may have. |

| Instance Requests | |
|---|---|
| Area (aTheme, aNumericFieldName, numberOfClasses) | Classifies using equal areas. |
| Clone : aLegend | Clones a legend. |
| ColumnChart (aTheme, aListOfFieldNames, aListOfSymbols, aBackgroundSymbol) | Establishes a column chart legend. |
| Copy (srcLegend) | Copies a legend to another. |
| DisplayNoDataClass (aBoolean) | The no-data class is drawn if set to true. |
| Dot (aFieldName, dotValue, aMarkerSymbol, aShadeSymbol) | Creates a dot density legend. |
| Draw (aDisplay, aTextSymbol, aRect, isClippedBoolean) | Draws a legend inside the specified rectangle. |
| GetClassifications : aList | Gets the list of classifications. |
| GetClassType : LegendClassTypeEnum | Returns the type of classification. |
| GetDensity : aNumber | Returns the dot's value in a dot density legend. |
| GetExtension (aLegendExtensionClass) : aLegendExtension | Gets the legend extension object for the specified class. |
| GetFieldNames : fieldNamesList | Gets the legend fields. |
| GetIndex (fieldNamesList) : aNumber | Returns the index of the class related to the specified field names. |
| GetLegendType : LegendTypeEnum | Returns the type of legend. |
| GetLineOffset : aNumber | Returns the offset in points for line symbols. |
| GetNormFieldName : aString | Returns the normalization field name. |
| GetNormType : LegendNormTypeEnum | Returns the type of normalization. |
| GetNullSymbol : aSymbol | Gets the symbol for values outside the legend's range. |

| | |
|---|---|
| GetNullValue (aFieldName) : aValue | Returns the null value of a field. |
| GetNumClasses : aNumber | Returns the number of classifications. |
| GetNumStdDev : aNumber | Returns the number of standard deviations per class. |
| GetPrecision : aNumber | Returns the precision displayed in a legend. |
| GetRefScale : aNumber | Returns the reference scale for scaling symbols. |
| GetRotationFieldName : fieldName | Returns the field name for symbol rotation. |
| GetSizeFieldName : fieldName | Returns the field name for sizing pie charts. |
| GetSymbol (fielNameList, selectSymbolBoolean) : aSymbol | Gets the symbol related to the specified fields. |
| GetSymbols : aSymbolList | Gets the list of symbols used in the classification. |
| Interval (aTheme, aNumericFieldName, numberOfClasses) | Sets a legend to be equal interval. |
| IsNoDataClassDisplayed : aBoolean | Returns true when no-data class is displayed. |
| IsScaled : aBoolean | Returns true when symbols are scaled to the view's scale. |
| Load (aFileName, LegendLoadTypeEnum) : aBoolean | Loads a legend file and returns true if successful. |
| Natural (aTheme, aNumericFieldName, numberOfClasses) | Creates a natural-break legend. |
| PieChart (aTheme, fieldNamesList, symbolsList, backgroundSymbol, sizeFieldName) | Creates a pie chart legend. |
| Quantile (aTheme, aNumericField, numberOfClasses) | Sets a quantile legend. |
| RemoveExtension (aLegendExtensionClass) | Removes the extension object for the specified class. |
| ResetClasses (aTheme, aFieldName) | Resets a legend to only the no-data class. |
| RestoreSymbols | Restores the symbols of a legend after drawing. |
| ReturnBestDensity (aTheme, fieldName, aMarkerSymbol) : aNumber | Suggests a dot value based on the size of the specified maker. |
| ReturnClassInfo (anIndex) : aList | Returns classification data in a list. |
| ReturnDefaultSymbol : aSymbol | Returns the default symbol. |
| ReturnFieldNames : fieldNamesList | Returns field names needed by GetSymbol and GetIndex requests. |
| Save (aFileName) : aBoolean | Saves the legend to a file name. |
| ScaleSymbols (aDisplay) | Scales the symbols in the legend based on the scale and units of the specified display. |

| | |
|---|---|
| SetClassInfo (classIndex, aList) | Sets the specifications of the indicated class to the information given in the list. |
| SetDensity (aNumber) | Sets the dot value for a dot density legend. |
| SetExtension (aLegendExtension) | Sets the legend extension object of a legend. |
| SetFieldNames (fieldNameList) | Sets the classification fields for a legend. |
| SetLegendType (LegendTypeEnum) | Sets the legend type. |
| SetLineOffset (aNumber) | Sets the offset in points for drawing line symbols. |
| SetNormFieldName (fieldName) | Sets the field used in normalization. |
| SetNormType (LegendNormTypeEnum) | Sets the type of normalization. |
| SetNullSymbol (aSymbol) | Sets the symbol for values outside the legend's range. |
| SetNullValue (fieldName, aValue) | Sets the null value for a field. |
| SetPrecision (aNumber) | Sets the number of decimal points displayed in a legend. |
| SetRefScale (aNumber) | Sets the reference scale for scaling legend symbols. |
| SetRotationFieldName (fieldName) | Sets the field used in rotating legend symbols. |
| SetScaleSymbols (aBoolean) | If set to true, legend symbols are scaled with the view. |
| SetSizeFieldName (fieldName) | Sets the field name used for sizing legend symbols. |
| SingleSymbol | Sets the legend to its single symbol mode. |
| StdDev (aTheme, fieldName, numberOfStdDev) | Creates a standard deviation legend. |
| Unique (aTheme, fieldName) : aBoolean | Creates a unique value legend and returns true if successful. |

## Enumerations

### LegendClassTypeEnum

| | |
|---|---|
| #LEGEND_CLASSTYPE_EQUALAREA | Equal area classification. |
| #LEGEND_CLASSTYPE_INTERVAL | Equal interval classification. |
| #LEGEND_CLASSTYPE_NATURAL | Natural breaks classification. |
| #LEGEND_CLASSTYPE_QUANTILE | Quantile classification. |
| #LEGEND_CLASSTYPE_STDDEV | Standard deviation classification. |

### LegendLoadTypeEnum

| | |
|---|---|
| #LEGEND_LOADTYPE_ALL | Load all legend types. |

| #LEGEND_LOADTYPE_CLASSES | Load classes. |
|---|---|
| #LEGEND_LOADTYPE_CLASSESANDSYMBOLS | Load classes and symbols. |
| #LEGEND_LOADTYPE_SYMBOLS | Load symbols. |

## LegendNormTypeEnum

| #LEGEND_NORMTYPE_FIELD | Normalize by field. |
|---|---|
| #LEGEND_NORMTYPE_NONE | No normalization. |
| #LEGEND_NORMTYPE_PERCENT | Normalize by percent of total. |

## LegendTypeEnum

| #LEGEND_TYPE_CHART | Chart Legend. |
|---|---|
| #LEGEND_TYPE_COLOR | Graduated Color Legend. |
| #LEGEND_TYPE_DOT | Dot Density Legend. |
| #LEGEND_TYPE_OLD | ArcView 2 legend. |
| #LEGEND_TYPE_SIMPLE | Single Symbol Legend. |
| #LEGEND_TYPE_SYMBOL | Graduated Symbol Legend. |
| #LEGEND_TYPE_UNIQUE | Unique Value Legend. |

## StatTypeEnum

| #STATTYPE_COUNT | Count. |
|---|---|
| #STATTYPE_SUM | Sum. |
| #STATTYPE_MIN | Minimum. |
| #STATTYPE_MAX | Maximum. |
| #STATTYPE_MEAN | Mean. |
| #STATTYPE_STDDEV | Standard deviation. |

## SymbolEnum

| #SYMBOL_CHART | Chart symbol. |
|---|---|
| #SYMBOL_FILL | Fill symbol. |
| #SYMBOL_MARKER | Marker symbol. |
| #SYMBOL_PEN | Pen symbol. |
| #SYMBOL_TEXT | Text symbol. |

***See also:*** Classification; Field; FTab; FTheme; Symbol

***Code example:*** page 487

# LegendExtension

This abstract class supports the addition of extension specific information to the Legend class.

**Inherits from** ExtensionObject

| Instance Requests | |
|---|---|
| OverrideAdvancedDialog : aBoolean | Overrides the advanced properties editor dialog box used in the legend editor. Returns true if successful. |
| ShowAdvancedDialog (theOwningTheme) | Shows the advanced dialog box to edit the legend extension properties. |

*See also:* Doc; Legend

# LegendFrame

This class implements placement of a legend in a layout document.

**Inherits from** Frame

| Class Requests | |
|---|---|
| Make (aRect) : aLegendFrame | Creates a legend frame at the size of the specified rectangle. |

| Instance Requests | |
|---|---|
| CanSimplify : aBoolean | Always returns true. |
| Draw | Draws a legend frame on its associated display. |
| Edit (aListofGraphics) : aBoolean | Starts the graphic editor. |
| EditSizeAndPos : aBoolean | Starts the size and position editor. |
| GetFillObject : anObj | Gets the fill object of a legend frame. |
| GetViewFrame : aViewFrame | Gets the view frame associated with a legend frame. |
| IsFilled : aBoolean | Returns true when the legend frame is filled. |
| IsFilledBy (aClass) : aBoolean | Returns true when objects of the specified class can fill the legend frame. |
| ReturnSymbols : aList | Returns a list of symbols. |

| | |
|---|---|
| SetBounds (aRect) | Sets the extent of a legend frame. |
| SetDisplay (aDisplay) | Sets the display object of a legend frame. |
| SetFillObject (anObj) | Sets the fill object. |
| SetSymbols (aSymbolList) | Sets the symbols. |
| SetViewFrame (aViewFrame) | Sets the view frame for a legend frame. |

***See also:*** Display; GraphicList; Template; View

# Librarian

This class implements the ArcStorm database.

**Inherits from** Obj

| Class Requests | |
|---|---|
| Make (aLibrarianName) : aLibrarian | Creates a librarian object. |
| ReturnLibrarians : aList | Returns a list of accessible librarians. |

| Instance Requests | |
|---|---|
| GetName : aString | Gets the name of a librarian. |
| GetPath : aString | Gets the path directory of a librarian. |
| ReturnLibraries : aList | Returns a list of libraries in a librarian. |

***See also:*** Layer; Library; SrcName

# Library

This class contains layers in an ArcStorm or ARC/INFO map library.

**Inherits from** Obj

| Class Requests | |
|---|---|
| Exists (aLibraryName) : aBoolean | Returns true if the specified library exists. |
| GetLocation (aLibraryName) : aFileName | Gets the directory path where the specified library exists. |
| Make (aLibraryName, aLibrarianName) : aLibrary | Creates a library object from the specified parameters. |

| Instance Requests | |
|---|---|
| GetIndexPath : aString | Gets the path to the index coverage for a library. |
| GetLibrarianName : aString | Gets the name of an ArcStorm librarian containing a library. |
| GetName : aString | Gets a library's name. |
| GetPath : aString | Gets the directory path to a library. |
| IsArcStorm : aBoolean | Returns true for an ArcStorm library. |
| ReturnIndexSrcNames : aSrcNamelist | Returns a list of source names for the index coverage of a library. |
| ReturnLayers : aList | Returns a list of layers in a library. |

***See also:*** Layer; Librarian; SrcName

# *Line*

This class implements the line geometric shape.

### **Inherits from** Shape

| Class Requests | |
|---|---|
| Make (aStartPoint, anEndPoint) : aLine | Creates a line object. |
| MakeNull : aLine | Creates a line with null values for start and end points. |

| Instance Requests | |
|---|---|
| = anObj : aBoolean | Returns true when a line is the same as another object. |
| AsMultiPoint : aMultiPoint | Returns a multi-point shape equivalent to a line. |
| AsPolyLine : aPolyLine | Returns a polyline shape equivalent to a line. |
| AsString : aString | Returns start and end points in a string format. |
| Contains (aShape) : aBoolean | Returns true if the specified shape is contained within a line. |
| GetDimension : aNumber | Always returns 1. |
| Intersects (aShape) : aBoolean | Returns true when the specified shape intersects a line. |
| IsNull : aBoolean | Returns true if the start or end point is null. |
| Move (moveX, moveY) : aLine | Moves a line by the specified amounts. |

| ReturnCenter : aPoint | Returns the center point of a line. |
|---|---|
| ReturnClipped (aRect) : aLine | Returns the line resulting from a line clipped to the specified rectangle. |
| ReturnEnd : aPoint | Returns the end point. |
| ReturnMerged (aShape) : aPolyLine | Merges the specified shape into a line. |
| ReturnOffset (aNumber) : aLine | Returns a new line at the given offset in points. |
| ReturnProjected (aPrj) : aShape | Projects a line. |
| ReturnStart : aPoint | Returns the starting point of a line. |
| ReturnUnprojected (aPrj) : aShape | Unprojects a line shape. |
| Snap (aShape, snapDistance) : aBoolean | Returns true when a line is snapped to the vertices of a shape. |

**See also:** Circle; Line; Oval; Point; Polygon

# LinearLookup

This class implements linear image look-up tables.

**Inherits from** ImageLookup

| Class Requests | |
|---|---|
| Make (pointList) : aLinearLookup | Creates a linear look-up based on the values provided as points. |
| MakeLine (aStartPoint, anEndPoint) : aLinearLookup | Creates a linear look-up between the values represented by start and end points. |

| Instance Requests | |
|---|---|
| AddPoint (aPoint) | Inserts a point into a linear look-up. |
| Clone : aLinearLookup | Clones a linear look-up. |
| GetPoints : aList | Gets a list of points in a look-up table. |
| Lookup (aValue) : aNumber | Finds a value in a linear look-up table. |

**See also:** ImageLegend; ISrc

# LineFile

This class implements disk files with lines as data elements.

## Inherits from File

| Class Requests | |
|---|---|
| Make (aFileName, FilePermEnum) : aLineFile | Creates or opens a disk file according to the specified parameters. |

| Instance Requests | |
|---|---|
| GetPos : aNumber | Returns the position of a file pointer in number of lines. |
| GetSize : aNumber | Returns the size of a file in number of lines. |
| GotoBeg | Positions the file pointer at the beginning. |
| GotoEnd | Positions the file pointer after the last line. |
| IsAtEnd : aBoolean | Returns true if the file pointer is positioned after the last line. |
| Read (lineList, numberOfLines) | Adds to the specified list the specified number of lines from the position of the file pointer. |
| ReadElt : aString | Returns the line at the position of the line file pointer. |
| SetPos (lineNumber) | Sets the position of the file pointer. |
| Write (lineList, numberOfLines) | Writes the specified number of lines from the list to a file. |
| WriteElt (aString) | Writes the specified line to a line file. |

## Enumerations

| FilePermEnum | |
|---|---|
| #FILE_PERM_APPEND | Opens an existing or new file to append elements. |
| #FILE_PERM_CLEARMODIFY | Overwrites an existing file or creates a new file for reading and writing. |
| #FILE_PERM_MODIFY | Opens an existing or new file for reading and writing. |
| #FILE_PERM_READ | Opens an existing file for reading. |
| #FILE_PERM_WRITE | Overwrites an existing file or creates a new file for writing. |

**See also:** FileName; TextFile

**Code example:** page 353

# List

Objects of this class hold other objects.

**Inherits from** Collection

**Inherited by** Clipboard, GraphicList, GraphicSet, SymbolList, and Template

| Class Requests | |
| --- | --- |
| Make : aList | Creates an empty list. |

| Instance Requests | |
| --- | --- |
| + anotherCollection : aList | Returns a new list by adding another container to a list. |
| - anotherCollection : aList | Subtracts a list from another list. |
| Add (anObj) : aList | Adds the specified object to a list. |
| AsList : aList | Converts the words in a string to elements of a list. |
| Clone : aList | Shallow clones a list by copying only the list object. |
| Count : aNumber | Returns the number of objects in a list. |
| DeepClone : aList | Deep clones a list by copying the elements of the list. |
| Empty | Clears a list. |
| Find (anObj) : aNumber | Returns the list index that contains the specified object. |
| FindByClass (aClass) : aNumber | Finds the first element of the list that has the specified class. |
| FindByValue (anObj) : aNumber | Finds an object by its value and returns the list index. |
| Get (aIndexNumber) : anObj | Gets the object at the specified list index. |
| HasKindOf (aClass) : aBoolean | Returns true if the list contains objects of the specified class. |
| Insert (anObj) | Places an object at the top of the list. |
| Merge (anotherCollection) : aList | Merges the specified container with a list while avoiding duplicate objects. |
| Remove (anIndexNumber) | Removes the object at the list index from a list. |
| RemoveDuplicates | Removes any duplicate object from a list. |
| RemoveObj (anObj) | Removes the specified object from a list. |

| Set (anIndexNumber, anObj) | Places the specified object at the specified index number of a list. |
|---|---|
| Shuffle (anObj, anIndexNumber) | Moves the specified object from its current position to the specified new position in a list. |
| Sort (ascendingBoolean) | Sorts the literals in a list. |

**See also:** Dictionary; String

**Code example:** page 389

# ListDisplay

This class accumulates a list of graphic primitives from a display so that it can be merged with a graphic list.

**Inherits from** Display

| Class Requests | |
|---|---|
| Make : aListDisplay | Creates an empty list display. (Not available in 2.0.) |

| Instance Requests | |
|---|---|
| DrawIcon (aRect, anIcon) | Draws the specified icon in the specified rectangle on a list display. |
| DrawLine (aLine, aSymbol) | Draws a line on a list display with the specified symbol. |
| DrawMultiPoint (aMultiPoint, aSymbol) | Draws a multi-point shape on a list display with the specified symbol. |
| DrawOval (anOval, aSymbol) | Draws an oval on a list display with the specified symbol. |
| DrawPoint (aPoint, aSymbol) | Draws a point on a list display with the specified symbol. |
| DrawPolygon (aPolygon, aSymbol, showOutlineBoolean) | Draws a polygon on a list display with the specified symbol. |
| DrawPolyLine (aPolyLine, aSymbol) | Draws a polyline on a list display with the specified symbol. |
| DrawRect (aRect, aSymbol) | Draws a rectangle on a list display with the specified symbol. |
| DrawText (aPoint, aString, aSymbol) | Draws a string at the specified location with the specified symbol. |

| DrawTextPolyLine (aPolyLine, aString, aSymbol) | Draws text along a polyline using the specified symbol. |
|---|---|
| HookupSym (aSymbol) | Hooks a symbol to the scale of a display. |
| ListEnd : aListOfGraphics | Closes the list display and returns a list of graphics. (Not available in 2.0.) |

*See also:* Display; GraphicList

# *LocateDialog*

This class implements the window for matching a single address.

### Inherits from ModalDialog

| Class Requests | |
|---|---|
| GetLastAddress : aString | Gets the last address used in locate dialog. (Not available in 2.0.) |
| GetLastZone : aString | Gets the last zone used in locate dialog. (Not available in 2.0.) |
| GetMatchPref : aMatchPref | Gets the match preference object. |
| SetMatchPref (aMatchPref) | Sets the match preference object. |
| Show (aMatchableTheme) : aShape | Places a shape at the match location. |
| ShowUsingDefault (aTheme, addressString, zoneString) : aShape | Returns a shape at the location of the specified address and zone. (Not available in 2.0.) |

# *LockMgr*

This class manages multi-user editing of shapes and tables.

### Inherits from Obj

| Class Requests | |
|---|---|
| Make (aMachineName, aServerID, aVersionNo) : aLockMgr | Creates a lock manager based on an RPC link. |
| MakeFromFile : aLockMgr | Creates a lock manager based on an RPC link established from the parameters in the default.flm file. |
| The : aLockMgr | Returns the current lock manager object. |

| Instance Requests | |
|---|---|
| GetErrorMsg : aString | Returns the error message of a lock manager. |
| GetRefreshRate : aNumber | Returns the refresh rate of a lock manager. |
| IsActive : aBoolean | Returns true when lock manager is active. |
| IsConnected : aBoolean | Returns true when lock manager is connected to a server. |

*See also:* RPCClient; RPCServer

# Mac

This class provides access to Macintosh system functions. *Mac* is used only in the Macintosh environment.

## Inherits from Obj

| Class Requests | |
|---|---|
| BringAVToFront | Brings the ArcView application to the front. |
| GetAboutLabel : aString | Gets Apple menu's *About...* label. |
| GetAboutScript : aString | Gets Apple menu's *About...* item script. |
| GetVoice : aString | Gets the current voice. |
| IsGUIVisible : aBoolean | Returns true if GUI is visible. |
| IsSpeechAvailable : aBoolean | Returns true if the Speech Manager is present. |
| IsValidType (extensionList, AFileName) : aBoolean | Returns true if the file type of the specified file matches any file type corresponding to the specified file extensions. |
| QueryDocClass (aFileName, aString) | Populates the specified string object with the document class of the specified file name. |
| QueryTypes (extensionList, fileTypeList) | Populates the specified string object with the document class name of the specified file name. |
| ResetTypes | Resets the default mapping between extensions and Macintosh file types. |
| ReturnVoices : stringList | Returns a list of available voices. |
| SetAboutLabel (aString) | Sets the Apple menu's *About...* label. |
| SetAboutScript (aString) | Sets the Apple menu's *About...* script. |
| SetDocInfo (aFilename, ADocSubClass) | Sets the document information, file type, and icon for the specified file name. |

| | |
|---|---|
| SetFileIconDoc (aFileName, aDocSubClass) | Sets the finder icon for the specified file name. |
| SetFileIconID (aFileName, iconResourceID) | Sets the finder icon of the given file name to the specified icon. |
| SetGUIVisible (aBoolean) | Makes the GUI visible if the Boolean value is true. |
| SetTypes (extensionList, FileTypeList) | Maps file name extensions and Macintosh file types. |
| SetVoices (aString) | Sets the voice for the Speak request. |
| Speak (aString) | Speaks the specified string. |

***See also:*** System

# *MapDisplay*

This class implements areas of a screen that can draw a map using map coordinates.

**Inherits from** Display

| *Instance Requests* | |
|---|---|
| BeginClip : aBoolean | Initial request to start drawing on a display. |
| DrawIcon (aRect, anIcon) | Draws the specified icon within the specified rectangle on a map display. |
| DrawLine (aLine, aSymbol) | Draws the specified line using the specified symbol on a map display. |
| DrawMultiPoint (aMultiPoint, aSymbol) | Draws the specified multi-point shape using the specified symbol on a map display. |
| DrawOval (aOval, aSymbol) | Draws the specified oval using the specified symbol on a map display. |
| DrawPoint (aPoint, aSymbol) | Draws the specified point using the specified symbol on a map display. |
| DrawPolygon (aPolygon, aSymbol, drawOutlineBoolean) | Draws the specified polygon using the specified symbol on a map display. If the specified Boolean value is true it also draws the polygon outline. |
| DrawPolyLine (aPolyLine, aSymbol) | Draws the specified polyline shape using the specified symbol on a map display. |
| DrawRect (aRect, aSymbol) | Draws the specified rectangle using the specified symbol on a map display. |
| DrawText (aPoint, aString, aSymbol) | Draws the specified text string at the specified point using the specified symbol on a map display. |

| | |
|---|---|
| DrawTextPolyLine (aPolyLine, aString, aSymbol) | Draws a text string along the polyline using the symbol on a map display. |
| EndClip | Final request to stop drawing on a map display. |
| ExportEnd | Stop exporting the graphics of a map display. |
| Flush | Forces drawing of buffer to a map display. |
| GetResolution : DisplayResEnum | Gets the resolution quality of a map display. |
| HookUpSymbol (aSymbol) | Hooks a symbol to the display's scale. |
| Invalidate (eraseBoolean) | Invalidates a map display. If the specified Boolean is true, the display area is erased prior to redrawing. |
| InvalidateRect (aRect) | Invalidates the specified rectangular area of a map display. |
| Pan | Allows user to pan a map display area in an apply event. |
| ReturnPixmap (aRect) | Returns a pixmap contained in the specified rectangle. |
| ReturnUserPoint : aPoint | Returns location of the mouse in an apply event. |
| SetResolution (DisplayResEnum) | Sets the resolution quality of a map display. |
| Validate | Validates a map display and stops redraw. |
| ZoomToRect (aRect) | Changes the display extent to the specified rectangle. |

## Enumerations

| DisplayResEnum | |
|---|---|
| #DISPLAY_RES_LOW | Low resolution. |
| #DISPLAY_RES_NORMAL | Normal resolution. |
| #DISPLAY_RES_HIGH | High resolution. |

***See also:*** PageDisplay; SrcName; View

# Marker

This class implements the symbology to draw a point on the screen.

**Inherits from** Symbol

**Inherited by** BasicMarker and CompositeMarker

| Instance Requests | |
|---|---|
| GetType : SymbolEnum | Gets the #SYMBOL_MARKER symbol type. |

## Enumerations

| SymbolEnum | |
|---|---|
| #SYMBOL_CHART | Chart symbol. |
| #SYMBOL_FILL | Fill symbol. |
| #SYMBOL_MARKER | Marker symbol. |
| #SYMBOL_PEN | Pen symbol. |
| #SYMBOL_TEXT | Text symbol. |

*See also:* Fill; Pen; Text

# MatchCand

This class implements candidate records for matching an address event.

**Inherits from** Obj

| Instance Requests | |
|---|---|
| GetMatchSourceRec : aNumber | Gets the feature theme's record number for a match candidate. |
| GetMatchSourceRec2 : aNumber | Gets the record number from the second feature table of an intersection candidate. |
| GetPosition : percentNumber | Gets the position along the address range. |
| GetScore : aNumber | Gets the match score of a match candidate. |
| GetSide : aString | Returns L or R as the side of the match candidate for which an event has been found. |
| GetValue (aMatchField) : aString | Gets the match field value for a match candidate. |
| IsIntersection : aBoolean | Returns true if a match candidate is an intersection. |

*See also:* MatchCase; MatchField; MatchKey; MatchSource

# MatchCase

Objects of this class contain a list of matching candidate records for a single event.

**Inherits from** Obj

| Class Requests | |
| --- | --- |
| Make (aMatchSource, aMatchKey) : aMatchCase | Creates an empty match case based on the specified parameters. |

| Instance Requests | |
| --- | --- |
| AllowIntersections (aMatchSource, aMatchKey) | Allows intersection matching in a match case. |
| GetBestCand : aMatchCand | Gets the best match candidate after scoring. |
| GetNthCand (aNumber) : aMatchCand | Gets the nth best candidate. |
| GetNumCands : aNumber | Gets the number of match candidates. |
| ScoreCandidates : aBoolean | Scores candidates and returns true if successful. |

*See also:* MatchCand; MatchKey; MatchSource

# MatchField

This class implements the field within the matching candidate record used for the matching process.

**Inherits from** Obj

| Instance Requests | |
| --- | --- |
| GetField : aField | Gets the associated field from the feature theme. |
| GetName : aString | Gets the name of a match field. |
| IsRequired : aBoolean | Returns true when an associated feature theme field is required for a match field. |

*See also:* MatchCand; MatchCase; MatchSource

# MatchKey

This class standardizes the event string in an address matching operation.

**Inherits from** Obj

| Class Requests | |
|---|---|
| Make (filenameString) : aMatchKey | Creates a match key based on the specified standardization file. |

| Instance Requests | |
|---|---|
| AllowIntersections (filenameString, delimiterString) | Allows a match key to have intersection events. |
| CanDoIntersections : aBoolean | Returns true if a match key includes intersections. |
| GetFieldLabel (matchKeyFieldName) : aString | Returns the field label of a match key field. |
| GetKey : aString | Gets the key value of a match key. |
| GetName : aString | Gets the name of a match key. |
| GetNthField (aFieldNumber) : aString | Gets the name of the field at the specified position. |
| GetNumFields : aNumber | Gets the number of fields in a match key. |
| GetValue (aMatchKeyFieldName) : aString | Gets the value of the specified field in a match key. |
| GetXName : aString | Gets the name of an intersection match key. |
| GetZoneKey : aString | Gets the zone value of a match key. |
| HasIntersection : aBoolean | Returns true when a match key contains an intersection event. |
| SetAliasTable (aliasVTab) | Sets the alias table for a match key to associate addresses with place names. |
| SetKey (anEventString) | Standardizes the specified string based on a match key. |
| SetValue (aMatchKeyFieldName, valueString) | Sets the value of the field in a match key to the specified string. |
| SetZoneKey (zoneValueString) | Sets the zone value in a match key. |

*See also:* AddressStyle; MatchPref; MatchSource

# *MatchPref*

The match preference class implements the matching criteria for creating a geo-coded dataset.

**Inherits from** Obj

| Class Requests | |
|---|---|
| Make : aMatchPref | Creates a default match preference object. |

| Instance Requests | |
|---|---|
| GetPref (MatchPrefEnum) : aBoolean | Gets the value of a preference Boolean element. |
| GetPrefVal (MatchPrefEnum) : aNumber | Gets the value of a preference numeric element. |
| Reset | Resets a match preference to its default values. |
| SetPref (MatchPrefEnum, aBoolean) | Sets the value of a match preference Boolean element. |
| setPrefVal (MatchPrefEnum, aNumber) | Sets the value of a match preference numeric element. |

## *Enumerations*

| MatchPrefEnum | |
|---|---|
| #MATCHPREF_NOREVIEW | If true, matching candidates are not reviewed. |
| #MATCHPREF_MULTBEST | If true, candidates with the same highest score are reviewed. |
| #MATCHPREF_MULTGOOD | If true, candidates with passing scores are reviewed. |
| #MATCHPREF_NOCAND | If true, all candidates are reviewed when none attain the minimum score. |
| #MATCHPREF_SPELLWEIGHT | Acceptable score for spelling. |
| #MATCHPREF_MINMATCHSCORE | Minimum acceptable match score. |
| #MATCHPREF_MINCANDSCORE | Minimum acceptable candidate score. |

**See also:** GeoName, MatchPrefDialog, MatchSource

# MatchPrefDialog

The match preference dialog class allows editing of matching preferences.

**Inherits from** ModalDialog

| Class Requests | |
|---|---|
| Show (aMatchPref) : aBoolean | Displays the values of a match preference object in an edit window. |

*See also:* MatchPref

# *MatchSource*

An object of this class locates address events on a matchable feature theme.

**Inherits from** Obj

| Class Requests | |
|---|---|
| Make (anAddressStyle, aTheme, aFieldList) : aMatchSource | Creates a match source based on the specified address style and list of fields from the specified theme. |

| Instance Requests | |
|---|---|
| BatchMatch (aGeoName, geoNameFTab) | Performs geocoding in batch mode. |
| EndMatch | Ends matching after a search request. |
| GetDefAliasTable : aVTab | Gets the default alias table for associating place names with addresses. |
| GetDefOffset : aNumber | Gets the default value of the offset from the street line for marking the matched location. |
| GetDefSqueeze : aNumber | Gets the default percentage of squeeze from an intersection for marking a matched location. |
| GetMatchFields : aFieldList | Gets the match fields of a match source. |
| GetMaxScore : aNumber | Gets the maximum possible score for a match source. |
| GetName : aString | Gets the address style name of a match source. |
| GetPoint (aMatchCand) : aPoint | Gets the location of a match candidate. |
| GetScore (recordNumber) : aNumber | Returns the match score. |
| GetStanRules : aFileNameString | Gets the file name with the standardization rules. |
| GetXDelimiter : aString | Gets the intersection delimiter. |
| GetXMatchFields : aFieldList | Gets the intersection match fields of a match source. |

| | |
|---|---|
| GetXMaxScore : aNumber | Gets the maximum score possible for matching against intersections. |
| GetXStanRules : aFileNameString | Gets the file name with the intersection standardization rules. |
| HasIntersections : aBoolean | Returns true when a match source can match against intersections. |
| HasValidIndex : aBoolean | Returns true when the index is valid for the current search. |
| HasZone : aBoolean | Returns true if a match source uses a zone for matching. |
| InitGeoTheme (aGeoName) : aVTab | Initializes the specified geocoded data set. |
| InvalIndex | Invalidates the index of a match source. |
| IsMatched (recordNumber) : aBoolean | Returns true if a match is found for the specified record. |
| OpenIndex | Opens the index to locate an event but does not create a geocoded theme. |
| Search (aMatchKey, SpellWeight, aMatchCase) : aNumber | Performs a search on a match source based on the specified parameters and returns the number of candidates found. |
| SetDefAliasTable (aVTab) | Sets the default table to associate place names with addresses. |
| SetDefOffset (aNumber) | Sets the default offset for placing an event marker from the street centerline. |
| SetDefSqueeze (aPercentNumber) | Sets the default percentage for placing a marker from an intersection. |
| SetOffset (aNumber) | Sets the offset value for placing an event marker from the street centerline. |
| SetSqueeze (aPercentNumber) | Sets the percentage for placing a marker from an intersection. |
| SetXDelimiter (delimiterString) | Sets the intersection delimiter. |
| ShowCandidate (aMatchCand) | Flashes the specified candidate. |
| SingleMatch (aMatchPref, addressString, zoneString) : aPoint | Matches the specified address and returns its geocoded point. |
| WriteMatch (recordNumber, aMatchKey, aMatchCand) | Places a match marker in the specified record number. |
| WriteUnMatch (recordNumber, aMatchKey) | Places a null marker in the specified record number. |

**See also:** AddressStyle; GeoName; MatchCand; MatchCase; MatchField; Theme

# Menu

This class implements the menu control that appears on the menu bar.

**Inherits from** ControlSet

| Class Requests | |
| --- | --- |
| Make : aMenu | Creates a menu control. |

| Instance Requests | |
| --- | --- |
| FindByLabel (labelString) : aChoice | Finds the first menu choice with the specified label. |
| GetLabel : aString | Gets the label of a menu. |
| Remove (aChoice) | Removes a menu item from the menu. |
| SetLabel (aString) | Sets the menu label. |

***See also:*** Choice; MenuBar; Space

***Code example:*** page 290

# MenuBar

This class implements the top row of a document GUI.

**Inherits from** ControlSet

| Instance Requests | |
| --- | --- |
| FindByLabel (labelString) : aMenu | Finds the first menu with the specified label. |
| FindByScript (scriptName) : aChoice | Finds the first choice that is associated with the specified script. |

***See also:*** ButtonBar; Choice; DocGUI; Menu; ToolBar

***Code example:*** page 326

# Mercator

This class implements the Mercator projection.

**Inherits from** Prj

| Class Requests | |
| --- | --- |
| Make (aRect) : aMercator | Creates a projection bounded by the specified rectangle. |

| Instance Requests | |
|---|---|
| ProjectPt (aPoint) : aBoolean | Projects a point and returns true if successful. |
| Recalculate | Calculates the derived constants. |
| ReturnCentralMeridian : aNumber | Returns the central meridian in decimal degrees. |
| ReturnLatitudeOfTrueScale : aNumber | Returns latitude of true scale in decimal degrees. |
| SetCentralMeridian (aLongitude) | Sets the central meridian to the specified degrees. |
| SetLatitudeOfTrueScale (aLatitude) | Sets the latitude of true scale to the specified degrees. |
| UnProjectPt (aPoint) : Boolean | Unprojects a point and returns true if successful. |

*See also:* CoordSys; Spheroid

# Miller

This class implements the Miller projection.

**Inherits from** Prj

| Class Requests | |
|---|---|
| Make (aRect) : aMiller | Creates a projection bounded by the specified rectangle. |

| Instance Requests | |
|---|---|
| ProjectPt (aPoint) : aBoolean | Projects a point and returns true if successful. |
| Recalculate | Calculates the derived constants. |
| ReturnCentralMeridian : aNumber | Returns the central meridian in decimal degrees. |
| SetCentralMeridian (aLongitude) | Sets the central meridian to the specified degrees. |
| UnProjectPt (aPoint) : aBoolean | Unprojects a point and returns true if successful. |

*See also:* CoordSys; Spheroid

# ModalDialog

This class implements modal dialog windows in ArcView.

**Inherits from** Obj

**Inherited by** AddDBThemeDialog, AnalysisPropertiesDialog, AreaOf-InterestDialog, AutoLabelDialog, DensitySurfaceDialog, EventDialog, ExtensionWin, FileDialog, FocalStatisticsDialog, GeocodeDialog, GridLegendWindow, IconMgr, InterpolationDialog, LocateDialog, Match-PrefDialog, NorthArrowMgr, PageSetupDialog, ProjectionDialog, ReclassEditor, ReMatchDialog, ScriptMgr, SourceDialog, SourceMan-ager, StanEditDialog, SummaryDialog, TemplateMgr, and ThemeOn-ThemeDialog

| Class Requests | |
|---|---|
| Show (libFileName, libraryName, moduleName, resourceName) : aString | Displays a modal window specified by the resource name stored in the module of the specified library. |

*See also:* MsgBox; Window

# Mollweide

This class implements the Mollweide projection.

**Inherits from** Prj

| Class Requests | |
|---|---|
| Make (aRect) : aMollweide | Creates a Mollweide projection. |
| **Instance Requests** | |
| ProjectPt (aPoint) : aBoolean | Projects the given point and returns true if successful. |
| Recalculate | Applies the changes in the projection parameters. |
| ReturnCentralMeridian : aNumber | Returns the central meridian longitude in degrees. |
| SetCentralMeridian (aNumber) | Sets the central meridian to the specified degrees of longitude. |
| UnProjectPt (aPoint) : aBoolean | Unprojects a point and returns true if successful. |

*See also:* CoordSys; Spheroid

# MovieWin

Apple's QuickTime movie can be displayed using objects of this class. *MovieWin* can be used only in the Macintosh environment.

**Inherits from** Window

| Class Requests | |
|---|---|
| Make (aFileName, titleString) : aMovieWin | Creates a movie window. |

| Instance Requests | |
|---|---|
| GetDuration : aNumber | Gets the duration of movie in a movie window. |
| GetFileName : aFileName | Gets the file name associated with a movie window. |
| GetStatus : MovieStatusEnum | Gets the status of the movie in a movie window. |
| GetTime : aNumber | Gets the current time of the movie. |
| IsModal : aBoolean | Returns true if a movie window is modal. |
| Open | Opens a movie window. |
| Pause | Pauses the running movie in a movie window. |
| Play | Runs the movie in a movie window. |
| SetFileName (aFileName) | Associates the specified file with a movie window. |
| SetModal (aBoolean) | Makes a movie window modal if the Boolean value is true. |
| SetTime (aNumber) | Sets the current time of the movie. |
| Stop | Stops the movie in a movie window. |

## Enumerations

| MovieStatusEnum | |
|---|---|
| #MOVIESTATUS_PAUSED | Movie is paused. |
| #MOVIESTATUS_RUNNING | Movie is currently running. |
| #MOVIESTATUS_STOPPED | Movie is stopped. |

# MsgBox

This class provides a predefined set of modal dialogs for interaction with the user.

## Inherits from Obj

| Class Requests | |
| --- | --- |
| AllYesNo (aMsg, aTitle, defaultIsYes) : aBoolean | Returns true for yes, false for no, and nil for yes-all. |
| Banner (anImageFileName, durationInSeconds, titleString) : aBoolean | Displays an image in a mode-less window. |
| Choice (aList, messageString, titleString) : anObj | Displays object names in a list for user selection in a drop-down control. |
| ChoiceAsString (aList, messageString, titleString) : anObj | Displays objects in a list by applying the AsString request. |
| Error (messageString, titleString) | Displays an error dialog box. |
| ErrorHelp (messageString, titleString, helpTopicName) | Displays an error dialog box with a help button. |
| GetErrorReporting : aBoolean | Returns true if ArcView errors are displayed in a modal dialog window and not redirected to a log file. |
| Info (messageString, titleString) | Displays an information dialog box. |
| Input (messageString, titleString, defaultString) : aString | Accepts a single input from the user. |
| List (aList, messageString, titleString) : anObj | Displays object names in a list for user selection. |
| ListAsString (aList, messageString, titleString) : anObj | Displays objects in a list by applying the AsString request. |
| LongYesNo (messageString, titleString, defaultIsYesBoolean) : aBoolean | Displays a large-window yes-no message box. |
| MiniYesNo (messageString, defaultIsYesBoolean) : aBoolean | Displays a dialog box to obtain a yes or no answer from the user. |
| MultiInput (messageString, titleString, labelList, defaultList) : aList | Displays a dialog box with multiple input fields. |
| MultiList (aList, messageString, titleString) : aList | Displays object names in a list for multiple selection. |
| MultiListAsString (aList, messageString, titleString) : aList | Displays the string representation of objects in a list for multiple selection. |
| Password : aString | Displays a password dialog box. |
| Report (aLongMessageString, titleString) | Displays a report dialog window. |
| SaveChanges (messageString, titleString, DefaultIsYesBoolean) : anObj | Displays a dialog box to save the changes to the project. |
| SetErrorReporting (aBoolean) | If set to true, errors are reported through a dialog box. |

| | |
|---|---|
| Warning (messageString, titleString) | Displays a warning dialog box. |
| YesNo (messageString, titleString, defaultIsYesBoolean) : aBoolean | Displays a yes/no dialog box. |
| YesNoCancel (messageString, titleString, defaultIsYesBoolean) : anObj | Displays a yes/no dialog box with a cancel button. |

***See also:*** FileDialog; SourceDialog; ScriptMgr; TextWin

***Code example:*** page 383

# *MultiBandLegend*

This class implements a legend for multi-band images.

**Inherits from** ImageLegend

| Class Requests | |
|---|---|
| Make (aRedImageLookup, aBlueImageLookup, aGreenImageLookup) : MultiBandLegend | Creates a multi-band legend by applying red, blue, and green look-up tables to bands 0, 1, and 2. |

| Instance Requests | |
|---|---|
| Clone : aMultiBandLegend | Clones a multi-band legend. |
| GetBlueBand : aNumber | Gets the band number displayed as blue. |
| GetBlueImageLookup : anImageLookup | Gets the image look-up table used in displaying the blue band. |
| GetGreenBand : aNumber | Gets the band number displayed as green. |
| GetGreenImageLookup : anImageLookup | Gets the image look-up table used in displaying the green band. |
| GetRedBand : aNumber | Gets the band number displayed as red. |
| GetRedImageLookup : anImageLookup | Gets the image look-up table used in displaying the red band. |
| SetBlueBand (bandNumber) | Sets the blue band number in a multi-band legend. |
| SetBlueImageLookup (anImageLookup) | Sets the blue band image look-up table. |
| SetGreenBand (bandNumber) | Sets the green band number in a multi-band legend. |
| SetGreenImageLookup (anImageLookup) | Sets the green band image look-up table. |
| SetRedBand (bandNumber) | Sets the red band number in a multi-band legend. |
| SetRedImageLookup (anImageLookup) | Set the red image look-up table. |

***See also:*** ImageLookup; ISrc; SingleBandLegend

# *MultiPoint*

This class implements a shape composed of several points.

**Inherits from** Shape
**Inherited by** PolyLine

| Class Requests | |
|---|---|
| Make (pointList) : aMultiPoint | Creates a multi-point shape. |
| MakeNull : aMultiPoint | Creates a null multi-point shape. |

| Instance Requests | |
|---|---|
| = aMultiPoint : aBoolean | Returns true if two multi-point objects are the same. |
| AsList : pointList | Returns the points in a multi-point shape. |
| AsMultiPoint : aMultiPoint | Clones a multi-point. |
| AsString : aString | Returns the string representation of the points in a multi-point shape. |
| Contains (aShape) : aBoolean | Returns true if a shape is contained within a multi-point shape. |
| Count : aNumber | Returns the number of points in a multi-point shape. |
| GetDimension : aNumber | Always returns zero. |
| Intersects (aShape) : aBoolean | Returns true if the specified shape intersects a multi-point shape. |
| IsEmpty : aBoolean | Returns true if a multipoint is empty. |
| IsNull : aBoolean | Returns true if no points have been defined for a multi-point shape. |
| Move (moveX, moveY) : aMultiPoint | Moves a multi-point by the specified amounts. |
| PointIntersection (anotherShape) : aMultiPoint | Intersects a shape and a multipoint. |
| ReturnBuffered (aDistance) : aPolygon | Creates a polygon by applying a buffer to a multipoint. |
| ReturnCenter : aPoint | Returns the center of a multi-point shape's extent. |
| ReturnMerged (aShape) : aMultiPoint | Merges the specified shape into a multi-point shape. |
| ReturnProjected (aPrj) : aShape | Projects a multi-point shape. |
| ReturnUnprojected (aPrj) : aShape | Unprojects a multi-point shape. |
| Snap (aShape, snapDistance) : aBoolean | Returns true if a multi-point can be snapped to vertices of the specified shape. |

***See also:*** Circle; Line; Oval; Point; Polygon; PolyLine, Rect

# NameDictionary

This class implements a specialized dictionary where keys are the object names.

**Inherits from** Collection

| Class Requests | |
|---|---|
| Make (hashSize) : aNameDictionary | Creates a name dictionary based on the specified hash size. |

| Instance Requests | |
|---|---|
| + anotherCollection : aCollection | Returns a new name dictionary or list by combining another collection object with a name dictionary. |
| Add (anObj) : aBoolean | Adds the specified object to a name dictionary and returns true if successful. |
| AsList : aList | Returns a list of objects stored in a name dictionary. |
| Count : aNumber | Returns the number of objects stored in a name dictionary. |
| Empty | Clears all objects from a name dictionary. |
| ForceMerge (anotherCollection) : aCollection | Merges a name dictionary with the specified collection object. |
| Get (aName) : anObj | Returns the object with the specified name from a name dictionary. |
| GetSize : aNumber | Gets the hash size of a name dictionary. |
| Merge (anotherCollection) : aCollection | Combines another collection with a name dictionary. |
| Remove (aName) : aBoolean | Removes the object with the specified name and returns true if successful. |
| ReturnKeys : aList | Returns a list of object names in a name dictionary. |
| Set (anObj) | Adds the specified object to a name dictionary. Objects are replaced with the same name. |
| SetSize (newHashSize) | Changes the hash size for a name dictionary. |

***See also:***   Dictionary; List

# NbrHood (Spatial Analyst)

This class defines a cell's neighborhood.

**Inherits from** Obj

| Class Requests | |
| --- | --- |
| Make : aNbrHood | Makes a default 3 by 3 cells rectangular neighborhood. |
| MakeAnnulus (innerRadius, outerRadius, inMapUnitsBoolean) : aNbrHood | Creates a donut shape neighborhood. |
| MakeCircle (aRadius, inMapUnitsBoolean) : aNbrHood | Creates a circular neighborhood. |
| MakeIrregular (aListOf NumberLists) : aNbrHood | Makes an irregular neighborhood. |
| MakeRectangle (aWidth, aHeight, inMapUnitsBoolean) : aNbrHood | Makes a rectangular neighboorhood. |
| MakeWedge (aRadius, startAngle, endAngle, inMapUnitsBoolean) : aNbrHood | Creates a wedge shape neighborhood. |

***See also:*** Grid

# Nearside

This class implements the Vertical Near-Side Perspective projection.

**Inherits from** Perspective

| Class Requests | |
| --- | --- |
| CanDoCustom : aBoolean | Always returns true. |
| Make (aRect) : aNearside | Creates a projection bounded by the specified rectangle. |

| Instance Requests | |
| --- | --- |
| ReturnHeight : aNumber | Returns height of view point. |
| SetHeight (aNumber) | Sets the height of view point. |

***See also:*** CoordSys; Spheroid

# *NetCostField (Network Analyst)*

This class represents the cost field in solving network problems.

**Inherits from** Obj

| Instance Requests | |
|---|---|
| GetFields : aFieldList | Returns the cost fields. |
| GetInputUnits : NetUnitsEnum | Returns the input units for a cost field. |
| GetLabel : aString | Returns the label for a cost field. |
| GetReportUnits : NetUnitsEnum | Returns the reporting units. |
| IsDir : aBoolean | Returns true when a cost field object has two directional fields. |
| SetReportUnits (NetUnitsEnum) | Sets the reporting units. |

## *Enumeration*

| NetUnitsEnum | |
|---|---|
| #NETUNITS_LINEAR_UNKNOWN | Unknown units of length. |
| #NETUNITS_LINEAR_INCHES | Inches. |
| #NETUNITS_LINEAR_FEET | Feet. |
| #NETUNITS_LINEAR_YARDS | Yards. |
| #NETUNITS_LINEAR_MILES | Miles. |
| #NETUNITS_LINEAR_MILLIMETERS | Millimeters. |
| #NETUNITS_LINEAR_CENTIMETERS | Centimeters. |
| #NETUNITS_LINEAR_METERS | Meters. |
| #NETUNITS_LINEAR_KILOMETERS | Kilometers. |
| #NETUNITS_LINEAR_NAUTICALMILES | Nautical miles. |
| #NETUNITS_LINEAR_DEGREES | Decimal degrees. |
| #NETUNITS_TIME_SECONDS | Seconds. |
| #NETUNITS_TIME_MINUTES | Minutes. |
| #NETUNITS_TIME_HOURS | Hours. |
| #NETUNITS_TIME_HMS | Time reported in hh:mm:ss. |
| #NETUNITS_TIME_UNKNOWN | Unknown units of time. |
| #NETUNITS_UNKNOWN | Unknown units of time or length. |

***See also:*** Field; NetDef; NetUnits; Network

# *NetDef (Network Analyst)*

Objects of network definition class manage the fields in the line theme feature table and in the turntable that define travel costs and obstacles.

**Inherits from** Obj

| Class Requests | |
|---|---|
| CanMakeFromFTab (anFTab) : aBoolean | Returns true if a network definition can be created from a feature table. |
| CanMakeFromTheme (aTheme) : aBoolean | Returns true if a network definition can be created from the specified theme. |
| Make (anFTab) : aNetDef | Makes a network definition from the specified feature table. |
| NetworkDBExists (anFTab) : aBoolean | Returns true if a network index directory exists for the given FTab. |

| Instance Requests | |
|---|---|
| BuildTopology | Builds node numbers for the edge table. |
| Close | Closes network definition to free memory. |
| GetCostFields : aFieldList | Gets the cost fields. |
| GetEdgeFTab : aFTab | Gets the line feature table for edges. |
| GetElevFields : aFieldList | Gets the elevation fields. |
| GetNetworkDBFN : aFileName | Gets the name of the network index directory. |
| GetNodeFields : aFieldList | Gets the node fields. |
| GetNodeVTab : aVTab | Gets the node table. |
| GetOneWayField : aField | Gets the one-way field. |
| GetSchema : aDictionary | Gets the network dictionary. |
| GetTurnCosts : aFieldList | Gets the cost fields in a turntable. |
| GetTurnVTab : aVTab | Gets the turntable. |
| HasError : aBoolean | Returns true if the network index direcory contains an error or does not have write permission. |
| HaveTopology : aBoolean | Returns true if node topology exists. |
| IsBeingUpdated : aBoolean | Returns true if the network index directory is currently being updated. |
| IsCurrent : aBoolean | Returns true if the network index directory is up to date with the network data. |
| IsOpen : aBoolean | Returns true if the internal network files are open. |

| | |
|---|---|
| IsWritable : aBoolean | Returns true if the network index directory is writable. |
| Open : aBoolean | Creates and updates the network index directory. |
| SetSchema (aDictionary) | Sets the network dictionary. |
| SetTurnVTab (aVTab) : aBoolean | Sets the given turntable for a shapefile. |
| StartUpdate | Opens the network index directory for editing. |
| StopUpdate | Saves edits to the network index directory. |
| UpdateCosts | Updates the network index directory with a selection from the line feature table. |

**See also:**  NetCostField; NetUnits; Network

# NetUnits (Network Analyst)

This class provides services for transforming units of time and length useful for maniuplating and reporting the cumulative cost to traverse a path.

**Inherits from** Obj

| *Class Requests* | |
|---|---|
| ConvertMeasEnum (UnitsLinearEnum) : NetUnitsEnum | Returns the corresponding network unit. |
| ConvertNetUnitsEnum (NetUnitsEnum) : UnitsLinearEnum | Returns the corresponding linear unit. |
| GetDomain (NetUnitsEnum) : NetUnitsDomainEnum | Returns unit's domain. |
| GetEnum (aName) : NetUnitsEnum | Returns the network unit enumeration from the given input or report name. |
| GetFullUnitString (NetUnitsEnum) : aString | Returns the name of the specified network unit enumeration. |
| GetInputNames (NetUnitsEnum) : aStringList | Returns the allowed conversions for the domain of the given enumeration. |
| GetReportNames (NetUnitsEnum) : aStringList | Returns the allowed forward conversions for the domain of the given enumeration. |
| GetUnitString (NetUnitsEnum) : aString | Returns the abbreviated name of the specified enumeration. |
| GetUnknownName (NetUnitsDomainEnum) : aString | Gets the name of the unknown unit in the given domain. |
| IsUnknown (NetUnitsEnum) : aBoolean | Returns true if the enumeration indicates an unknown unit. |

| Make (inputNetUnitsEnum, reportNetUnitsEnum) : aNetUnits | Creates a network unit object. |
| VelocityToTime (fromLinearMeasure, fromNetUnitsEnum, velocityLength, velocityNetUnitsEnum, velocityTimeNetUnitsEnum) : aNumber | Calculates from linear measure to seconds. |

| Instance Requests | |
| --- | --- |
| Convert (aValuej) : aValue | Converts a value from input to report unit. |
| GetDefNumber : aNumber | Gets the default number in formatting network units. |
| GetInputUnits : NetUnitsEnum | Gets the input unit enumeration. |
| GetReportUnits : NetUnitsEnum | Gets the report unit enumeration. |
| HasErr : aBoolean | Returns true when a conversion causes error. |
| SetHMSFormat (formatString, amString, pmString) | Sets the format for reading and writing hh:mm:ss data. |
| SetInputUnits (NetUnitEnum) | Sets the input unit. |
| SetReportUnits (NetUnitEnum) | Sets the report unit. |
| SetStartMeasure (NetUnitEnum, anObj) | Sets a starting value for reported values. |

## Enumerations

| NetUnitsEnum | |
| --- | --- |
| #NETUNITS_LINEAR_UNKNOWN | Unknown units of length. |
| #NETUNITS_LINEAR_INCHES | Inches. |
| #NETUNITS_LINEAR_FEET | Feet. |
| #NETUNITS_LINEAR_YARDS | Yards. |
| #NETUNITS_LINEAR_MILES | Miles. |
| #NETUNITS_LINEAR_MILLIMETERS | Millimeters. |
| #NETUNITS_LINEAR_CENTIMETERS | Centimeters. |
| #NETUNITS_LINEAR_METERS | Meters. |
| #NETUNITS_LINEAR_KILOMETERS | Kilometers. |
| #NETUNITS_LINEAR_NAUTICALMILES | Nautical miles. |
| #NETUNITS_LINEAR_DEGREES | Decimal degrees. |
| #NETUNITS_TIME_SECONDS | Seconds. |
| #NETUNITS_TIME_MINUTES | Minutes. |
| #NETUNITS_TIME_HOURS | Hours. |

| #NETUNITS_TIME_HMS | Time reported in hh:mm:ss. |
| #NETUNITS_TIME_UNKNOWN | Unknown units of time. |
| #NETUNITS_UNKNOWN | Unknown units of time or length. |

| **NetUnitsDomainEnum** | |
| --- | --- |
| #NETUNITS_DOMAIN_TIME | Time domain. |
| #NETUNITS_DOMAIN_LINEAR | Linear domain. |
| #NETUNITS_DOMAIN_UNKNOWN | Unknown domain. |

**See also:** NetCostField; NetDef; Network; Units

# NetWork (Network Analyst)

This class is used in solving network problems.

**Inherits from Obj**

| **Class Requests** | |
| --- | --- |
| Make (aNetDef) : aNetwork | Creates a network object. |

| **Instance Requests** | |
| --- | --- |
| ClearClosestFacResult | Clears the closest facility result. |
| ClearPathResult | Clears the path result. |
| ClearServiceAreaResult | Clears the service area result. |
| FindClosestFac (aOriginPointList, aFacilityPointList, numberOfFacilities, cutOffNumber, ToFacilityBoolean) : aNumberList | Solves closest facility problem. |
| FindPath (aPointList, FindBestOrderBoolean, ReturnToOriginBoolean) : aNumber | Solves the best path problem. |
| FindPathCost (aPointList, FindBestOrderBoolean, ReturnToOriginBoolean) : aNumber | Gets a paths cost by solving for the best path. |
| FindServiceArea (aPointList, aListOfCostLists, FromPointsBoolean, CompactAreaBoolean) : aBoolean | Solves the service area problem. |
| GetClosestFacDirections (resultFTab) : aString | Returns the direction string between each origin and its closest facilities. |
| GetClosestFacIndex (originIndex, nthClosestFacility) : aNumber | Returns the index of the nth closest facility. |

| | |
|---|---|
| GetClosestFacPathCost (originIndex, nthClosestFacility) : aNumber | Returns the cost of travel to the nth closest facility. |
| GetDefaultSearchTol : aNumber | Gets the default search tolerance. |
| GetNetDef : aNetDef | Gets the network definition object of a network. |
| GetPathDirections (resultFTab) : aString | Returns the path's direction. |
| GetSearchTol : aNumber | Gets the search tolerance. |
| HasClosestFacResult : aBoolean | Returns true if the closest facility problem is solved succcessfully. |
| HasPathResult : aBoolean | Returns true if the path problem is solved succcessfully. |
| HasServiceAreaResult : aBoolean | Returns true if the service area problem is solved succcessfully. |
| IsPointOnNetwork (aPoint) : aBoolean | Returns true if the specified point is on the network within the search tolerance. |
| ResetSearchTol | Sets the search tolerance to the network's default value. |
| ReturnClosestFacShape (originIndex, nthClosestFac) : aShape | Returns the path shape to the nth closest facility. |
| ReturnPathShape : aShape | Returns the result of a path problem. |
| SetCostField (aCostField) | Sets the cost field. |
| SetDirCostField (aCostField) | Sets the field for reporting travel cost. |
| SetDirLandMarks (aPointTheme, aLabelField) | Sets the field in a point theme for reporting land marks as part of directions. |
| SetDirStreetNameFields (aFieldList) | Sets the fields used for street names when generating directions. |
| SetSearchTol (aNumber) | Sets the search tolerance. |
| WriteClosestFac (aFileName) : aBoolean | Saves the closest facility paths to a shape file. |
| WriteClosestFacLong (aFileName) : aBoolean | Saves the closest facility paths in their long format to a shape file. |
| WriteClosestFacPath (aFileName, originIndex, nthClosestFac) : aBoolean | Saves the path for the specified pair to a shape file. |
| WriteClosestFacPathLong (aFileName, ithOrigin, jthClosestFac) : aBoolean | Saves the path in long format for the specified pair to a shape file. |
| WritePath (aFileName) : aBoolean | Writes the network paths to a file. |
| WritePathLong (aFileName) : aBoolean | Writes the network paths in long format to a file. |
| WriteServiceArea (aNetFileName, anAreaFileName) | Writes service network shapefile and service area shapefile to the specified files. |

***See also:*** NetCostField; NetDef; NetUnits; NetworkSrc

# NetWorkSrc (Network Analyst)

This class models the storage and retrieval for network objects associated with an FTab.

**Inherits from** SrcExtension

| Class Requests | |
|---|---|
| Make (aNetwork) : aNetworkSrc | Creates a network source object. |

| Instance Requests | |
|---|---|
| GetNetwork : aNetwork | Gets the network object. |

*See also:* Network

# NetworkWin (Network Analyst)

This abstract class models network problem definition dialog boxes.

**Inherits from** Window

**Inherited by** ClosestFacilityWin, ServiceAreaWin, and ShortestPath-Win

| Class Requests | |
|---|---|
| GetInvalidSymbol : aSymbol | Gets the symbol used for invalid points. |
| GetValidSymbol : aSymbol | Gets the symbol used for valid points. |
| SetInvalidSymbol (aSymbol) | Sets the invalid symbol. |
| SetValidSymbol (aSymbol) | Sets the valid symbol. |

| Instance Requests | |
|---|---|
| AddPoint (aFTab, aLabel, aPoint) | Adds the specified point to a network window. |
| CanSolve : aBoolean | Returns true if a network window contains enough information to solve its query. |
| Close | Closes a network window. |
| GetFTab : aFTab | Gets the feature table associated with a network window. |
| GetNetwork : aNetwork | Returns the network associated with a network window. |

| GetNextPointNum : aNumber | Returns the next point number used for labeling a new graphic pick. |
|---|---|
| GetNumPoints : aNumber | Returns the number of points in a network window. |
| GetSourceTheme : aFTheme | Gets the network source theme. |
| GetViewSearchTol : aNumber | Gets the search tolerance based on the given View's extent. |
| IsOpen : aBoolean | Returns true if a network window is open. |
| Refresh | Refreshes the display of a network window. |
| Solve : aBoolean | Processes the query of a network window. |

*See also:* Network; NetworkWinSrc

# NetworkWinSrc (Network Analyst)

This class stores and retrieves the network window object.

### Inherits from ThemeExtension

| Class Requests | |
|---|---|
| Make (aNetworkWin) : aNetworkWinSrc | Creates a network window source. |

| Instance Requests | |
|---|---|
| GetNetworkWin : aNetworkWin | Gets the network window. |

*See also:* NetworkWin

# NewZealand

This class implements the Cauchy-Rieman based general conformal projection used for the New Zealand National Grid.

### Inherits from Prj

| Class Requests | |
|---|---|
| CanDoCustom : aBoolean | Always returns false. |
| CanDoSpheroid : aBoolean | Always returns true. |
| Make (aRect) : aNewZealand | Creates a projection bounded by the specified rectangle. |

| *Instance Requests* | |
|---|---|
| CalcHorizon | Calculates the horizon polygon. |
| ProjectPt (aPoint) : aBoolean | Projects a point and returns true if successful. |
| Recalculate | Calculates derived constants. |
| UnProjectPt (aPoint) : aBoolean | Unprojects a point and returns true if successful. |

**See also:** CoordSys; Spheroid

# Nil

This class implements an object without a value.

### Inherits from Value

| *Instance Requests* | |
|---|---|
| = anObj : aBoolean | Returns true when the specified object is nil. |
| AsString : aString | Returns the word nil. |

**See also:** Boolean; Number; String

# NorthArrow

This class implements north arrows.

### Inherits from GraphicGroup

| *Class Requests* | |
|---|---|
| Make (aRect) : aNorthArrow | Creates a north arrow bounded by the specified rectangle. |

| *Instance Requests* | |
|---|---|
| CanSimplify : aBoolean | Always returns true. |
| Edit (aListofGraphics) : aBoolean | Displays the north arrow manager dialog box. |
| EditSizeAndPos : aBoolean | Displays the size and position editor. |
| GetAngle : aNumber | Gets the rotation angle of a north arrow in decimal degrees. |
| IsGroup : aBoolean | Always returns true. |

| | |
|---|---|
| Offset (aPoint) | Moves a north arrow by the specified amount. |
| SetAngle (aNumber) | Sets the rotation angle for a north arrow. |
| SetArrow (aGraphic) | Sets a north arrow to the specified graphic object. |
| SetBounds (aRect) | Sets a north arrow's bounds to the specified rectangle. |

***See also:*** Display; GraphicList; Layout; Template

# NorthArrowMgr

This class implements the dialog window for managing north arrow objects.

**Inherits from** ModalDialog

| Class Requests | |
|---|---|
| Add (aGraphic) | Adds the graphic object to the list of north arrows. |
| Merge (aListofGraphics) | Adds a list of graphic objects as north arrows. |

***See also:*** Graphic; NorthArrow

# Number

This class implements numerical values.

**Inherits from** Value

| Class Requests | |
|---|---|
| GetDefFormat : aString | Gets the default format. |
| GetEuler : aNumber | Gets Euler number. |
| GetPi : aNumber | Gets the pi value. |
| MakeNull : aNumber | Creates a null number. |
| MakeRandom (min, max) : aNumber | Generates a random number. |
| SetDefFormat (aFormatString) | Sets the default format. |

| Instance Requests | |
|---|---|
| .. anotherNumber : anInterval | Creates an interval. |
| + anotherNumber : aNumber | Returns the sum of two numbers. |

| | |
|---|---|
| - anotherNumber : aNumber | Returns the result of a subtraction. |
| * anotherNumber : aNumber | Returns the multiplication result of two numbers. |
| / anotherNumber : aNumber | Returns the result of a division. |
| ^ anotherNumber : aNumber | Returns the result of a number to the power of another. |
| = anotherNumber : aBoolean | Returns true when two number objects have the same value. |
| @ yNumber : aPoint | Creates a point at the specified y and x. |
| Abs : aNumber | Returns the absolute value of a number. |
| ACos : aNumber | Returns the arc cosine of a number. |
| AsChar : aString | Returns the equivalent ASCII character of a number. |
| AsDate : aDate | Returns a date from a number in YYYYMMDD format. |
| AsDays : aDuration | Creates a duration object. |
| AsDegrees : aNumber | Converts radians to degrees. |
| AsGrid : aGrid | Creates a uniform value grid. |
| AsHexString : aString | Converts decimal numbers to hexadecimal. |
| AsHours : aDuration | Creates a duration object. |
| ASin : aNumber | Returns the arc sine of a number. |
| AsMinutes : aDuration | Creates a duration object. |
| AsRadians : aNumber | Converts degrees to radians. |
| AsSeconds : aDuration | Creates a duration object. |
| AsString : aString | Converts a number to its string format. |
| AsYears : aDuration | Creates a duration object. |
| ATan : aNumber | Returns the inverse tanget in radians of a number. |
| Ceiling : aNumber | Rounds an integer up. |
| Cos : aNumber | Returns cosine of a number in radians. |
| Floor : aNumber | Rounds a number down. |
| GetFormat : aString | Gets the current format string for a number. |
| IsInfinity : aBoolean | Returns true if a number is infinity. |
| IsNull : aBoolean | Returns true when the value of a number object is null. |
| Ln : aNumber | Returns the natural logarithm of a number. |
| Log (baseNumber) : aNumber | Returns logarithm of a number. |

| | |
|---|---|
| Max anotherNumber : aNumber | Returns the maximum of two numbers. |
| Min anotherNumber : aNumber | Returns the minimum of two numbers. |
| Mod anotherNumber : aNumber | Returns the remainder of a division operation. |
| Negate : aNumber | Negates a number. |
| Round : aNumber | Rounds a number to the nearest integer. |
| SetFormat (aFormatString) : aNumber | Sets the format of a number to the specified format string. |
| SetFormatPrecision (umberOfDecimalPoints) : aNumber | Sets the displayed precision of a number. |
| Sin : aNumber | Returns sine of a number in radians. |
| Sqrt : aNumber | Returns the square root of a positive number. |
| Tan : aNumber | Returns the tangent of a number in radian. |
| Truncate : aNumber | Truncates the decimal portion of a number. |

***See also:*** Date; Interval; Script; String

# *Obj*

This is the abstract superclass for all other classes in ArcView.

**Inherits from** Obj

**Inherited by** AddressStyle, Application, AttrUpdate, BandStatistics, ChartDisplay, ChartPart, Classification, Codepage, Collection, Color, ColorMap, Control, CoordSys, Coverage, Date, DBICursor, DDEClient, DDEServer, Digit, Display, DLL, DLLProc, Doc, DocGUI, DocWin, ExtensionObject, Field, File, FileName, Font, FontManager, Graphic, Grid, Help, Icon, ImageLegend, ImageLookup, ImgSrc, INFODir, Interp, Labeler, Layer, Legend, Librarian, Library, LockMgr, MatchCand, MatchCase, MatchField, MatchKey, MatchPref, MatchSource, Modal-Dialog, MsgBox, NbrHood, NetCostField, NetDef, NetUnits, NetWork, ODB, PageManager, Palette, Pixmap, Printer, Prj, Radius, RPCClient, RPCServer, Script, SDEDataset, SDEFeature, SDELayer, SDELog, Shape, Spheroid, SQLCon, SrcName, Stipple, Symbol, SymbolWin, System, TextComposer, TextPositioner, TGraphic, Theme, Threshold, TOC, Units, Value, VTab, and Window

| Class Requests | |
| --- | --- |
| GetClassName : aString | Gets the name of a class. |
| HasInstances : aBoolean | Returns true when at least one instance of a class exists. |
| IsSubclassOf (anotherClass) : aBoolean | Returns true if a class is the subclass of a specified class. |

| Instance Requests | |
| --- | --- |
| <> anotherObj : aBoolean | Returns true when two objects are not the same. |
| = anotherObj : aBoolean | Returns true when two objects are the same. |
| AsString : aString | Returns the object name. |
| Clone : anObj | Clones an object. |
| GetClass : aClass | Returns the class of an object. |
| GetName : aString | Gets the object name. |
| Is (aClass) : aBoolean | Returns true when an object is from the specified class. |
| SetName (aName) | Sets an object's name. |

*See also:* ODB

# ObliqueMercator

This class implements the Oblique Mercator projection.

**Inherits from** Prj

| Class Requests | |
| --- | --- |
| CanDoSpheroid : aBoolean | Always returns true. |
| Make (aRect) : anOblmerc | Creates a projection bounded by the specified rectangle. |

| Instance Requests | |
| --- | --- |
| ProjectPt (aPoint) : aBoolean | Projects a point and returns true when successful. |
| Recalculate | Computes derived constants. |
| ReturnAzimuthOfCentralLine : aNumber | Returns the azimuth angle. |
| ReturnCentralLatitude : aNumber | Returns the latitude in degrees. |
| ReturnCentralLongitude : aNumber | Returns the longitude in degrees. |
| ReturnFalseEasting : aNumber | Returns the value of X at the central meridian. |

| ReturnFalseNorthing : aNumber | Returns the value of Y at the central meridian. |
|---|---|
| ReturnScale : Number | Returns the scale along the central line. |
| SetAzimuthOfCentralLine (anAzimuthNumber) | Sets the azimuth of a line. |
| SetCentralLatitude (aLatitude) | Sets the latitude of the central point. |
| SetCentralLongitude (aLongitude) | Sets the longitude of the central point. |
| SetFalseEasting (aNumber) | Sets the value of X at the central meridian. |
| SetFalseNorthing (aNumber) | Sets the value of Y at the central meridian. |
| SetScale (scaleNumber) | Sets the scale along the central meridian. |
| UnProjectPt (aPoint) : aBoolean | Unprojects a point and returns true when successful. |

*See also:* CoordSys; Spheroid

# *ODB*

An object database is a file containing a collection of objects.

**Inherits from** Obj

**Inherited by** Extension

| Class Requests | |
|---|---|
| Make (aFileName) : anODB | Creates an ODB object. |
| Open (anODBFilename) : anODB | Opens an existing ODB file. |
| OpenAndInstall (aFileName) : anODB | Loads and installs an extension. |

| Instance Requests | |
|---|---|
| Add (anObj) | Adds the object to an ODB. |
| Clear | Removes all objects from an ODB. |
| Commit : aBoolean | Saves changes to an ODB. |
| Count : aNumber | Returns the number of root objects in an ODB. |
| Get (aListIndexNumber) : anObj | Gets an object from an ODB. |
| GetCodepage : aNumber | Returns the codepage used for an object database. |
| GetDependencies : filenameList | Returns an ODB's dependency. |
| GetFileName : aFileName | Gets ODB's filename. |
| GetVersion : aNumber | Returns ODB's version number. |

*See also:* Application; File; Project

*Code example:* page 461

# Orthographic

This class implements the Orthographic projection.

**Inherits from** Perspective

| Class Requests | |
|---|---|
| CanDoCustom : aBoolean | Always returns true. |
| Make (aRect) : anOrtho | Creates a projection bounded by the specified rectangle. |

**See also:** CoordSys; Spheroid

# Oval

This class implements an oval geometric shape.

**Inherits from** Shape

| Class Requests | |
|---|---|
| Make (anOriginPoint, aSizePoint) : anOval | Creates an oval bounded by the rectangle created from the origin and size points. |
| MakeNull : anOval | Creates a null oval. |

| Instance Requests | |
|---|---|
| = anObj : aBoolean | Returns true when the object is the same as an oval. |
| AsArc (startAngle, endAngle) : aPolyLine | Returns a polyline for the specified arc of an oval. |
| AsMultiPoint : aMultiPoint | Returns the multi-point equivalent of an oval. |
| AsPolygon : aPolygon | Returns the polygon equivalent of an oval. |
| AsString : aString | Returns the description of an oval. |
| Contains (aShape) : aBoolean | Returns true if the specified shape is contained in an oval. |
| ExpandBy (distanceNumber) : anOval | Enlarges an oval's axes by the specified length. |
| GetDimension : aNumber | Always returns 2. |
| InsetBy (distanceNumber) : anOval | Shrinks an oval's axes by the specified amount. |
| Intersects (aShape) : aBoolean | Returns true if the shape intersects an oval. |
| IsNull : aBoolean | Returns true if an oval is null. |
| Move (moveX, moveY) : anOval | Moves an oval by the specified amount. |

| | |
|---|---|
| ReturnCenter : aPoint | Returns an oval's center point. |
| ReturnDifference (aShape) : aPolygon | Returns the difference between an oval and the given shape. |
| ReturnIntersection (aShape) : aPolygon | Returns the intersection of an oval and the given shape. |
| ReturnMerged (aShape) : aPolygon | Merges the shape with an oval. |
| ReturnProjected (aPrj) : aShape | Projects an oval. |
| ReturnUnion (aShape) : aPolygon | Returns the union of an oval and the given shape. |
| ReturnUnprojected (aPrj) : aShape | Unprojects an oval. |
| SetOrigin (aPoint) : anOval | Sets the origin point of an oval. |
| SetSize (aPoint) : anOval | Sets the width and height of an oval. |
| Snap (aShape, snapDistance) : aBoolean | Returns true if the shape can snap to an oval within the specified tolerance. |

**See also:** Circle; Line; Point; Polygon; Rect

# PageDisplay

This class implements the layout display.

**Inherits from** Display

| Instance Requests | |
|---|---|
| BeginClip : aBoolean | First request before drawing on a page display. |
| DrawIcon (aRect, anIcon) | Draws the specified icon in the specified rectangle on a page display. |
| DrawLine (aLine, aSymbol) | Draws the specified line using the specified symbol on a page display. |
| DrawMultiPoint (aMultiPoint, aSymbol) | Draws the specified multi-point shape using the specified symbol on a page display. |
| DrawOval (aOval, aSymbol) | Draws the specified oval using the specified symbol on a page display. |
| DrawPoint (aPoint, aSymbol) | Draws the specified point using the specified symbol on a page display. |
| DrawPolygon (aPolygon, aSymbol, drawOutlineBoolean) | Draws the specified polygon using the specified symbol on a page display. |
| DrawPolyLine (aPolyLine, aSymbol) | Draws the specified polyline shape using the specified symbol on a page display. |
| DrawRect (aRect, aSymbol) | Draws the specified rectangle using the specified symbol on a page display. |

| | |
|---|---|
| DrawText (aPoint, aString, aSymbol) | Draws the specified text string at the specified point and using the specified symbol on a page display. |
| DrawTextPolyLine (aPolyLine, aString, aSymbol) | Draws the specified text string along the specified polyline using the specified symbol on a page display. |
| EndClip | Last request after drawing on a page display. |
| ExportEnd | Ends exporting of graphics on a page display. |
| Flush | Draws all graphics in the buffer on a page display. |
| GetResolution : DisplayResEnum | Gets the resolution quality. |
| HookupSymbol (aSymbol) | Hooks a symbol to a page display's scale. |
| Invalidate (eraseBoolean) | Invalidates a page display if the specified Boolean value is true. The page display is erased before redrawing. |
| InvalidateRect (aRect) | Invalidates the area of a page display indicated by the specified rectangle. |
| IsGridActive : aBoolean | Returns true when grid is active on a page display. |
| IsGridVisible : aBoolean | Returns true when grid is visible on a page display. |
| IsMarginVisible : aBoolean | Returns true when margin is visible on a page display. |
| IsUsingPrinterMargins : aBoolean | Returns true when a page display is using printer's margin. |
| IsUsingPrinterPageSize : aBoolean | Returns true when a page display is using printer's page size. |
| Pan | Pans a display during the apply event. |
| ReturnPixmap (aRect) | Returns a pixmap contained in the specified rectangle. |
| ReturnUserPoint : aPoint | Returns the location of the mouse during the apply event. |
| SetGridActive (isActiveBoolean) | Sets a grid's state to the specified Boolean value. |
| SetGridMesh (aPoint) | Sets the size of a grid to the specified X and Y values. |
| SetGridVisible (isVisibleBoolean) | Sets the visibility state of a grid to the specified Boolean value. |
| SetMargin (aRect) | Sets the margin to the specified value. |
| SetMarginVisible (isVisibleBoolean) | Sets the visibility state of margin lines to the specified Boolean value. |

| SetPageSize (aPoint) | Sets the page size of a page display to the specified X and Y values. |
|---|---|
| SetResolution (DisplayResEnum) | Sets the resolution quality of a page display. |
| SetUsingPrinterMargins (isUsedBoolean) | Uses the printer margin if the specified Boolean value is true. |
| SetUsingPrinterPageSize (isUsedBoolean) | Uses the printer's page size if the specified Boolean value is true. |
| UndoZoom | Reverts the last zoom or pan on a page display. |
| Validate | Validates a page display. |
| ZoomToActual | Zooms to a one-to-one scale. |
| ZoomToPage | Shows the entire page display. |
| ZoomToRect (aRect) | Zooms to a page display of the area specified by the specified rectangle. |

## *Enumerations*

| **DisplayResEnum** | |
|---|---|
| #DISPLAY_RES_LOW | Low resolution. |
| #DISPLAY_RES_NORMAL | Normal resolution. |
| #DISPLAY_RES_HIGH | High resolution. |

***See also:*** MapDisplay; PageSetupDialog

# *PageManager*

This class implements standard page sizes for printer and plotter.

**Inherits from** Obj

| **Class Requests** | |
|---|---|
| FindSize (aPoint) : PageManagerSizeEnum | Returns a page size enumeration that matches the specified width and height. |
| GetUnits (PageManagerSizeEnum) : UnitsLinearEnum | Gets the measurement unit used in a standard page size. |
| ReturnSize (PageManagerSizeSize) : aPoint | Returns the width and height of a page size. |

## Enumerations

| PageManagerSizeEnum | |
|---|---|
| #PAGEMANAGER_SIZE_A0 | 84.1 x 118.9 centimeters. |
| #PAGEMANAGER_SIZE_A1 | 59.4 x 84.1 centimeters. |
| #PAGEMANAGER_SIZE_A2 | 42.0 x 59.4 centimeters. |
| #PAGEMANAGER_SIZE_A3 | 29.7 x 42.0 centimeters. |
| #PAGEMANAGER_SIZE_A4 | 21.0 x 29.7 centimeters. |
| #PAGEMANAGER_SIZE_C | 17.0 x 22.0 inches. |
| #PAGEMANAGER_SIZE_CAMERA | 7.3 x 11.0 inches. |
| #PAGEMANAGER_SIZE_CUSTOM | A user defined size. |
| #PAGEMANAGER_SIZE_D | 22.0 x 34.0 inches. |
| #PAGEMANAGER_SIZE_E | 34.0 x 44.0 inches. |
| #PAGEMANAGER_SIZE_LEGAL | 8.5 x 14.0 inches. |
| #PAGEMANAGER_SIZE_LETTER | 8.5 x 11.0 inches. |
| #PAGEMANAGER_SIZE_PRINTER | Printer's default size. |
| #PAGEMANAGER_SIZE_TABLOID | 11.0 x 17.0 inches. |

| UnitsLinearEnum | |
|---|---|
| #UNITS_LINEAR_CENTIMETERS | Centimeters. |
| #UNITS_LINEAR_DEGREES | Decimal degrees. |
| #UNITS_LINEAR_FEET | Feet. |
| #UNITS_LINEAR_INCHES | Inches. |
| #UNITS_LINEAR_KILOMETERS | Kilometers. |
| #UNITS_LINEAR_METERS | Meters. |
| #UNITS_LINEAR_MILES | Miles. |
| #UNITS_LINEAR_MILLIMETERS | Millimeters. |
| #UNITS_LINEAR_NAUTICALMILES | Nautical miles. |
| #UNITS_LINEAR_UNKNOWN | Unknown. |
| #UNITS_LINEAR_YARDS | Yards. |

# PageSetupDialog

This class implements the modal dialog box for setting page sizes.

**Inherits from** ModalDialog

| Class Requests | |
|---|---|
| Show (aDisplay) | Shows the page setup modal window for a specified display object. |

*See also:* Display; Layout

# *Palette*

This class manages ArcView's symbols.

**Inherits from** Obj

| Class Requests | |
|---|---|
| Make : aPalette | Creates an empty palette. |
| MakeFromFile (aFile name) : aPalette | Creates a palette from a file. |

| Instance Requests | |
|---|---|
| Clear (PaletteListEnum) | Clears the specified symbol list in a palette. |
| GetList (PaletteListEnum) : aSymbolList | Gets the requested symbol list from a palette. |
| Import : aBoolean | Shows the file window to load an icon and adds it to the marker symbol list of a palette. |
| ImportAlLineSet (aFile name) : aBoolean | Loads ARC/INFO's line symbol set from the specified file and adds it to the pen symbol list of a palette. |
| ImportAlShadeSet (aFile name) : aBoolean | Loads ARC/INFO's shade symbol set from the specified file and adds it to the fill symbol list of a palette. |
| ImportFromFile (aFile name) : aBoolean | Loads an icon from the specified file and adds it to the marker symbol list of a palette. |
| Load (PaletteListEnum) : aBoolean | Shows the file window to load and adds the specified symbol list to a palette. |
| LoadFromFile (PaletteListEnum, aFile name) : aBoolean | Loads the specified symbol list from the specified file and adds it to a palette. |
| Save (PaletteListEnum) | Shows the file window to save the specified symbol list of a palette. |
| SaveToFile (PaletteListEnum, aFile name) | Saves the specified symbol list of a palette to the specified file. |
| SetList (PaletteListEnum, aSymbolList) | Sets the specified symbol list of a palette to the specified symbol list object. |

## Enumerations

| PaletteListEnum | |
|---|---|
| #PALETTE_LIST_ALL | All symbol lists. |
| #PALETTE_LIST_FILL | Shade or fill symbol list. |
| #PALETTE_LIST_PEN | Line or pen symbol list. |
| #PALETTE_LIST_MARKER | Marker symbol list. |
| #PALETTE_LIST_COLOR | Color symbol list. |

**See also:** SymbolList

# Pattern

Objects of this class represent string objects with wildcard characters.

**Inherits from String**

| Class Requests | |
|---|---|
| Make (aString) : aPattern | Makes a pattern object from the specified string. Use * and ? as wildcards. |

| Instance Requests | |
|---|---|
| = anObj : aBoolean | Returns true if comparison is between two identical patterns or a pattern matches a string object. |

# Pen

This class symbolizes lines.

**Inherits from Symbol**

**Inherited by BasicPen, CompositePen, and VectorPen**

| Instance Requests | |
|---|---|
| GetBgColor : aColor | Always returns nil. |
| GetOffset : aNumber | Gets a pen's offset value. |
| GetType : SymbolEnum | Always returns #SYMBOL_PEN. |

| SetBgColor (aColor) | Does not set anything because lines lack background color. |
|---|---|
| SetOffset (aNumber) | Sets a pen's offset value. |

*See also:* Fill; Marker

# Perspective

This class implements the General Vertical Perspective projection.

**Inherits from** Prj

**Inherited by** Gnomonic, Nearside, Ortho, and Stereo

| Class Requests | |
|---|---|
| CanDoCustom : aBoolean | Always returns false. |
| Make (aRect) : aPerspective | Creates a projection bounded by the specified rectangle. |

| Instance Requests | |
|---|---|
| CalcHorizon | Determines the boundary of the visible area. |
| Recalculate | Computes derived constants. |
| ReturnCentralMeridian : aNumber | Returns the central meridian in decimal degrees. |
| ReturnReferenceLatitude : aNumber | Returns the reference latitude in decimal degrees. |
| SetCentralMeridian (aLongitude) | Sets the central meridian to the specified decimal degrees. |
| SetReferenceLatitude (aLatitude) | Sets the reference latitude to the specified decimal degrees. |
| ProjectPt (aPoint) : aBoolean | Projects the specified point and returns true if successful. |
| UnProjectPt (aPoint) : aBoolean | Unprojects the specified point and returns true if successful. |

*See also:* CoordSys; Spheroid

# PictureFrame

This class provides the ability to display graphic files on layout documents.

**Inherits from** Frame

**Inherited by** DocFrame

| Class Requests | |
|---|---|
| Make (aRect) : aPictureFrame | Creates an empty picture frame. |
| MakeClipboard (aRect) : aPictureFrame | Creates a picture frame to display the contents of the system clipboard. |

| Instance Requests | |
|---|---|
| CanSimplify : Boolean | Always returns true. |
| Draw | Draws a picture frame. |
| Edit (aListofGraphics) : aBoolean | Shows the property editor. |
| GetFile name : aFile name | Gets the disk file associated with a picture frame. |
| GetFillObject : anObj | Gets the fill object of a picture frame. |
| IsFilled : aBoolean | Returns true if a picture frame is filled. |
| IsFilledBy (aClass) : aBoolean | Returns true if a picture frame is filled by the objects of the specified class. |
| SetFile name (aFile name) | Associates the disk file with a picture frame. |
| SetFillObject (anObj) | Fills a picture frame with the specified object. |

*See also:* Display; GraphicList; Template

# PieChartSymbol

This class models the graphic symbology used to draw a pie chart.

**Inherits from** ChartSymbol

| Class Request | |
|---|---|
| Make : aPieChartSymbol | Creates a pie chart symbol. |

| Instance Requests | |
|---|---|
| IsFirstValueRadius : aBoolean | Returns true when the first value of a pie chart symbol's value list is used to set the radius. |
| SetFirstValueRadius (aBoolean) | If set to true then the first value of a pie chart symbol's value list is used to set the radius. |

*See also:* ChartSymbol; ColumnChartSymbol

# Point

This class implements a location through X and Y, or latitude and longitude values.

**Inherits from** Shape

| Class Requests | |
|---|---|
| Make (anX, aY) : aPoint | Creates a point based on the specified coordinate. |
| MakeNull : aPoint | Creates a point without coordinates. |

| Instance Requests | |
|---|---|
| / (anotherPoint) : aPoint | Returns a new point by dividing the coordinates of two points. |
| - (anotherPoint) : aPoint | Returns a new point by subtracting the coordinates of the specified point from a point. |
| + (anotherPoint) : aPoint | Returns a new point by adding the coordinates of the specified point to a point. |
| * (anotherPoint) : aPoint | Returns a new point by multiplying the coordinates of the specified point by a point. |
| = anObj : aBoolean | Returns true when the specified object is the same as a point. |
| AsMultiPoint : aMultiPoint | Converts a point to a multi-point shape. |
| AsString : aString | Returns the string representation of a point. |
| GetDimension : aNumber | Always returns zero. |
| GetX : aNumber | Gets the X or longitude value of a point. |
| GetY : aNumber | Gets the Y or latitude value of a point. |
| Intersects (aShape) : aBoolean | Returns true when the specified shape intersects a point. |
| IsNull : aBoolean | Returns true if point coordinates are null. |
| Move (moveX, moveY) : aPoint | Moves a point by the specified amount. |
| ReturnCenter : aPoint | Returns a new point with the same coordinates. |
| ReturnMerged (aShape) : aMultiPoint | Returns a multi-point shape after merging the shape with a point. |
| ReturnProjected (aPrj) : aShape | Projects a point. |
| ReturnUnprojected (aPrj) : aShape | Unprojects a point. |
| SetX (aNumber) | Sets the X or longitude value of a point. |

| SetY (aNumber) | Sets the Y or latitude value of a point. |
| Snap (aShape, snapDistance) : aBoolean | Returns true if a point is snapped to the vertices of a shape within the specified distance. |

***See also:*** Circle; Line; Oval; PointTextPositioner; Polygon; PolyLine; Rect

# PointTextPositioner

This class implements placing a text string relevant to a point.

**Inherits from** TextPositioner

| Class Requests ||
|---|---|
| Make : aPointTextPositioner | Creates a default point text positioner that places text to the right of a point. |

| Instance Requests ||
|---|---|
| Calculate (anchorPoint, textExtentPoint, pointMarkerSize, nil) | Computes the position of a text block around the anchor point based on the text extent. The last parameter is not used. |

***See also:*** GraphicText; Point

# Polygon

This class implements polygon shapes.

**Inherits from** PolyLine

**Inherited by** Annotation

| Class Requests ||
|---|---|
| Make (aListofPointsList) : aPolygon | Creates rings from points specified in each list. |
| MakeNull : aPolygon | Creates a polygon without points. |

| Instance Requests ||
|---|---|
| AsPolygon : aPolygon | Clones a polygon object. |
| AsPolyLine : aPolyLine | Converts a polygon to polyline. |
| Clean : aPolygon | Removes a polygon's intersections with itself. |
| Contains (aShape) : aBoolean | Returns true if the specified shape is contained within a polygon. |

| | |
|---|---|
| Explode : polygonList | Returns a list of single-ring polygons for a multi-ring polygon. |
| GetDimension : aNumber | Always returns a 2. |
| Intersects (aShape) : aBoolean | Returns true if the specified shape intersects a polygon. |
| IsEmpty : aBoolean | Returns true for an empty polygon. |
| IsNull : aBoolean | Returns true if a polygon object has no points. |
| LineIntersection (anotherShape) : aPolyLine | Returns the intersection of a polygon and a shape. |
| PointIntersection (anotherShape) : MultiPoint | Returns the intersection of a polygon and a shape. |
| ReturnBuffered (aDistance) : aPolygon | Returns a new polygon by setting a buffer. |
| ReturnCenter : aPoint | Returns the centroid of a polygon. |
| ReturnClipped (aRect) : aShape | Clips a polygon by the specified rectangle. |
| ReturnDifference (aShape) : aPolygon | Subtracts the specified shape from a polygon. |
| ReturnIntersection (aShape) : aPolygon | Intersects the specified shape with a polygon. |
| ReturnMerged (aShape) : aPolygon | Merges a shape and polygon. |
| ReturnProjected(aPrj) : aShape | Projects a polygon. |
| ReturnUnion (aShape) : aPolygon | Returns the union of the given shape and a polygon. |
| ReturnUnprojected(aPrj) : aShape | Unprojects a polygon. |
| Split (aPolyLine) : polygonsList | Splits a polygon with the specified polyline. |

***See also:*** Circle; Line; MultiPoint; Oval; Point; Rect; PolygonTextPositioner

# *PolygonTextPositioner*

This class implements placing a text string relevant to a polygon.

**Inherits from** TextPositioner

| Class Requests | |
|---|---|
| Make : PolygonTextPositioner | Creates a default polygon text positioner. |

| Instance Requests | |
|---|---|
| Calculate (polygonAnchor, textExtentPoint, 0, alternateAnchorPoint) | Computes the position of a text block around the center of the specified polygon or alternate anchor point (if not nil) based on the text extent. The third parameter is not used. |

***See also:*** GraphicText; Polygon

# *PolyLine*

This class implements polyline shapes.

**Inherits from** MultiPoint

**Inherited by** Polygon

| Class Requests | |
|---|---|
| Make (aListofPointsList) : aPolyLine | Creates a polyline from points specified in each list. |
| MakeNull : aPolyLine | Creates an empty polyline. |

| Instance Requests | |
|---|---|
| = anObj : aBoolean | Returns true when the specified object is the same as a polyline. |
| Along (aPercentageNumber) : aPoint | Returns a point on a polyline at the specified percentage. |
| AsLine : aLine | Creates a line from the polyline. |
| AsList : aList | Returns a list of polyline points. |
| AsMultiPoint : aMultiPoint | Converts a polyline to a multi-point shape. |
| AsPolyLine : aPolyLine | Clones a polyline. |
| AsString : aString | Returns a string representation of a polyline. |
| Clean : aPolyLine | Optimizes the number of nodes in a polyline. |
| Contains (aShape) : aBoolean | Returns true if the specified shape is contained in a polyline. |
| CountParts : aNumber | Returns the number of line segments in a polyline. |
| Explode : polylineList | Returns a list of single-part polylines from a multi-part polyline. |
| Flip : aPolyLine | Flips the order of vertices in a polyline. |
| FlipPart (aPartNumber) : aPolyLine | Flips the order of vertices in the specified part of a polyline. |
| GetDimension : aNumber | Always returns 1. |
| Intersects (aShape) : aBoolean | Returns true if the specified shape intersects a polyline. |
| IsEmpty : aBoolean | Returns true for empty polylines. |
| IsNull : aBoolean | Returns true if a polyline is empty. |

| | |
|---|---|
| LineIntersection (anotherShape) : aPolyLine | Returns the intersection of a polyline with the specified shape. |
| PointIntersection (anotherShape) : aMultiPoint | Returns the intersection of a polyline with the specified shape. |
| PointPosition (aPoint) : aNumber | Returns the position of the point along a polyline in a percentage of the polyline's length. |
| QueryPointDistance (aPoint, aDistance) : aNumber | Returns the distance of a point from a polyline. |
| ReShape (anotherPolyLine) : aPolyLine | Returns a new polyline by replacing paths. |
| ReturnBuffered (aDistance) : aPolygon | Returns a polygon by applying a buffer distance to a polyline. |
| ReturnCenter : aPoint | Returns the point at 50% of a polyline's length. |
| ReturnClipped (aRect) : aShape | Clips a polyline based on the specified rectangle. |
| ReturnConnected : aPolyLine | Optimizes the number of paths in a polyline. |
| ReturnDensified (stepSizeLength) : aPolyLine | Adds points to a polyline at the specified increments. |
| ReturnMerged (aShape) : aPolyLine | Merges the specified shape with a polyline. |
| ReturnOffset (anOffset) : aPolyLine | Creates a new polyline at the given offset distance. |
| ReturnProjected (aPrj) : aShape | Projects a polyline. |
| ReturnUnprojected (aPrj) : aShape | Unprojects a polyline. |
| Snap (aShape, snapDistance) : aBoolean | Returns true if a polyline can be snapped to a shape within the specified snap distance. |
| Split (aShape) : polylineList | Splits a polyline with the specified shape. |
| SplitLines (anotherPolyLine) : polylineLists | Splits a polyline with the specified polyline. |

**See also:** Point; PolyLine; PolyLineTextPositioner; Shape

# PolyLineTextPositioner

This class implements placing a text string relevant to a polyline.

**Inherits from** TextPositioner

| Class Requests | |
|---|---|
| Make : aPolyLineTextPositioner | Creates a default polyline text positioner. |

| Instance Requests | |
|---|---|
| Calculate (polyLineAnchor, textExtentPoint, 0, alternateAnchorPoint) | Computes the position of a text block based on the bounding rectangle of the specified polyline or alternate anchor point (if not nil) based on the text extent. The third parameter is not used. |

*See also:* GraphicText; PolyLine

# Popup

This class models Pop-up menus.

### Inherits from Menu

| Class Requests | |
|---|---|
| Make : aPopup | Creates a pop-up menu. |

*See also:* PopupSet

# PopupSet

This class manages all pop-up menus associated with a document GUI.

### Inherits from ControlSet

| Instance Requests | |
|---|---|
| FindByLabel (aLabel) : aPopup | Finds a pop-up by its label. |
| GetActive : aPopup | Gets the active pop-up menu in a pop-up set. |
| SetActive (aPopup) | Sets the given pop-up menu as the active pop-up. |

*See also:* Popup

# Printer

This class represents the hardware independent printer device.

### Inherits from Obj

| Class Requests | |
|---|---|
| The : aPrinter | Creates an object for the current system printer. |

| Instance Requests | |
|---|---|
| Edit (aListOfStrings) : aNumber | Displays a print dialog box for selecting an item to print from a list. It returns an index to the list. |
| GetFile name : aString | Gets the print output file name. |
| GetFormat : PrinterFormatEnum | Gets the printer's print format. |
| GetJobName : aString | Gets the name of the current print job. |
| IsReady : aBoolean | Returns true if printer is ready to print. |
| ReturnMargins : aRect | Returns printer's margins. |
| ReturnPageSize : aPoint | Returns the width and height of printer's page. |
| SetFile name (aString) | Sets the printer output file name. |
| SetFormat (PrinterFormatEnum) | Sets the printer's print format. |
| SetJobName (aString) | Sets the name of the print job. |
| Setup | Displays the printer's setup dialog box. |

## Enumerations

| PrinterFormatEnum | |
|---|---|
| #PRINTER_FORMAT_CGM | CGM print format. |
| #PRINTER_FORMAT_POSTSCRIPT | PostScript print format. |

**See also:** Doc

# Prj

Subclasses of Prj implement specific projections.

**Inherits from** Obj

**Inherited by** Albers, Cassini, EqualAreaAzimuthal, EqualAreaCylindrical, EquidistantAzimuthal, EquidistantConic, EquidistantCylindrical, Hammer, Lambert, Mercator, Miller, Mollweide, NewZealand, ObliqueMercator, Perspective, Robinson, RSO, Sinusoid, and TransverseMercator

| Class Requests | |
|---|---|
| CanDoCustom : aBoolean | Returns true when a projection can be customized. Projections with fixed parameters return false. |
| CanDoSpheroid : aBoolean | Returns true if a projection supports ellipsoids. Projections that support only spheres return false. |

| | |
|---|---|
| Make (aRect) : aPrj | Creates a projection bounded by the specified rectangle. |
| MakeNull : aPrj | Creates an empty projection. |
| ReturnPrjName : aString | Returns the full name of a projection subclass. |

### Instance Requests

| | |
|---|---|
| = (anotherPrj) : aBoolean | Returns true when two projections are the same. |
| CalcHorizon | Calculates the visible horizon. |
| FixExtent (anExtent) | Adjusts the extent to fall within the acceptable bounds of a projection. |
| GetDisplayQuality : PrjDisplayQualityEnum | Gets the projection display quality. |
| GetSpheroid : aSpheroid | Gets the associated ellipsoid. |
| GetHorizon : aPolygon | Gets the visible horizon. |
| IsNull : aBoolean | Returns true if the projection is empty. |
| IsUnProjectable (aPoint) : aBoolean | Returns true if the point can be unprojected. |
| ProjectPt (aPoint) : aBoolean | Projects a point and returns true if successful. |
| ProjectRect (aRect,  PrjRectQualityEnum) | Projects the specified rectangle. |
| Recalculate | Computes derived constants. |
| ReturnDescription : aString | Returns description of a projection. |
| ReturnShift : aNumber | Returns the prime meridian in decimal degrees. |
| SetDescription (aSytring) | Sets the description for a projection. |
| SetDisplayQuality (PrjDisplayQualityEnum) | Sets the projection display quality. |
| SetSpheroid  (SpheroidEnum) | Sets the ellipsoid of a projection. |
| SetShift (aLongitude) | Sets the prime meridian to the specified decimal degrees. |
| UnProjectPt (aPoint) : aBoolean | Unprojects a point and returns true if successful. |
| UnProjectRect (aRect,  PrjRectQualityEnum) | Unprojects the specified rectangle. |

## Enumerations

### PrjDisplayQualityEnum

| | |
|---|---|
| #PRJ_DISPLAYQUALITY_CLEAN | Accurate projection but slow. |
| #PRJ_DISPLAYQUALITY_CLIPPED | The most accurate projection. (Not implemented in version 2.1.) |
| #PRJ_DISPLAYQUALITY_FAST | Fast projection but not fully checked for accuracy. |

| PrjEnum | |
|---|---|
| #PRJ_ALBERS | Albers equal area conic projection. |
| #PRJ_AZIMUTHALEQUALAREA | Lambert azimuthal equal area projection. |
| #PRJ_CASSINI | Cassini projection. |
| #PRJ_CYLINDRICALEQUALAREA | Cylindrical equal area projection. |
| #PRJ_EQUICYLINDRICAL | Equidistant cylindrical projection. |
| #PRJ_EQUIDISTANTAZIMUTHAL | Equidistant azimuthal projection. |
| #PRJ_EQUIDISTANTCONIC | Equidistant conic projection. |
| #PRJ_HAMMERAITOFF | Hammer-Aitoff projection. |
| #PRJ_LAMBERT | Lambert conformal conic projection. |
| #PRJ_MERCATOR | Mercator projection. |
| #PRJ_MILLER | Miller cylindrical projection. |
| #PRJ_MOLLWEIDE | Mollweide projection. |
| #PRJ_NEWZEAL | Cauchy-Rieman based conformal projection for New Zealand National Grid. |
| #PRJ_NONE | No projection. |
| #PRJ_OBLIQUEMERCATOR | Oblique Mercator projection. |
| #PRJ_PERSPECTIVE | Vertical perspective projection. |
| #PRJ_ROBINSON | Robinson projection. |
| #PRJ_RSO | Rectified Skew Orthomorphic for Malaysia only. |
| #PRJ_SINUSOIDAL | Sinusoidal projection. |
| #PRJ_TRNMERC | Transverse Mercator projection. |

| PrjRectQualityEnum | |
|---|---|
| #PRJ_RECTQUALITY_FINE | Projects points on the rectangle's perimeter. |
| #PRJ_RECTQUALITY_ROUGH | Projects only the four corners. |

| SpheroidEnum | |
|---|---|
| #SPHEROID_AIRY | Airy 1830. |
| #SPHEROID_AUSTRALIAN | Australian National 1965. |
| #SPHEROID_BESSEL | Bessel 1841. |
| #SPHEROID_CLARKE1866 | Alexander Ross Clarke 1866. |
| #SPHEROID_CLARKE1880 | Alexander Ross Clarke 1880. |
| #SPHEROID_EVEREST | Everest 1830. |

| #SPHEROID_FIRST | For ArcView's internal use only. |
|---|---|
| #SPHEROID_GRS80 | Geodetic Reference System 1980. |
| #SPHEROID_INTERNATIONAL1909 | John Fillmore Hayford 1909 (IUGG 1924). |
| #SPHEROID_KRASOVSKY | Krasovsky 1940. |
| #SPHEROID_LAST | For ArcView's internal use only. |
| #SPHEROID_OTHER | Unknown. |
| #SPHEROID_SPHERE | The world as a sphere. |
| #SPHEROID_WGS72 | World Geodetic System 1972. |
| #SPHEROID_WGS84 | World Geodetic System 1984. |

*See also:* CoordSys; Spheroid

*Code example:* page 513

# Project

This class implements the ArcView project.

**Inherits from** Doc

| Class Requests | |
|---|---|
| Make : aProject | Creates a new project. |
| Open (projectFile name) : aProject | Opens an existing project. |

| Instance Requests | |
|---|---|
| AddDoc (aDoc) | Adds a document object to the project. |
| AddGUI (aDocGUI) : aDocGUI | Adds a document user interface to the project. |
| AddScript (aScript) | Adds a script object to the project. |
| Close | Closes the project. |
| CloseAll | Closes all documents within a project. |
| Customize | Shows the customization dialog box. |
| Edit | Shows the project's property dialog box. |
| EncryptScripts | Encrypts and embeds the scripts. |
| FindDoc (aDocName) : aDoc | Finds a document by its name. |
| FindGUI (aGUIName) : aDocGUI | Finds a document user interface by its name. |
| FindGUIsFor (aDocSubClass) : docGUIList | Returns a list of DocGUIs associated with the specified class in a project. |

| | |
|---|---|
| FindScript (aScriptName) : aScript | Finds a script by its name. |
| GetButtons : labelButtonList | Gets a list of buttons that appear on the project window. |
| GetDependencies : extensionList | Returns a list of extension dependencies. |
| GetDocs : docList | Gets a list of all documents in a project. |
| GetDocsWithGroupGUI (aDocGUI) : docList | Gets a list of ArcView documents that have the specified DocGUI as their group DocGUI. |
| GetDocsWithGUI (aDocGUI) : docList | Gets a list of ArcView documents that are associated with the specified DocGUI in a project. |
| GetFile name : aFile name | Gets a project file name. |
| GetGUIs : docGUIList | Gets the document GUI objects in a project. |
| GetScripts : aNameDictionary | Gets the embedded scripts in a project. |
| GetSelColor : aColor | Gets the color used in highlighting a selection. |
| GetSelectedDocs : docList | Gets a list of documents selected in the project window of a project. |
| GetSelectedGUI : aDocGUI | Gets the DocGUI selected in the project window of a project. |
| GetSerialNumber : aString | Gets the ArcView serial number that created this project. |
| GetShutDown : aString | Gets the shutdown script name. |
| GetStartUp : aString | Gets the start-up script name. |
| GetVisibleGUIsWidth : numberOfPixels | Returns the width of DocGUI list in the project window. |
| GetWorkDir : aFile name | Gets a project's work directory. |
| HasDoc (aDoc) : aBoolean | Returns true if the specified document exists in a project. |
| Import (projectFile name) : aDoc | Imports the specified project into a project. |
| IsClosing : aBoolean | Returns true when a project is closing. |
| IsModified : aBoolean | Returns true if the project has been modified since the last save. |
| MakeDocName (aGUI) : aString | Creates a unique name for a document of the specified GUI class. |
| MakeFile name (prefixString, extensionString) : aFile name | Creates a unique file name in the working directory of a project. |
| MakeGUIName (baseDocGUIName) : aString | Creates a unique name for a DocGUI. |
| MakeSysDefault (aFile name, encryptedBoolean) | Creates a system default project. |

| | |
|---|---|
| MakeUserDefault | Sets the user default to a project. |
| RemoveDoc (aDoc) | Removes a document from a project. |
| RemoveGUI (GUIName) : aBoolean | Removes a GUI from a project. |
| RemoveScript (aScriptName) | Removes the specified script from a project. |
| ResetGUIs | Restores the default document GUIs. |
| Save : aBoolean | Saves the project and returns true if successful. |
| SetFile name (aFile name) | Sets a project's file name. |
| SetModified (aBoolean) | Sets the modified flag of a project. |
| SetSelColor (aColor) | Sets the selection color in a project. |
| SetSelectedDoc (aDoc, addToSelectionBoolean) | Selects the specified document in the project window. |
| SetSelectedGUI (aDocGUI) | Selects the specified document GUI from the project window. |
| SetShutDown (aScriptName) | Sets the shutdown script in a project. |
| SetStartUp (aScriptName) | Sets the start-up script in a project. |
| SetVisibleGUIsWidth (numberOfPixels) | Sets the width of DocGUI list in a project window. |
| SetWorkDir (aFile name) | Sets the working directory of a project. |
| ShowCodepage | Shows a project's codepage. |
| Update | Refreshes the list of DocGUIs. |

***See also:*** Application; DocGUI; ODB; Script

***Code examples:*** page 330, page 405

# ProjectionDialog

This class implements a projection dialog box.

**Inherits from** ModalDialog

| *Class Requests* | |
|---|---|
| Show (aView, UnitsLinearEnum) : aPrj | Displays the projection dialog box for a view. |

***See also:*** Prj; View

# QueryWin

This class implements the query builder window.

**Inherits from** Window

| Class Requests | |
|---|---|
| Make (aVTab) : aQueryWin | Creates a query builder window for the specified virtual table. |
| ReturnCalculation (aVTab, aField) : aString | Displays a modal dialog box to build a calculation expression for the specified field in the specified virtual table. |
| ReturnQuery (aVTab, defaultExpressionString) : aString | Displays a modal dialog box to build a query expression. |

| Instance Requests | |
|---|---|
| Close | Closes a query window. |
| Open | Displays a query window. |

***See also:*** Table; VTab

# Radius (Spatial Analyst)

Defines the search radius for interpolators that are based on a local moving average.

**Inherits from** Obj

| Class Requests | |
|---|---|
| Make : aRadius | Makes a variable radius with a sample count of 12 and no maximum search distance. |
| MakeFixed (searchDistance, minimumSampleCount) : aRadius | Creates a fixed radius. |
| MakeVariable (aSampleCount, maxSearchDistance) : aRadius | Creates a variable radius. |

***See also:*** Grid; Interp

# RasterFill

This class fills polygons with a raster template.

**Inherits from** Fill

| Class Requests | |
| --- | --- |
| Make : RasterFill | Creates a default raster fill object. |

| Instance Requests | |
| --- | --- |
| = anObj : aBoolean | Returns true if a raster fill and the object are the same. |
| Copy (anotherRasterFill) | Copies the attributes of the specified raster fill into a raster fill. |
| GetBgColor : aColor | Gets the background color of a raster fill. |
| GetStipple : aStipple | Gets the stipple object used in a raster fill. |
| GetStyle : RasterFillStyleEnum | Gets a raster fill's style. |
| SetBgColor (aColor) | Sets the background color for a raster fill. |
| SetStipple (aStipple) | Sets the stipple object for a raster fill. |
| SetStyle (RsterFillStyleEnum) | Sets a raster fill's style. |
| UnHook | Unhooks the scale symbol. |

## Enumerations

| RasterFillStyleEnum | |
| --- | --- |
| #RASTERFILL_STYLE_EMPTY | No shading. |
| #RASTERFILL_STYLE_OPAQUESTIPPLE | Stipple shading with solid background. |
| #RASTERFILL_STYLE_SOLID | Solid background. |
| #RASTERFILL_STYLE_STIPPLE | Stipple shading on a transparent background. |

**See also:** CompositeFill; Stipple; VectorFill

# ReclassEditor (Spatial Analyst)

This class models the dialog box for specifying reclassification parameters.

**Inherits from** ModalDialog

| Class Requests | |
| --- | --- |
| Show (aTheme) : aList | Displays the reclassification dialog box and returns specified values. |

**See also:** Classification; Grid

# *Rect*

This class implements the rectangular shape.

**Inherits from** Shape

| Class Requests | |
|---|---|
| Make (originPoint, sizePoint) : aRect | Creates a rectangle defined by the starting point and the specified width and height in a point object. |
| MakeEmpty : aRect | Creates a rectangle with undefined origin and size. |
| MakeNull : aRect | Creates a rectangle with undefined origin and size. |
| MakeXY (x1, y1, x2, y2) : aRect | Creates a rectangle that extends to the specified coordinates. |

| Instance Requests | |
|---|---|
| = anObj : aBoolean | Returns true when the object and a rectangle are the same. |
| AsMulti-point : aMulti-point | Converts a rectangle to a multi-point shape. |
| AsPolygon : aPolygon | Converts a rectangle to a polygon. |
| AsString : aString | Returns the string representation of a rectangle. |
| ExpandBy (aNumber) : aRect | Moves each side outward by the specified amount. |
| GetBottom : aNumber | Gets the lower Y coordinate. |
| GetDimension : aNumber | Always returns 2. |
| GetHeight : aNumber | Gets the height of a rectangle. |
| GetLeft : aNumber | Gets the lower X coordinate. |
| GetRight : aNumber | Gets the higher X coordinate. |
| GetTop : aNumber | Gets the higher Y coordinate. |
| GetWidth : aNumber | Gets the width of a rectangle. |
| InsetBy (aNumber) : aRect | Moves inward on each side of a rectangle by the specified amount. |
| Intersects (aShape) : aBoolean | Returns true if the specified shape intersects a rectangle. |
| IntersectWith (anotherRect) : aRect | Returns a rectangle from the intersection of two rectangles. |
| IsEmpty : aBoolean | Returns true if a rectangle's attributes have no values. |

| | |
|---|---|
| IsNull : aBoolean | Returns true if a rectangle's attributes have no values. |
| Move (xNumber, yNumber) : aRect | Moves a rectangle by the specified amount. |
| ReturnCenter : aPoint | Returns the center of a rectangle. |
| ReturnClipped (anotherRect) : aShape | Returns a shape after clipping a rectangle. |
| ReturnDifference (aShape) : aPolygon | Subtracts the specified shape from a rectangle. |
| ReturnIntersection (aShape) : aPolygon | Intersects the specified shape with a rectangle. |
| ReturnMerged (aShape) : aPolygon | Merges a rectangle with a shape. |
| ReturnOrigin : aPoint | Returns the starting point of a rectangle. |
| ReturnProjected (aPrj) : aShape | Projects a rectangle using the specified projection. |
| ReturnSize : sizePoint | Returns the width and height of a rectangle. |
| ReturnUnion (aShape) : aPolygon | Returns the union of a the specified shape and a rectangle. |
| ReturnUnprojected (aPrj) : aShape | Unprojects a rectangle. |
| Scale (ratioNumber) : aRect | Scales a rectangle from its center by the specified amount. |
| SetOrigin (aPoint) : aRect | Sets the starting point of a rectangle. |
| SetSize (sizePoint) : aRect | Sets the width and height of a rectangle. |
| Snap (aShape, snapDistance) : aBoolean | Returns true if the center of a rectangle can snap to a shape within the snap distance. |
| UnionWith (anotherRect) : aRect | Returns a rectangle from the overlapping area of two rectangles. |

***See also:*** Display; GraphicShape

***Code example:*** page 484

# *RematchDialog*

This class models the modal dialog box for rematching a geocoded theme.

**Inherits from** ModalDialog

| *Class Requests* | |
|---|---|
| Show (geocodedFTheme) : aBoolean | Shows the rematch dialog box. |

***See also:*** GeoName

# *Robinson*

This class implements the Robinson projection.

**Inherits from P r j**

| Class Requests | |
|---|---|
| Make (aRect) : aRobinson | Creates a Robinson projection bounded by the specified rectangle. |

| Instance Requests | |
|---|---|
| ProjectPt (aPoint) : aBoolean | Projects a point and returns true if successful. |
| Recalculate | Computes derived constants. |
| ReturnCentralMeridian : aNumber | Returns the central meridian in decimal degrees. |
| SetCentralMeridian (aLongitude) | Sets the central meridian. |
| UnProjectPt (aPoint) : aBoolean | Unprojects a point and returns true if successful. |

***See also:*** CoordSys; Spheroid

# *RPCClient*

This class implements the client side of a Remote Procedure Call (RPC).

**Inherits from Obj**

| Class Requests | |
|---|---|
| GetMachineName : aString | Returns the name of the local machine. |
| HasServer (hostnameString, serverNumber, versionNumber) : aBoolean | Returns true if the specified host machine is running the specified RPC server. |
| IsMachine (hostnameString) : aBoolean | Returns true if the current process can connect to the specified workstation. |
| Make (hostnameString, serverNumber, versionNumber) : anRPCClient | Creates an RPC client connected to the specified server. |

| Instance Requests | |
|---|---|
| Close | Closes the RPC connection. |
| Execute (serviceIDNumber, commandString, returnClass) : anObj | Executes a service with the specified command and returns an object of the specified class. |
| GetErrorID : aNumber | Returns the last RPC error code. |

| | |
|---|---|
| GetErrorMsg : aString | Returns the description of the last RPC error. |
| GetTimeout : aNumber | Returns the time-out value for an execution to complete in seconds. |
| HasError : aBoolean | Returns true if the last RPC request contained errors. |
| SetCodepage (aBoolean, codepageName) | Sets the codepage for an RPC session. |
| SetTimeout (aNumber) | Sets the maximum allowable number of seconds for an execution to complete. |

***See also:*** DDEClient; DDEServer; RPCServer

# RPCServer

This class implements the server side of a Remote Procedure Call (RPC). The *RPC-Server* class is used only in the UNIX environment.

**Inherits from** Obj

| Class Requests | |
|---|---|
| AddUser (userNameString, workstationName) | Allows a user or users to access the server if an asterisk (*) is specified on a given workstation. |
| GetControlledAccess : aBoolean | Returns true if authorized users have limited access to the server. |
| RemoveUser (userNameString, workstationName) | Removes a user from the authorized users list. |
| ReturnUsers : stringList | Returns a list of user and workstation names. |
| SetControlledAccess (aBoolean) | If set to true, access is allowed only to authorized users. |
| Start (serverNumber, versionNumber) | Starts ArcView's RPC server identified by the server and version numbers. |
| Stop | Stops ArcView's RPC server. |

***See also:*** DDEClient; DDEServer; RPCClient

# RSO

This class implements the Rectified Skew Orthomorphic projection for the Malayan Grid, which is used in Malaysia, Singapore, and Brunei.

### Inherits from Prj

| Class Requests | |
| --- | --- |
| CanDoCustom : aBoolean | Always returns false. |
| CanDoSpheroid : aBoolean | Always returns true. |
| MakeBrunei : anRSO | Creates a projection for the Brunei area. |
| MakeMalaysian : anRSO | Creates a projection for the Malaysian area. |

| Instance Requests | |
| --- | --- |
| CalcHorizon | Computes the visible horizon. |
| ProjectPt (aPoint) : aBoolean | Projects a point and returns true if successful. |
| Recalculate | Calculates the derived constants. |
| UnProjectPt (aPoint) : aBoolean | Unprojects a point and returns true if successful. |

***See also:*** CoordSys; Spheroid

# ScaleBarFrame

This class implements the placement of a scale bar on a layout.

### Inherits from Graphic

| Class Requests | |
| --- | --- |
| Make (aRect) : ScaleBarFrame | Creates a scale bar frame based on the origin and the width of the specified rectangle. |

| Instance Requests | |
| --- | --- |
| CanSimplify : aBoolean | Always returns true. |
| Draw | Draws a scale bar frame. |
| Edit (scaleBarFramesList) : aBoolean | Displays the editor dialog box for all scale bar frames in the layout. |
| GetDivisions : aNumber | Gets the number of divisions in the first interval of a scale bar frame. |
| GetFillObject : anObj | Gets the fill object. |
| GetInterval : aNumber | Gets the interval of a scale bar frame. |
| GetIntervals : aNumber | Gets the number of intervals to the right of zero. |

| | |
|---|---|
| GetPreserveInterval : aBoolean | Retruns true if a scale bar preserves its interval. |
| GetStyle : ScaleBarFrameStyleEnum | Gets the style of a scale bar frame. |
| GetTextRatio : aNumber | Gets the ratio in the text style scale bar. |
| GetUnits : UnitsLinearEnum | Gets the measurement unit in a scale bar frame. |
| GetViewFrame : aViewFrame | Gets the view frame associated with a scale bar frame. |
| IsFilled : aBoolean | Returns true if a scale bar frame is filled. |
| IsFilledBy (aClass) : aBoolean | Returns true if a scale bar frame is filled by the objects of the specified class. |
| ReturnSymbols : symbolsList | Returns a list of symbols used in a scale bar frame. |
| SetBounds (aRect) | Establishes boundaries for a scale bar frame. |
| SetDisplay (aDisplay) | Associates the display object to a scale bar frame. |
| SetDivisions (aNumber) | Sets the number of divisions to the left of zero. |
| SetFillObject (anObj) | Fills a scale bar frame with the specified object. |
| SetInterval (aNumber) | Sets the value of each interval on a scale bar frame. |
| SetIntervals (aNumber) | Sets the number of intervals to the right of zero. |
| SetPreserveInterval (preserveBoolean) | Preserves a scale bar's interval if set to true. |
| SetStyle (ScaleBarFrameStyleEnum) | Sets the style of a scale bar frame. |
| SetSymbols (symbolList) | Sets the symbols used in a scale bar frame. |
| SetTextRatio (aNumber) | Sets the ratio value in the text style scale bar frame. |
| SetUnits (UnitsLinearEnum) | Sets the measurement unit of a scale bar frame. |
| SetViewFrame (aViewFrame) | Associates the view frame with a scale bar frame. |

## Enumerations

| ScaleBarFrameStyleEnum | |
|---|---|
| #SCALEBARFRAME_STYLE_ALTFILLED | Filled alternate halves of bands. |
| #SCALEBARFRAME_STYLE_FILLED | Filled alternate bands. |
| #SCALEBARFRAME_STYLE_HOLLOW | Hollow bands. |
| #SCALEBARFRAME_STYLE_LINED | Lines down the middle of alternate bands. |
| #SCALEBARFRAME_STYLE_TEXT | Ratio. |

| UnitsLinearEnum | |
|---|---|
| #UNITS_LINEAR_UNKNOWN | Unknown. |
| #UNITS_LINEAR_INCHES | Inches. |

| #UNITS_LINEAR_FEET | Feet. |
|---|---|
| #UNITS_LINEAR_YARDS | Yards. |
| #UNITS_LINEAR_MILES | Miles. |
| #UNITS_LINEAR_MILLIMETERS | Millimeters. |
| #UNITS_LINEAR_CENTIMETERS | Centimeters. |
| #UNITS_LINEAR_METERS | Meters. |
| #UNITS_LINEAR_KILOMETERS | Kilometers. |
| #UNITS_LINEAR_NAUTICALMILES | Nautical miles. |
| #UNITS_LINEAR_DEGREES | Decimal degrees. |

*See also:* GraphicList; Layout; Template; ViewFrame

# Script

This class implements Avenue programs.

**Inherits from** Obj

**Inherited by** EncryptedScript

| Class Requests | |
|---|---|
| Make (scriptString) : aScript | If no syntax errors exist, creates a compiled script from the specified string. |
| The : aScript | Returns the currently executing script. |

| Instance Requests | |
|---|---|
| AsEncrypted : anEncryptedScript | Encrypts a script. |
| AsString : aString | Returns a script's source code. |
| BreakExists (characterPositionNumber) : aBoolean | Returns true if a break exists at the specified character position. |
| ClearAllBreaks | Clears all break points in a script. |
| DoIt (anObj) : anObj | Executes a script and passes the specified object. |
| GetDateFormat : aString | Gets the date format of a script. |
| GetErrorMsg : aStr | Gets the description of the last error. |
| GetErrorPos : aNumber | Gets the token number where the last error occurred. |
| GetErrorTopic : aString | Gets the help topic for the last error. |
| GetNumberFormat : aString | Gets the number format of a script. |
| HasError : aBoolean | Returns true when compile or run-time errors occur. |

| | |
|---|---|
| IsCompiled : aBoolean | Returns true when a script is compiled. |
| SetDynamicBreak | Sets a break point on the next executable statement. |
| SetDateFormat (aString) | Sets the date format for a script. |
| SetNumberFormat (aString) | Sets the number format for a script. |

***See also:*** SEd; Application;. Project

***Code examples:*** page 315, page 414

# ScriptMgr

This class implements the dialog box that displays available scripts.

**Inherits from** ModalDialog

| Class Requests | |
|---|---|
| GetFromSubset (patternString) : aScript | Displays the script manager dialog box with the script names matching the specified pattern. |
| Show : aScript | Displays the script manager dialog box. |

***See also:*** Script; SEd

***Code example:*** page 435

# SDEDataSet (Database Theme)

Objects of this class represent datasets in Spatial Database Engine (SDE).

**Inherits from** Obj

| Class Requests | |
|---|---|
| FindActiveDataset : anSDEDataset | Returns the active SDE data set. |
| GetError : SDEErrorEnum | Returns the error message for the last SDE request. |
| GetErrorMsg (SDEErrorEnum) : aString | Returns description of an SDE request error. |
| IsUsingServerExtent : aBoolean | Returns true when SDE is using the server's data set for determining the extent. |
| Make (aServerName, aDatasetName, aUserName, aPassword) : anSDEDataset | Creates an SDE data set with the specified name on the specified server. |
| UseServerExtent (isUsingServerExtentBoolean) | If set to true, server's data set is used to determine the extent. |

| Instance Requests | |
|---|---|
| AddLayer (newSDELayer, numberOfFeatures, averagePointsPerFeature, gridSizeInDistanceUnits) : aBoolean | Makes a new empty layer and returns true if sucessful. |
| AdjustPriority (adjustByNumber) : aNumber | Lowers the processing priority. |
| ClearSpatialFilter : aBoolean | Removes the current spatial filter and returns true when sucessful. |
| CommitTransaction : aBoolean | Commits the changes in the transaction. |
| DeleteLayer (anSDELayer, featuresOnlyBoolean) : aBoolean | Deletes the features from the specified layer, the layer is also deleted if the boolean is set to false. |
| FreeAllLocks : aBoolean | Releases all locks within a data set. |
| GetAllLayerNames : nameList | Returns the names of layers in an SDE data set. |
| GetMaxLayers : aNumber | Returns the maximum number of layers allowed in a data set. |
| GetMaxPoints : aNumber | Returns the maximum number of points each feature is allowed to have. |
| GetName : aString | Returns the name of an SDE data set. |
| GetPriority : aNumber | Returns the priority number on an SDE server for processing. |
| GetResolution : aNumber | Retruns the smallest distance that can be measured in an SDE data set. |
| GetUnits : UnitsLinearEnum | Returns measuring unit in a data set. |
| IsConnected : aBoolean | Returns true when an SDE data set is connected to an SDE server. |
| RollbackTransaction : aBoolean | Rollsback the changes in a transaction. |
| SetBuffers (minBufferSizeInBytes, maxBufferSizeInBytes, numberOfObjects) : aBoolean | Sets the data transfer parameters. |
| SetConnected (connectBoolean) : aBoolean | Connects or disconnects a data set from the server. |
| SetSpatialFilter (anSDEFeature, SDELayerSearchEnum, expectedResultBoolean) : aBoolean | Sets the spatial filter and returns true if successful. |
| SetSpatialFilterByID (aLayerID, aFeatureID, SDELayerSearchEnum, expectedResultBoolean) : aBoolean | Sets the spatial filter and returns true if successful. |
| StartTransaction : aBoolean | Starts a new transaction. |

## Enumerations

| SDEErrorEnum | |
|---|---|
| #SDEERR_ALL_SLIVERS | Result features are all sliver polygons. |
| #SDEERR_ATTRIBUTE_CONVERT_ERROR | Invalid data type. |
| #SDEERR_ATTRIBUTE_NOTEXIST | Field does not exist. |
| #SDEERR_ATTRIBUTE_TABLE_EXISTS | Duplicate attribute table. |
| #SDEERR_ATTRIBUTES_EXIST | Duplicate attribute record. |
| #SDEERR_AUTOLOG_NOTSET | The auto log mask has not been set. |
| #SDEERR_BLOBSIZE_TOOLARGE | Maximum BLOB size has been exceeded. |
| #SDEERR_BUFFER_OUTOF_RANGE | Exceeds the valid buffer range. |
| #SDEERR_DATASET_EXISTS | Data set already exists. |
| #SDEERR_DATASET_NOTFOUND | Dataset table does not exist. |
| #SDEERR_DATASET_NOTEXIST | Data set does not exist. |
| #SDEERR_DB_IOERROR | Database level I/O error. |
| #SDEERR_DUPLICATE_ARC | Duplicate arc exists. |
| #SDEERR_DUPLICATE_GROUP | Duplicate group name. |
| #SDEERR_DUPLICATE_MEMBER | Duplicate group member. |
| #SDEERR_DUPLICATE_USER | Duplicate user. |
| #SDEERR_ENTITYTYPE_NOTALLOWED | Invalid feature type. |
| #SDEERR_FEATURE_INTEGRITY_ERROR | Feature integrity error. |
| #SDEERR_FEATURE_UNCHANGED | Feature was unchanged. |
| #SDEERR_FEATUREID_EXISTS | Duplicate feature ID number. |
| #SDEERR_FEATUREID_NOTEXIST | Missing feature ID number. |
| #SDEERR_FILE_IOERROR | Error writing to or creating an output text file. |
| #SDEERR_GENERAL_FAILURE | General SDE failure. |
| #SDEERR_GRIDSIZE_TOOSMALL | Grid size too small. |
| #SDEERR_GROUP_NOTEXIST | Group does not exist. |
| #SDEERR_INDEX_EXISTS | Duplicate attribute index. |
| #SDEERR_INDEX_NOTEXIST | Invalid attribute index. |
| #SDEERR_INVALID_ATTRIBUTEMASK_MODE | Invalid attribute mask mode. |
| #SDEERR_INVALID_AUTOLOG_MASK | Current auto log mask is invalid. |
| #SDEERR_INVALID_DATASETNAME | Invalid data set name. |
| #SDEERR_INVALID_END_POINT | Invalid end point. |
| #SDEERR_INVALID_ENTITYTYPE_MASK | Invalid feature typefor a mask. |

| | |
|---|---|
| #SDEERR_INVALID_ENTITYTYPE | Invalid feature type. |
| #SDEERR_INVALID_EXTENT | Invalid extent. |
| #SDEERR_INVALID_FEATURE_VERSION | Feature being replaced has been modified. |
| #SDEERR_INVALID_FEATURE | Invalid SDE feature. |
| #SDEERR_INVALID_FIELDTYPE | Field type is invalid. |
| #SDEERR_INVALID_FIELDNAME | Field name is invalid. |
| #SDEERR_INVALID_GRIDSIZE | Grid size too small or too large. |
| #SDEERR_INVALID_LAYERNUMBER | Invalid layer ID number. |
| #SDEERR_INVALID_LAYERKEYWORD | Invalid layer configuration keyword. |
| #SDEERR_INVALID_LAYERNAME | Invalid layer name. |
| #SDEERR_INVALID_LOCKMODE | Invalid lock mode. |
| #SDEERR_INVALID_MEMBER | Invalid group member. |
| #SDEERR_INVALID_MIDLE_POINT | Invalid mid-point. |
| #SDEERR_INVALID_PARAMATER | Invalid parameter. |
| #SDEERR_INVALID_PASSWORD | Invalid password. |
| #SDEERR_INVALID_POINTER | Invalid data pointer. |
| #SDEERR_INVALID_RADIUS | Radius too small to perform the operation. |
| #SDEERR_INVALID_RELEASE | Invalid release version of the SDE server. |
| #SDEERR_INVALID_SEARCHMETHOD | Invalid search method. |
| #SDEERR_INVALID_SECURITYCLASS | Invalid security class. |
| #SDEERR_INVALID_SECURITYMASK | Invalid security mask. |
| #SDEERR_INVALID_SERVER | Server machine was not found. |
| #SDEERR_INVALID_SQL | Invalid SQL where clause. |
| #SDEERR_INVALID_STATISTICS_TYPE | Invalid field type for statistics. |
| #SDEERR_INVALID_TRANSACTIONID | Invalid internal SDE Transaction ID. |
| #SDEERR_INVALID_UNIQUE_TYPE | Invalid field type for unique values. |
| #SDEERR_INVALID_USER | Invalid user name. |
| #SDEERR_IOMANAGER_NOTAVAILABLE | IO manager is not started on the SDE server. |
| #SDEERR_LAYER_EXISTS | Layer already exists. |
| #SDEERR_LAYER_INUSE | Layer in use by another user. |
| #SDEERR_LAYER_LOCKED | Layer is already locked. |
| #SDEERR_LAYER_NOTEXIST | Layer does not exist. |
| #SDEERR_LICENSE_EXPIRED | The runtime license has expired. |
| #SDEERR_LOADONLY_LAYER | Layer in LOAD_ONLY mode. |
| #SDEERR_LOCK_CONFLICT | Object locked by another user. |

| | |
|---|---|
| #SDEERR_LOG_IOERROR | Error in logfile I/O. |
| #SDEERR_LOG_NOACCESS | Error accessing logfile. |
| #SDEERR_LOG_NOTEXIST | Logfile does not exist. |
| #SDEERR_LOG_NOTOPEN | Logfile is not open. |
| #SDEERR_LOGIN_NOTALLOWED | No login allowed. |
| #SDEERR_MEMBER_NOTEXIST | Group member does not exist. |
| #SDEERR_NETWORK_FAILURE | Network connection failure. |
| #SDEERR_NETWORK_TIMEOUT | Connection time out. |
| #SDEERR_NO_ACCESS | Access to data set denied. |
| #SDEERR_NO_ATTRIBUTE_TABLE | Layer does not have an associated attribute table. |
| #SDEERR_NO_ATTRIBUTE_DATA | Missing attribute record or invalid attribute mask. |
| #SDEERR_NO_ATTRIBUTE_MASK | Missing attribute mask for the layer. |
| #SDEERR_NO_FEATURES | No features found. |
| #SDEERR_NO_LOCKS | Layer has no locks. |
| #SDEERR_NO_NODE_TABLE | Node table does not exist. |
| #SDEERR_NO_PERMISSIONS | Permission denied. |
| #SDEERR_NO_SEARCHFEATURE_SET | Search feature has not been set. |
| #SDEERR_NOT_IMPLEMENTED_YET | Not implemented yet. |
| #SDEERR_NOT_NETWORK | Network operation only. |
| #SDEERR_ONLY_2D_OPERATION | Two-dimensional features operation. |
| #SDEERR_OUTOF_CLIENT_MEMORY | Client out of dynamic memory. |
| #SDEERR_OUTOF_CONTEXT | Illegal operation. |
| #SDEERR_OUTOF_LOCKS | The maximum number of locks is reached. |
| #SDEERR_OUTOF_SERVER_MEMORY | Server out of memory. |
| #SDEERR_POINT_NOTEXIST | Point does not exist. |
| #SDEERR_POINTS_NOT_ADJACENT | Points not adjacent. |
| #SDEERR_QUERY_FINISHED | Query stream finished successfully. |
| #SDEERR_QUERY_IN_PROGRESS | Query stream in progress. |
| #SDEERR_SDE_NOTSTARTED | SDE server process is not started. |
| #SDEERR_SEARCH_FINISHED | Search stream finished successfully. |
| #SDEERR_SEARCH_IN_PROGRESS | Search stream in progress. |
| #SDEERR_SERVICE_NOTFOUND | No SDE entry in the /etc/services file. |
| #SDEERR_SUCCESS | No error message has been generated. |

| #SDEERR_TEMPORARY_IOERROR | Temporary file permission errors. |
|---|---|
| #SDEERR_TOOMANY_POINTS | Too many points in the feature. |
| #SDEERR_TOOMANY_TASKS | Too many users loged in. |
| #SDEERR_TOPOLOGICAL_ERROR | Topological integrity violated. |
| #SDEERR_TRANSACTION_IN_PROGRESS | Open transaction. |
| #SDEERR_UNDEFINED_ERROR | Undefined SDE error. |
| #SDEERR_UNSUPPORTED_SHAPETYPE | Shape type is not supported. |
| #SDEERR_USER_NOTEXIST | User does not exist. |
| #SDEERR_VERSIONTABLE_EXISTS | A version table already exists. |

## SDELayerSearchEnum

| #SDELAYER_SEARCH_EXTENT | The extent of the search feature overlaps or touches the envelope of features in the layer. The fastest search method. |
|---|---|
| #SDELAYER_SEARCH_EXTENTBYGRID | The extent of the search feature overlaps or touches the envelope of features in the layer, guarantees that the layer features returned by the search are in spatial index order. |
| #SDELAYER_SEARCH_COMMONPOINT | The search feature and features in the layer share an identical vertex. Use to find adjacent features if topological integrity exists. |
| #SDELAYER_SEARCH_INTERSECTLINE | The search feature and features in the layer must intersect without sharing a vertex. Does not include one shape completely inside the other. Use to find errors in topological integrity. |
| #SDELAYER_SEARCH_INTERSECTLINEORPOINT | The search feature touches layer features. |
| #SDELAYER_SEARCH_COMMONLINE | The search feature shares a common line with layer features. |
| #SDELAYER_SEARCH_INTERSECTAREAORLINE | The search feature touches layer features or they share a common area. |
| #SDELAYER_SEARCH_INTERSECTAREA | The search feature and the layer features share a common area. Either the search feature or the layer features must be an area feature. Does not find adjacent features. |
| #SDELAYER_SEARCH_INTERSECTAREANOTOUCH | Either the search feature or the layer features are wholly inside the other and their boundaries do not touch. |

| | |
|---|---|
| #SDELAYER_SEARCH_CONTAINS | The search feature is completely contained by the layer features. If the layer features are area features, the search feature must be completely inside the layer features or their boundaries. If the search feature is a point, it must be one of the layer features' vertices. |
| #SDELAYER_SEARCH_CONTAINSNOTOUCH | The layer features must be area features. The search feature must be completely contained by the layer feature without touching its boundary. |
| #SDELAYER_SEARCH_ISWITHIN | The layer features are completely contained by the search feature. If the search feature is an area feature, the layer features must be completely inside the search feature or its boundary. If the layer features are point features, they must be vertices in the search feature. |
| #SDELAYER_SEARCH_ISWITHINNOTOUCH | The search feature must be an area feature. The layer features must be completely contained by the search feature without touching its boundary. |
| #SDELAYER_SEARCH_FIRSTPOINTIN | The layer features must be area features. The layer features contain the first point of the search feature |
| #SDELAYER_SEARCH_CONTAINSCENTROID | The search feature must be an area feature. The search feature contains the centroid of the layer features. |
| #SDELAYER_SEARCH_IDENTICAL | The search feature and the layer features are identical in both feature type and vertices. Use to find duplicate data. |

**See also:** SDELayer; SDELog; SDEFeature; DBTheme

# SDEFeature (Database Theme)

Objects of class SDEFeature model spatial features in instances of SDE-Layer.

**Inherits from** Obj

| Class Requests | |
|---|---|
| Make (aShape) : anSDEFeature | Makes an SDE feature from the specified shape. |
| MakeEmpty (numberOfPoints) : anSDEFeature | Creates an empty SDE feature and sets its maximum number of points. |
| MakeList (aShape) : listOfSDEFeatures | Creates SDE features from single and multiple connected shapes. |

| *Instance Requests* | |
|---|---|
| AddToLog (anSDELog) : aBoolean | Adds an SDE feature to the specified log and returns true if successful. |
| AsSDEFeature : anSDEFeature | Clones an SDE feature. |
| AsShape : aShape | Converts an SDE feature to a shape. |
| Buffer (aDistance, newSDEFeature) : aBoolean | Creates a new SDE feature by applying the specified buffer to an SDE feature. |
| ClipPoints (polygonSDEFeature) : aNumber | Clips the points of an SDE feature outside the specified SDE feature and returns the number of remaining points. |
| ClipToArc (aRect) : polylineList | Returns a list of the polyline shapes remained after clipping by a rectangle. |
| ClipToPoly (aRect) : polygomList | Returns a list of the polygon shapes remained after clipping by a rectangle |
| CopyGeometry (anSDELayer) : aNumber | Copies an SDEFeature's geometry to the specified SDELayer and returns the new SDEFeature's ID number. |
| Count : aNumber | Returns the number of vertices in an SDEFeature. |
| GetID : aNumber | Returns the ID number of an SDEFeature. |
| GetSDELayer : anSDELayer | Returns the SDELayer of an SDEFeature. |
| GetType : SDEFeatureTypeEnum | Returns an SDEFeature's type. |
| HasAttributes : aBoolean | Retruns true if attributes of an SDEFeature has been retrieved from the SDE server. |
| HasRelation (anotherSDEFeatureOrShape, SDEFeatureRelationEnum) : aBoolean | Returns true if the specified relationship exists. |
| Intersects (anotherSDEFeature) : pointList | Returns points of intersection between two SDE features. |
| Merge (anotherSDEFeature) : aBoolean | Merges two adjacent SDE features and returns true if successful. |
| MoveGeometry (anSDELayer) : aNumber | Moves the geometry part of an SDE feature to a new layer and returns its new ID. |
| Overlay (anotherSDEFeature, minAreaOfResult, maxSliverFactorOfResult) : listOfSDEFeaturesList | Overlays two SDE features. |
| Remove (anotherSDEFeature) : aBoolean | Removes the specified SDE feature from an SDE feature. |
| Resize (numberOfPoints) : aBoolean | Changes the maximum allowed number of points in an SDE feature. |

| | |
|---|---|
| ReturnArea : aNumber | Returns the area of an SDEFeature. |
| ReturnCenter : aPoint | Returns the centroid of an SDEFeature. |
| ReturnCommonLength (anotherSDEFeature) : aNumber | Returns the length of common boundaries between two SDE features. |
| ReturnConvexHull : anSDEFeature | Returns the smallest convex polygon containing all points of a multipoint SDE feature. |
| ReturnExtent : aRect | Returns the extent of an SDE feature. |
| ReturnLength : aNumber | Returns the length of an SDE feature. |
| ReturnValue (aField) : anObj | Returns the attribute value of the specified field. |
| SetGeometry (aShape) : aBoolean | Sets the geometry of an SDE feature and returns true if successful. |
| SetSDELayer (anSDELayer) | Assigns an SDE feature to the specified SDE layer. |
| SetValue (aField, anObj) | Sets the attribute value of an SDE feature. |
| Simplify (aDistance) : aBoolean | Reduces the number of vertices in an SDE feature. |
| Verify : aBoolean | Returns true when an SDE feature is an acceptable feature type. |

## Enumerations

| SDEFeatureRelationEnum | |
|---|---|
| #SDEFEATURE_RELATION_INTERSECTLINE | The two features intersect, but the point of intersection is not a shared vertex. |
| #SDEFEATURE_RELATION_COMMONPOINT | The two features have at least one common vertex. |
| #SDEFEATURE_RELATION_ COMMONBOUNDSAME | The two features have the same boundary, and all vertices are connected in the same direction. |
| #SDEFEATURE_RELATION_ COMMONBOUNDDIFF | The two features have the same boundary, but the vertices for the primary and secondary features are connected in opposite directions. |
| #SDEFEATURE_RELATION_IDENTICAL | The two features have the same feature type and coordinate description. |
| #SDEFEATURE_RELATION_INTERSECTAREA | At least one of the two features is an area feature type, and the other feature is wholly or partially contained inside it. |
| #SDEFEATURE_RELATION_CONTAINS | The primary feature must contain the secondary feature. |
| #SDEFEATURE_RELATION_ISWITHIN | The primary feature must be within the secondary feature. |

| SDEFeatureTypeEnum | |
|---|---|
| #SDEFEATURE_TYPE_POINT | Single point feature. |
| #SDEFEATURE_TYPE_MULTIPOINT | Multipoint feature. |
| #SDEFEATURE_TYPE_SPAGHETTI | Set of points connected in order by straight line segments taht can cross each other. |
| #SDEFEATURE_TYPE_LINESTRING | Set of points connected in order by straight line segments taht can not cross each other. |
| #SDEFEATURE_TYPE_RING | A loop feature. |
| #SDEFEATURE_TYPE_POLYGON | Polygon feature. |
| #SDEFEATURE_TYPE_DONUTPOLYGON | Donut shape feature. |
| #SDEFEATURE_TYPE_THREED | This feature type modifies an SDELayer so the features in it can have three dimensional coordinates. |
| #SDEFEATURE_TYPE_NETWORK | This feature type modifies an SDELayer so the linestring features in it connect at nodes. |

**See also:** SDELayer; SDELog; SDEDataset; DBTheme; Shape

# SDELayer (Database Theme)

This class models spatial layers of an SDE data set.

**Inherits from** Obj

| Class Requests | |
|---|---|
| ConnectLines (lineList, dissolveLinesBoolean, createLinesBoolean) : shapeList | Generates a polyline if dissolveLinesBoolean is set to false and createLinesBoolean is set to true. Generates a polygon if dissolveLinesBoolean and createLinesBoolean are set to false. Dissolves polygons if dissolveLinesBoolean is set to true and createLinesBoolean is set to false. |
| DeleteSDEFeatureByLog (anSDELogName) : aBoolean | Deletes the features in the specified SDE log from the current SDE data set and returns true if successful. |
| Make (anSDEDataset, anSDELayerName) : anSDELayer | Gets the specified SDE layer. |
| MakeFromLayerNumber (anSDEDataset, layerIDNumber) : anSDELayer | Gets the specified SDE layer. |

| | |
|---|---|
| MakeNew (uniqueLayerNumber, uniqueLayerName, featureTypeList) : anSDELayer | Creates a new SDE layer. |
| SetAutoSearch (aBoolean) | Sets the auto search property of an SDE layer. |

| Instance Requests | |
|---|---|
| AddFeaturesByLinearDissolve (sourceSDELayer) : aNumber | Creates new linear features for an SDE layer by dissolving all the linear features in the specified SDELayer and returns the number of new features created. |
| AddFeaturesByLinearDissolveLog (anSDELogName) : aNumber | Creates new linear features for an SDE layer by dissolving all the linear features in the specified SDELog and returns the number of new features created. |
| AddFeaturesByPolygonDissolve (sourceSDELayer) : aNumber | Creates new polygon features for an SDE layer by dissolving all the polygon features in the specified SDELayer and returns the number of new features created. |
| AddFeaturesByPolygonDissolveLog (anSDELogName) : aNumber | Creates new polygon features for an SDE layer by dissolving all the polygon features in the specified SDELog and returns the number of new features created. |
| AddFeaturesByPolygonize (sourceSDELayer) : aNumber | Creates new polygon features for an SDE layer from the polygon features in the specified SDELayer and returns the number of new features created. |
| AddFeaturesByPolygonizeLog (anSDELogName) : aNumber | Creates new polygon features for an SDE layer from the polygon features in the specified SDELog and returns the number of new features created. |
| AddFields (aFieldList) : aBoolean | Adds new attribute fields to an SDELayer and returns true if successful. |
| AddToLog (anSDELog, anSQLWhereClause) : aNumber | Adds the feature that satisfy the specified where clause to the given log. |
| BuildIslands (removeDonutsBoolean) : aBoolean | Builds donut shape polygons and removes the xisting donuts if the specified Boolean is set to true. |
| CancelQuery | Cancels a query stream. |
| CancelSearch | Cancels a search stream. |
| Clean (aRect, anSDELogName) : aNumber | Cleans the specified area of a log by enforcing topological relationships. |
| CreateAttributeTable (fieldList) : aBoolean | Creates an attribute table from the given fields for an SDE layer. |

| | |
|---|---|
| CreateIndex (isUniqueBoolean, isAscendingBoolean, indexName, fieldList) : aBoolean | Creates a composite index and returns true if successful. |
| DeleteAttributeTable : aBoolean | Deletes an SDE layer's attribute table and returns true if successful. |
| DeleteIndex (indexName) : aBoolean | Deletes the specified index for an SDE layer and returns true if successful. |
| DeleteSDEFeature (featureIDNumber, attributesOnlyBoolean) : aBoolean | Deletes the attributes or both attributes and geometry of the specified feature. |
| FindField (fieldName) : aField | Gets the specified attribute field for an SDE layer. |
| FindToLog (anSDELog, SDELayerSearchEnum, anSQLWhereClause) : aNumber | Finds SDE features satisfying the specified where clause and search method, and adds them to the given log. |
| GetCreator : aString | Returns the user ID that created an SDE layer. |
| GetFields : fieldList | Gets the attribute fields of an SDE layer. |
| GetGridSizes : numberList | Returns the sizes of the three spatial index for an SDE layer. |
| GetID : aNumber | Returns the ID number of an SDE layer. |
| GetName : aString | Returns the name of an SDE layer. |
| GetSDEDataset : aSDEDataset | Returns the SDE data set of an SDE layer. |
| GetSecurityClass : aNumber | Returns the security class number of an SDE layer. |
| HasType (SDEFeatureTypeEnum) : aBoolean | Returns true if the specified feature type exists in an SDE layer. |
| IsWritable : aBoolean | Returns true if attributes can be added or changed. |
| LockArea (aRect, canModifyBoolean) : aBoolean | Locks the features within the specified rectangle of an SDE layer. |
| LockLayer (canModifyBoolean) : aBoolean | Locks all features of an SDE layer and returns true if successful. |
| ModifyAreaLock (aRect, canModifyBoolean) : aBoolean | Changes the area of the area-lock. |
| Query (anSQLWhereClause, anSDEFeature) : aBoolean | Starts a query stream and returns true when no features is found. |
| QueryByDateRange (lastModifiedStartDate, lastModifiedEndDate, anSQLWhereCluase, anSDEFeature) : aBoolean | Starts a query stream selecting features last modified between the specified dates and satisfying the given where clause, returns true when no features is found. |
| QueryByIDRange (startFeatureID, endFeatureID, anSQLWhereClause, anSDEFeature) : aBoolean | Starts a query stream selecting features with ID within the specified range and satisfying the given where clause, returns true when no features is found. |

| | |
|---|---|
| QueryNext (anSDEFeature) : aBoolean | The next feature in the query stream is retrieved, returns true when no more features are found. |
| QuerySDEFeature (featureID, anSDEFeature) : aBoolean | Finds the feature with the specified feature id and returns true if successful. |
| ReturnExtent (anSQLWhereClause) : aRect | Returns the extent of features satisfying the specified where clause. |
| ReturnID (anSQLWhereClause) : aNumber | Starts a query stream retrieving feature IDs of features satisfying the specified where clause. |
| ReturnIDByDateRange (lastModifiedStartDate, lastModifiedEndDate, anSQLWhereClause) : aNumber | Starts a query stream returning feature IDs of features last modified between the specified dates and satisfying the given where clause, returns null when no features is found. |
| ReturnIDByIDRange (startFeatureID, endFeatureID, anSQLWhereClause) : aNumber | Starts a query stream returning feature IDs within the specified range and satisfying the given where clause, returns null when no features is found. |
| ReturnNextID : aNumber | The next feature ID in the query stream is retrieved, returns null when no more features are found. |
| ReturnSDEFeatureExtent (featureID) : aRect | Returns the extent of the specified feature in an SDE layer. |
| ReturnSDEFeatureValue (featureID, aField) : anObj | Returns the attribute value of the specified feature. |
| ReturnStatistics (fieldName, anSQLWhereClause) : numberList | Returns the basic statistical information from the values of the specified field for the features satisfying the given where clause. |
| ReturnUniqueValues (fieldName, anSQLWhereClause, maximumUniqueNumber) : listofTwoLists | Groups and counts the unique values in the specified field that satisfy the given where clause. |
| SearchByID (layerID, featureID) : aBoolean | Resets the current search feature and cancels any existing search streams. |
| SearchBySDEFeature (anSDEFeature) : aBoolean | Resets the current search feature and cancels any existing search streams. |
| SearchByShape (aShape) : aBoolean | Resets the current search feature to a shape and cancels any existing search streams. |
| SearchForID (SDELayerSearchEnum, anSQLWhereClause, maximumFeaturesRetrieved) : aNumber | Starts a search stream based on the specified where clause. |
| SearchForNextID : aNumber | Returns the next feature id in the search stream, returns null when no features are found. |
| SearchForNextSDEFeature (anSDEFeature) : aBoolean | Retrieves the next feature in the search stream, returns true when no more features are found. |

| | |
|---|---|
| SearchForSDEFeature (SDELayerSearchEnum, anSQLWhereClause, maximumFeatures, anSDEFeature) : aBoolean | Starts a search stream. |
| StoreSDEFeature (anSDEFeature, replaceBoolean, onlyValuesBoolean) : aBoolean | Stores an SDE feature in an SDE layer. |
| UnLock : aBoolean | Removes the locks for an SDE layer. |

## Enumerations

| SDEFeatureTypeEnum | |
|---|---|
| #SDEFEATURE_TYPE_POINT | Single point feature. |
| #SDEFEATURE_TYPE_MULTIPOINT | Multipoint feature. |
| #SDEFEATURE_TYPE_SPAGHETTI | Set of points connected in order by straight line segments that can cross each other. |
| #SDEFEATURE_TYPE_LINESTRING | Set of points connected in order by straight line segments that can not cross each other. |
| #SDEFEATURE_TYPE_RING | A loop feature. |
| #SDEFEATURE_TYPE_POLYGON | Polygon feature. |
| #SDEFEATURE_TYPE_DONUTPOLYGON | Donut shape feature. |
| #SDEFEATURE_TYPE_THREED | This feature type modifies an SDELayer so the features in it can have three dimensional coordinates. |
| #SDEFEATURE_TYPE_NETWORK | This feature type modifies an SDELayer so the linestring features in it connect at nodes. |

| SDELayerSearchEnum | |
|---|---|
| #SDELAYER_SEARCH_EXTENT | The extent of the search feature overlaps or touches the envelope of features in the layer. The fastest search method. |
| #SDELAYER_SEARCH_EXTENTBYGRID | The extent of the search feature overlaps or touches the envelope of features in the layer, guarantees that the layer features returned by the search are in spatial index order. |
| #SDELAYER_SEARCH_COMMONPOINT | The search feature and features in the layer share an identical vertex. Use to find adjacent features if topological integrity exists. |

| | |
|---|---|
| #SDELAYER_SEARCH_INTERSECTLINE | The search feature and features in the layer must intersect without sharing a vertex. Does not include one shape completely inside the other. Use to find errors in topological integrity. |
| #SDELAYER_SEARCH_INTERSECTLINEORPOINT | The search feature touches layer features. |
| #SDELAYER_SEARCH_COMMONLINE | The search feature shares a common line with layer features. |
| #SDELAYER_SEARCH_INTERSECTAREAORLINE | The search feature touches layer features or they share a common area. |
| #SDELAYER_SEARCH_INTERSECTAREA | The search feature and the layer features share a common area. Either the search feature or the layer features must be an area feature. Does not find adjacent features. |
| #SDELAYER_SEARCH_INTERSECTAREANOTOUCH | Either the search feature or the layer features are wholly inside the other and their boundaries do not touch. |
| #SDELAYER_SEARCH_CONTAINS | The search feature is completely contained by the layer features. If the layer features are area features, the search feature must be completely inside the layer features or their boundaries. If the search feature is a point, it must be one of the layer features' vertices. |
| #SDELAYER_SEARCH_CONTAINSNOTOUCH | The layer features must be area features. The search feature must be completely contained by the layer feature without touching its boundary. |
| #SDELAYER_SEARCH_ISWITHIN | The layer features are completely contained by the search feature. If the search feature is an area feature, the layer features must be completely inside the search feature or its boundary. If the layer features are point features, they must be vertices in the search feature. |
| #SDELAYER_SEARCH_ISWITHINNOTOUCH | The search feature must be an area feature. The layer features must be completely contained by the search feature without touching its boundary. |
| #SDELAYER_SEARCH_FIRSTPOINTIN | The layer features must be area features. The layer features contain the first point of the search feature |
| #SDELAYER_SEARCH_CONTAINSCENTROID | The search feature must be an area feature. The search feature contains the centroid of the layer features. |

| #SDELAYER_SEARCH_IDENTICAL | The search feature and the layer features are identical in both feature type and vertices. Use to find duplicate data. |
|---|---|

***See also:*** SDELog; SDEFeature; SDEDataset; DBTheme

# SDELog (Database Theme)

Instances of this class maintain collection of SDE features from an SDE data set.

### Inherits from Obj

| Class Requests | |
|---|---|
| Copy (sourceSDELogName, destinationSDELogName) : aBoolean | Copies the source SDE log to the destination and returns true if successful. |
| Delete (anSDELogName) : aBoolean | Deletes the specified log file and returns true if successful. |
| Make (anSDELogName, SDELogModeEnum) : anSDELog | Creates a new persistent SDE log. |
| MakeLogName : aString | Returns a unique SDE log name. |

| Instance Requests | |
|---|---|
| ChangeType (isTemporaryBoolean) : aBoolean | If set to true, makes the log file temporary. |
| Close : aBoolean | Closes a log file and returns true if successful. |
| Combine (anSDELogName, SDELogCombineTypeEnum) : aBoolean | Combines two SDE logs. |
| Count : aNumber | Returns the number of features in an SDE log. |
| GetName : aString | Returns the name of an SDE log. |
| Query (anSQLWhereClause, anSDEFeature) : aBoolean | Starts a query stream to retrieve features satisfying the specified where clause, returns true if none is found. |
| QueryNext (anSDEFeature) : aBoolean | Retrieves the next feature in the query stream, returns true when no feature is found. |
| ReturnExtent : aRect | Returns the extent of an SDE log. |
| ReturnID (anSQLWhereClause) : listOfTwoNumbers | Starts a query stream to return layer and feature ID numbers. |

| | |
|---|---|
| ReturnNextID : listOfTwoNumbers | Returns the next layer and feature ID in the query stream. |
| SetAutoLog (SDELogAutoEnum) | Begins the auto-log operation. |
| Sort : aNumber | Sorts by layer and feature IDs, removes duplicate features and returns the number of features remaining in the log. |

## Enumerations

| SDELogAutoEnum | |
|---|---|
| #SDELOG_AUTO_ALL | Combines #SDELOG_AUTO_READ and #SDELOG_AUTO_WRITE. |
| #SDELOG_AUTO_OFF | Stops auto-logging. |
| #SDELOG_AUTO_QUERY | All features retrieved by query streams are recorded. |
| #SDELOG_AUTO_READ | All features retrieved by query streams, search streams, or the QuerySDEFeature request are recorded. |
| #SDELOG_AUTO_SEARCH | All features retrieved by search streams are recorded. |
| #SDELOG_AUTO_WRITE | All features added to or replaced in an SDE layer using the StoreSDEFeature request are recorded. |

| SDELogCombineTypeEnum | |
|---|---|
| #SDELOG_COMBINETYPE_DIFFERENCE | Subtracts the SDE features. |
| #SDELOG_COMBINETYPE_INTERSECT | Intersects using the logical And operation. |
| #SDELOG_COMBINETYPE_SYMDIFF | Merges using the logical Xor operation. |
| #SDELOG_COMBINETYPE_UNION | Unites  using the logical Or operation. |

| SDELogModeEnum | |
|---|---|
| #SDELOG_MODE_READ | Read only mode. |
| #SDELOG_MODE_WRITE | Reading and writing mode, position at the top of the log. |
| #SDELOG_MODE_APPEND | Reading and writing mode, position at the end of the log. |
| #SDELOG_MODE_CREATE | Create a new log for reading and writing. |

***See also:*** DBTheme;  SDEDataset; SDEFeature; SDELayer

# SEd

This type of document object allows creation and manipulation of scripts.

**Inherits from** Doc

| Class Requests | |
|---|---|
| GetDefaultFontSize : SEdFontSizeEnum | Returns the current font size. |
| Make : anSEd | Creates a new script editor. |
| MakeFromSource (sourceString, nameString) : anSEd | Creates a script editor named as specified for the specified source code. |
| MakeWithGUI (aGUIName) : aSEd | Creates a new script editor with the specified DocGUI. |
| SetDefaultFontSize (SEdFontSizeEnum) | Sets the default font size. |

| Instance Requests | |
|---|---|
| AddBreak (characterPositionNumber) : aBoolean | Places a break at the statement near the specified position. |
| BreakExists (characterPositionNumber) : aBoolean | Returns true if a break exists at the specified position. |
| ClearAllBreaks | Removes all breaks from the source code in a script editor. |
| ClearBreak (characterPositionNumber) : aBoolean | Removes the break point at the specified position. |
| ClearSelected | Deletes the selected characters from the source code in a script editor. |
| Compile | Compiles the source code in a script editor. |
| CopySelected | Copies the selected text into the system clipboard. |
| CutSelected | Removes the selected text after copying it into the system clipboard. |
| Edit | Allows editing of the properties for a script editor. |
| GetCodepage: aNumber | Returns the codepage. |
| GetFontSize : SEdFontSizeEnum | Returns the font size for a script editor. |
| GetRemainActiveState : aBoolean | Returns true when a script editor remains active while executing. |
| GetScript : aScript | Gets the compiled source as a script object. |
| GetSearchString : aString | Gets the last search string. |
| GetSelected : aString | Gets the selected text. |

| | |
|---|---|
| GetSource : aString | Gets the source code in a script editor. |
| HasError : aBoolean | Returns true if a run-time error has occurred. |
| Insert (aString) | Inserts the specified string at the cursor position of a script editor. |
| IsCompiled : aBoolean | Returns true if the source code in a script editor is compiled. |
| KillLine | Deletes from the beginning of a line to the cursor. |
| Paste | Pastes the text in the system clipboard into a script editor. |
| Print | Sends the contents of a script editor to the printer. |
| ReturnInsertPos : characterPositionNumber | Returns the position of the cursor. |
| Run (anObj) | Executes or continues after a break point in a script. |
| Search (aString) : aBoolean | Selects and returns true if the specified string is found in the text between the cursor position and the end of the source text. |
| SelectAll | Selects all characters in the source code of a script editor. |
| SetCodepage (aNumber) | Sets the codepage. |
| SetInsertPos (characterPositionNumber) | Moves the cursor to the specified position of the source code in a script editor. |
| SetFontSize (SEdFontSizeEnum) | Sets the font size for a script editor. |
| SetName (nameString) | Sets the name of a script editor and its script. |
| SetRemainActiveState (aBoolean) | If set to true, the script editor remains active while executing. |
| SetSource (aString) | Sets the source of a script editor. |
| ShiftLeft | Moves the selected lines two spaces to the left. |
| ShiftRight | Moves the selected source code lines two spaces to the right. |
| ShowCodepage | Displays the codepage. |
| ShowHelpWin | Shows the help window. |
| ShowVariables | Allows examination of variables through a modal dialog box. |
| Step | Executes the next request. |
| Undo | Undoes the last change not saved. |

## *Enumerations*

| SEdFontSizeEnum | |
|---|---|
| #SED_FONTSIZE_SMALL | Small Courier. |
| #SED_FONTSIZE_MEDIUM | Medium Courier. |
| #SED_FONTSIZE_LARGE | Large Courier. |
| #SED_FONTSIZE_DEFAULT | Default size. |

***See also:*** Script

***Code example:*** page 338

# *ServiceAreaWin (Network Analyst)*

This class models the window used in solving service area problems.

**Inherits from** NetworkWin

| Class Requests | |
|---|---|
| Make (aView, aNetwork, serviceNetworkFTheme, serviceAreaFTheme, sourceNetworkFTheme) : aServiceAreaWin | Creates a service area window. |
| MakeNewResultAreaFTab (aNetwork, aFileName) : anFTab | Creates a service area result feature table. |
| MakeNewResultNetFTab (aNetwork, aFileName) : anFTab | Creates a service netwrok result feature table. |

| Instance Requests | |
|---|---|
| AddPoint (anFTab, labelString, aPoint) | Adds a point. |
| CanSolve : aBoolean | Returns true if sufficient information is provided on the window to solve the problem. |
| Open | Displays the service area window. |
| Refresh | Refreshes the service area window. |
| Solve : aBoolean | Solves the service area and network problem, returns true if successful. |

***See also:*** Network

# *Shape*

Shape subclasses implement planar geometric shapes.

**Inherits from** Obj

**Inherited by** Circle, Line, MultiPoint, Oval, Point, and Rect

| Class Requests | |
|---|---|
| SetCleanPreference (CleanPreferenceEnum) : CleanPreferenceEnum | Sets the clean preference and returns the previous preference. |

| Instance Requests | |
|---|---|
| AsMultiPoint : aMultiPoint | Converts the shape to a multi-point. |
| AsPolygon : aPolygon | Converts the shape to a polygon. |
| AsPolyLine : aPolyLine | Converts a shape to a polyline. |
| Clean : aShape | Cleans a shape. |
| Contains (anotherShape) : aBoolean | Returns true if the specified shape is contained within a shape. |
| ContainsCenter (anotherShape) : aBoolean | Returns true if the center of the specified shape is contained within a shape. |
| Distance (anotherShape) : aNumber | Returns the shortest distance between two shapes. |
| GetDimension : aNumber | Returns zero for points, one for lines, and two for polygons. |
| Intersects (anotherShape) : aBoolean | Returns true if two shapes have at least one intersection. |
| IsCenterContainedIn (anotherShape) : aBoolean | Returns true if the center of a shape is contained within the specified shape. |
| IsContainedIn (anotherShape) : aBoolean | Returns true if a shape is contained within the specified shape. |
| IsEmpty : aBoolean | Returns true if a shape is empty. |
| IsNull : aBoolean | Returns true if the attributes of a shape have not been defined. |
| IsWithin (anotherShape, aDistance) : aBoolean | Returns true if two shapes are within the specified distance. |
| LineIntersection (anotherShape) : aPolyLine | Intersects two shapes. |
| Move (xNumber, yNumber) : aShape | Moves a shape by the specified amount. |
| PointIntersection (anotherShape) : aMultiPoint | Intersects two shapes. |

| | |
|---|---|
| ReturnArea : aNumber | Returns the area of a shape. |
| ReturnBuffered (aDistance) : aPolygon | Applies a buffer to a shape. |
| ReturnCenter : aPoint | Returns the center point of a shape. |
| ReturnClipped (aRect) : aShape | Clips a shape by a rectangle. |
| ReturnDifference (anotherShape) : aShape | Subtracts the specified shape from a shape. |
| ReturnExtent : aRect | Returns the extent of a shape. |
| ReturnIntersection (anotherShape) : aShape | Intersects two shapes. |
| ReturnLength : aNumber | Returns length, perimeter, or circumference of a shape. |
| ReturnMerged (anotherShape) : aShape | Merges two shapes. |
| ReturnProjected (aPrj) : aShape | Projects a shape using the specified projection. |
| ReturnUnion (anotherShape) : aShape | Unites two shapes. |
| ReturnUnprojected (aPrj) : aShape | Unprojects a shape. |
| Snap (anotherShape, snapDistance) : aBoolean | Snaps vertices of a shape to a specified shape within a distance tolerance. |

## Enumerations

| CleanPreferenceEnum | |
|---|---|
| #SHAPE_CLEAN_FAST | Default setting; optimize cleaning for speed. |
| #SHAPE_CLEAN_HIGHEST_QUALITY | Cleans polygons to ensure that all rings are preserved, and that the ring topology is correct. |

***See also:*** GraphicShape; TextPositioner

# *ShortestPathWin (Network Analyst)*

This class models the window for solving shortest path problems.

**Inherits from** NetworkWin

| Class Requests | |
|---|---|
| Make (aView, aNetwork, anFTab, sourceNetworkFTheme) : aShortestPathWin | Creates a shortest path window. |
| MakeNewResultFTab (aNetwork, aFileName) : anFTab | Creates a result feature table. |

| Instance Requests | |
|---|---|
| AddPoint (anFTab, labelString, aPoint) | Adds a point. |
| CanSolve : aBoolean | Returns true if sufficient information is provided to solve the problem. |
| Open | Displays the window. |
| Refresh | Refreshes the window. |
| Solve : aBoolean | Solves the shortest path problem. |

*See also:* Network

# SingleBandLegend

This class implements the legend for a single band image.

**Inherits from** ImageLegend

| Class Requests | |
|---|---|
| Make (aBandNumber, anImageLookup, aColorMap) : aSingleBandLegend | Creates a single band legend from the specified parameters. |

| Instance Requests | |
|---|---|
| Clone : aSingleBandLegend | Creates a duplicate single band legend. |
| GetBand : aNumber | Gets the band number associated with a single band legend. |
| GetColorMap : aColorMap | Gets the colormap associated with a single band legend. |
| GetImageLookup : aImageLookup | Gets the image look-up table associated with a single band legend. |
| SetBand (bandNumber) | Sets the band number for a single band legend. |
| SetColorMap (aColorMap) | Associates a colormap with a single band legend. |
| SetImageLookup (anImageLookup) | Associates an image look-up table with a single band legend. |

*See also:* Color; ColorMap; ImageLookup

# Sinusoid

This class implements the Sinusoidal projection.

**Inherits from** Prj

| Class Requests | |
|---|---|
| Make (aRect) : aSinusoid | Creates a Sinusoidal projection bounded by the specified rectangle. |

| Instance Requests | |
|---|---|
| ProjectPt (aPoint) : aBoolean | Projects a point and returns true if successful. |
| Recalculate | Calculates the derived constants. |
| ReturnCentralMeridian : aNumber | Returns the central meridian in decimal degrees. |
| SetCentralMeridian (aLongitude) | Sets the central meridian. |
| UnProjectPt (aPoint) : aBoolean | Unprojects a point and returns true if successful. |

*See also:* CoordSys; Spheroid

# SourceDialog

This class implements the dialog box that searches for theme sources.

**Inherits from** ModalDialog

| Class Requests | |
|---|---|
| Show (titleString) : srcNameList | Displays the source dialog box and returns the selected sources. |
| ShowClass (titleString, firstDisplayClass) : srcNameList | Displays the source dialog box for image or feature sources. |
| ShowLibraries (titleString) | Displays the source dialog box for librarian. |

*See also:* SrcName; ISrc

# SourceManager

This class facilitates maintenance of data sets.

**Inherits from** ModalDialog

| Class Requests | |
|---|---|
| GetDataSet (sourceClass, titleString) : aFileName | Creates a source manager window for the specified source class to select a data set. |
| ManageDataSets (sourceClass, titleString) | Creates a source manager window for the specified source class. |

| PutDataSet (sourceClass, titleString, defaultFileName, doCheckBoolean) : aFileName | Creates a source manager window for the specified source class to save a data set. |
|---|---|
| ReturnDataSets (sourceClass, titleString) : fileNameList | Creates a source manager window for the specified source class to select multiple data sets. |

# Space

This class implements separators for user control objects.

**Inherits from** Control

| *Class Requests* | |
|---|---|
| Make : aSpace | Creates a user control space. |

***See also:*** ButtonBar; ControlSet; Menu; ToolBar

# Spheroid

This class implements various types of spheroid.

**Inherits from** Obj

| *Class Requests* | |
|---|---|
| Make (SpheroidEnum, aPrj) : aSpheroid | Creates a spheroid from the specified type and associates it with the specified projection. |

| *Instance Requests* | |
|---|---|
| = (anotherSpheroid) : aBoolean | Returns true when two spheroids are the same. |
| GetSpheroidName : aString | Gets the name associated with a spheroid type. |
| GetType : SpheroidEnum | Gets a spheroid's type. |
| GetUnits : UnitsLinearEnum | Gets the linear units of a spheroid. |
| ReturnEccentricity : aNumber | Returns the eccentricity of a spheroid. |
| ReturnRadius : aNumber | Returns the radius of a spheroid. |
| SetMajorAndMinorAxes (equatorialRadius, PolarRadius) | Sets the major and minor axes of a spheroid. |
| SetRadiusAndFlattening (equatorialRadius, flatteningNumber) | Sets the major axis and flattening factor of a spheroid. |
| SetType (SpheroidEnum) | Sets the type of a spheroid. |

| SetUnits (UnitsLinearEnum) | Sets the units for a spheroid. |
|---|---|

## *Enumerations*

| SpheroidEnum | |
|---|---|
| #SPHEROID_OTHER | Unknown. |
| #SPHEROID_FIRST | For internal use only. |
| #SPHEROID_AIRY | Airy 1830. |
| #SPHEROID_AUSTRALIAN | Australian National 1965. |
| #SPHEROID_BESSEL | Bessel 1841. |
| #SPHEROID_CLARKE1866 | Alexander Ross Clarke 1866. |
| #SPHEROID_CLARKE1880 | Alexander Ross Clarke 1880. |
| #SPHEROID_EVEREST | Everest 1830. |
| #SPHEROID_GRS80 | Geodetic Reference System 1980. |
| #SPHEROID_INTERNATIONAL1909 | John Fillmore Hayford 1909 (IUGG 1924). |
| #SPHEROID_KRASOVSKY | Krasovsky 1940. |
| #SPHEROID_SPHERE | The world as a sphere. Use this for world maps. |
| #SPHEROID_WGS72 | World Geodetic System 1972. |
| #SPHEROID_WGS84 | World Geodetic System 1984. |
| #SPHEROID_LAST | For internal use only. |

| UnitsLinearEnum | |
|---|---|
| #UNITS_LINEAR_CENTIMETERS | Centimeters. |
| #UNITS_LINEAR_DEGREES | Decimal degrees. |
| #UNITS_LINEAR_FEET | Feet. |
| #UNITS_LINEAR_INCHES | Inches. |
| #UNITS_LINEAR_KILOMETERS | Kilometers. |
| #UNITS_LINEAR_METERS | Meters. |
| #UNITS_LINEAR_MILES | Miles. |
| #UNITS_LINEAR_MILLIMETERS | Millimeters. |
| #UNITS_LINEAR_NAUTICALMILES | Nautical miles. |
| #UNITS_LINEAR_UNKNOWN | Unknown. |
| #UNITS_LINEAR_YARDS | Yards. |

***See also:*** CoordSys; Prj

# SQLCon

This class facilitates the connection to an SQL database.

**Inherits from** Obj

| Class Requests | |
|---|---|
| Find (aName) : anSQLCon | Establishes the connection identified by the ODBC name or database integrator. |
| GetConnections : anSQLConList | Gets a list of all SQL connections. |
| HasSQL : aBoolean | Returns true if an SQL connection can be created. |

| Instance Requests | |
|---|---|
| ExecuteSQL (aString) : aBoolean | Executes the string over an SQL connection. |
| HasError : aBoolean | Returns true if connection encountered errors. |
| IsLogin : aBoolean | Returns true if an SQL connection has logged in. |
| Login (aLoginString) | Logs in an SQL connection using the specified string. |
| Logout | Logs out an SQL connection. |

*See also:* DBICursor; SQLWin

# SQLWin

This class implements a window for using and managing an SQL connection.

**Inherits from** Window

| Class Requests | |
|---|---|
| Make : anSQLWin | Creates and displays an SQL connection window. |
| The : anSQLWin | Returns the SQL connection window. |

| Instance Requests | |
|---|---|
| Open | Opens an SQL connection window. |

*See also:* SQLCon; VTab

# SrcExtension

Instances of this class store extension-specific environments.

**Inherits from** Extension Object

# SrcName

This class implements the source of data for a theme.

**Inherits from** Obj

**Inherited by** DynName, GeoName, and XYName

| Class Requests | |
| --- | --- |
| Make (aSourceString) : aSrcName | Creates a source name from a shape file, image file, coverage, layer, or librarian. |

| Instance Requests | |
| --- | --- |
| GetDataSource : aString | Gets the data source for a source name. |
| GetFileName : aFileName | Gets the file name associated with a source name. |
| GetName : aString | Gets the name of a source name object. |
| GetSubName : aString | Gets the sub-name which is the feature class of a source name. |
| SetDataSource (aSourceString) | Sets the data source for a source string. |
| SetFileName (aFileName) | Associates the file name with a source name. |
| SetName (aString) | Sets the name of a source name object. |
| SetSubName (aString) | Sets the feature class of a source name. |

***See also:*** Coverage; FileName; FTab; ISrc; Layer; Theme

# Stack

This class implements a data collection based on the last-in-first-out structure.

**Inherits from** Collection

| Class Requests | |
| --- | --- |
| Make : aStack | Creates an empty stack. |

| Instance Requests | |
| --- | --- |
| Depth : aNumber | Returns the number of objects in the stack. |
| Empty | Clears out a stack. |
| IsEmpty : aBoolean | Returns true if a stack is empty. |
| Peek (indexNumber) : anObj | Returns the object at the specified position of stack. |

| Pop : anObj | Returns the object at the top of a stack and then removes it from the stack. |
|---|---|
| Push (anObj) | Places an object on the stack. |
| Size : aNumber | Returns the maximum number of objects a stack can accommodate. |
| Top : anObj | Returns the object on top of the stack without removing it from the stack. |

***See also:*** Dictionary; List

***Code example:*** page 401

# StanEditDialog

This class is the modal dialog box used to view and edit a geocoding event's standardization.

**Inherits from** ModalDialog

| Class Requests | |
|---|---|
| Show (aMatchKey) : aBoolean | Displays the dialog box. |

***See also:*** GeocodeDialog;  MatchKey

# Stereographic

This class implements the Stereographic projection.

**Inherits from** Perspective

| Class Requests | |
|---|---|
| CanDoCustom : aBoolean | Always returns true. |
| Make (aRect) : aStereo | Creates a projection bounded by the specified rectangle. |

***See also:*** CoordSys; Spheroid

# Stipple

A stipple is a monochrome image used to define fill patterns.

**Inherits from** Obj

| Class Requests | |
|---|---|
| Make (anIcon) : aStipple | Creates a stipple from the specified icon. |

| Instance Requests | |
|---|---|
| = anObj : aBoolean | Returns true if the object and stipple are the same. |
| GetWidth : aNumber | Gets the width of a stipple in pixels. |
| GetHeight : aNumber | Gets the height of a stipple in pixels. |

**See also:** BasicMarker; Icon; RasterFill

# String

This class implements the string data type.

**Inherits from** Value

**Inherited by** Pattern

| Class Requests | |
|---|---|
| MakeBuffer(numberOfSpaces) : aString | Creates a string with the specified number of blank characters. |

| Instance Requests | |
|---|---|
| + anotherString : aString | Concatenates two strings. |
| ++ anotherString : aString | Concatenates two strings with a space in between. |
| = anObj : aBoolean | Returns true if the object is the same as a string. |
| AsAscii : aNumber | Returns the ASCII value of the first character in a string. |
| AsDate : aDate | Converts a string to date using the default date format. |
| AsEnum : anEnumerationElt | Converts a string to an enumeration. |
| AsFileName : aFileName | Converts a string to a file name. |
| AsList : stringList | Converts the words in a string to string objects of a list. |

| | |
|---|---|
| AsNumber : aNumber | Converts a string to a number. |
| AsPattern : aPattern | Creates a pattern object from a string. |
| AsSrcName : aSrcName | Creates a source name from a string. |
| AsString : aString | Clones a string. |
| AsTokens (separatorString) : stringList | Creates a list of strings from the tokens of a string. |
| BasicProper (separatorString) : aString | Converts the first character of each token to upper case and remaining characters to lower case. |
| BasicTrim (leftTrimString, rightTrimString) : aString | Trims the left and right of a string from the specified characters. |
| Contains (anotherStringOrPattern) : aBoolean | Returns true if the specified string or pattern is found in a string. |
| Count : aNumber | Returns the number of characters in a string. |
| Extract (wordIndexNumber) : aString | Returns the word at the specified index. |
| IndexOf (aString) : aNumber | Returns the character index of the specified string within another string. |
| IsNull : aBoolean | Returns true if a string object has no value. |
| IsNumber : aBoolean | Returns true if a string object can be converted to a number. |
| IsQuoted : aBoolean | Returns true if the first and last characters are double quotations. |
| LCase : aString | Converts all characters of a string to lower case. |
| Left (aNumber) : aString | Returns the specified number of characters from the left side of a string. |
| Middle (offsetNumber, lengthNumber) : aString | Returns the specified number of characters from the specified position of a string. |
| Proper : aString | Converts the first character of every word to upper case and the rest to lower case. |
| Quote : aString | Places a string in quotations. |
| Right (aNumber) : aString | Returns the specified number of characters from the right side of a string. |
| Split (positionIndexList, insertString) : aString | Inserts the specified string at the specified positions. |
| Substitute (matchString, replaceString) : aString | Substitutes strings. |
| Translate (fromCharacterSetString, toCharacterSetString) : aString | Translates occurrences of one character to another. |
| Trim : aString | Trims blanks from the left and right side of a string. |

| UCase : aString | Converts all characters of a string to upper case. |
|---|---|
| Unquote : aString | Removes a string's quotations. |

***See also:*** Date; List; Number

***Code example:*** page 323

# SummaryDialog

This class implements the dialog box for specifying how summary values are computed.

**Inherits from** ModalDialog

| Class Requests | |
|---|---|
| Show (aVTab, aField) : summaryVTab | Returns a virtual table with the summary information. |

***See also:*** Field; VTab

# SVGram (Spatial Analyst)

SVGram is the abstract class for the semi-variogram used with Kriging interpolation.

**Inherits from** Obj

| Class Requests | |
|---|---|
| Make (pointFTab, aPrj, numericField, KrigingEnum, anInterval) : SVGram | Creates an SVGram from the specified point or multipoint feature table. |

| Instance Requests | |
|---|---|
| GetVTab : pointFTab | Gets the associated point or multipoint FTab. |

## *Enumerations*

| KringingEnum | |
|---|---|
| #KRINGING_CIRCULAR | Circular |
| #KRIGING_EXPONENTIAL | Exponential. |

| #KRIGING_GAUSSIAN | Gaussian. |
|---|---|
| #KRIGING_LINEAR | Linear with sill. |
| #KRIGING_SPHERICAL | Spherical. |
| #KRIGING_UNIVERSAL1 | Universal Kriging with linear drift. |
| #KRIGING_UNIVERSAL2 | Universal Kriging with quadratic drift |

*See also:* Grid; Interp

# Symbol

Subclasses of symbol provide the ability to draw a shape on the screen.

**Inherits from** Obj

*Inherited by* ChartSymbol; Fill; Marker; Pen; and TextSymbol

| Class Requests | |
|---|---|
| Make (SymbolEnum) : aSymbol | Creates a default symbol. |

| Instance Requests | |
|---|---|
| = anObj : aBoolean | Returns true if the object is the same as a symbol. |
| CanSetSize : aBoolean | Returns true when symbol's size can be changed. |
| Copy (anotherSymbol) | Copies the specified symbol into a symbol. |
| GetBgColor : aColor | Gets the background color. |
| GetColor : aColor | Gets the foreground color. |
| GetColorLock : aBoolean | Returns true if the color lock is set for a symbol. |
| GetOlColor : aColor | Gets the outline color. |
| GetSize : aNumber | Gets a symbol's size in points. |
| GetSizeInches : aNumber | Gets the size of a symbol in inches. |
| GetType : SymbolEnum | Gets a symbol's type. |
| SetBgColor (aColor) | Sets the background color. |
| SetColor (aColor) | Sets the foreground color. |
| SetColorLock (aBoolean) | Sets the color lock for a symbol. |
| SetOlColor (aColor) | Sets the outline color for a symbol. |
| SetSize (sizeNumber) | Sets the size of a symbol in points. |
| SetSizeInches (sizeNumber) | Sets the size of a symbol in inches. |
| UnHook | Unhook a symbol from its scale. |

## *Enumerations*

| SymbolEnum | |
|---|---|
| #SYMBOL_PEN | Pen symbol. |
| #SYMBOL_MARKER | Marker symbol. |
| #SYMBOL_FILL | Fill symbol. |
| #SYMBOL_TEXT | Text symbol. |

***See also:*** Graphic; Legend; SymbolWin

***Code example:*** page 375

# *SymbolList*

This class is a collection of symbols.

**Inherits from** List

| Class Requests | |
|---|---|
| FromList (aList) : aSymbolList | Creates a symbol list from a list of symbols. |
| GetPredefined (SymListTypeEnum) : symbolList | Gets the predefined symbols of the specified type. |
| Make : aSymbolList | Creates an empty symbol list. |

| Instance Requests | |
|---|---|
| RampColor (aStartColor, anEndColor) | Adds a colormap to a symbol list. |
| RandomColors | Adds random colors to a symbol list. |
| RampSavedColors (aSymbolList) | Creates a color ramp. |
| RampSizes (startSize, endSize) | Creates a size ramp. |
| RampSplitSavedColors (aSymbolList, splitIndex) | Creates a split color ramp. |
| RampSubset (startIndex, endIndex) | Creates a color ramp subset. |
| RandomSavedSymbols (aSymbolList) | Creates random symbol map. |
| RandomSymbols | Creates random symbols. |
| UniformColor (aColor) | Sets all symbols in a list to one color. |

## Enumerations

| SymListTypeEnum | |
|---|---|
| #SYMLIST_TYPE_COLORRAMP | Color ramps. |
| #SYMLIST_TYPE_COLORSCHEME | Color schemes. |
| #SYMLIST_TYPE_RANDOMMARKERS | Random marker symbols. |
| #SYMLIST_TYPE_RANDOMFILLS | Random fill symbols. |
| #SYMLIST_TYPE_RANDOMPENS | Random line symbols. |

*See also:* CompositeFill; CompositeMarker; CompositePen; Palette; Symbol; SymbolWin

*Code example:* page 492

# SymbolWin

This class implements the palette window for symbol selection.

**Inherits from** Obj

| Instance Requests | |
|---|---|
| Close | Closes the symbol window. |
| GetPalette : aPalette | Gets the symbol container of the symbol window. |
| GetPanel : SymbolWinPanelEnum | Gets the active panel. |
| IsOpen : aBoolean | Returns true if the symbol window is open. |
| Open | Displays the symbol window. |
| RefreshPalette (PaletteListEnum) | Refreshes the specified palette in the symbol window. |
| ReturnCurrentSymbol (SymbolEnum) : aSymbol | Returns the selected symbol from the specified type. |
| ReturnFrameSymbol : aSymbol | Returns the draft frame symbol. |
| SelectSymbol (aSymbol) | Selects a symbol from the symbol window. |
| SetPanel (SymbolWinPanelEnum) | Sets the current panel. |

## Enumerations

| PaletteListEnum | |
|---|---|
| #PALETTE_LIST_ALL | All symbol lists. |

| #PALETTE_LIST_FILL | Fill symbol list. |
| #PALETTE_LIST_PEN | Pen symbol list. |
| #PALETTE_LIST_MARKER | Marker symbol list. |
| #PALETTE_LIST_COLOR | Color symbol list. |

| *SymbolEnum* | |
| --- | --- |
| #SYMBOL_PEN | Pen. |
| #SYMBOL_MARKER | Marker. |
| #SYMBOL_FILL | Fill. |
| #SYMBOL_TEXT | Text. |

| *SymbolWinPanelEnum* | |
| --- | --- |
| #SYMBOL_WIN_PANEL_COLOR | Color panel. |
| #SYMBOL_WIN_PANEL_FILL | Fill panel. |
| #SYMBOL_WIN_PANEL_FONT | Font panel. |
| #SYMBOL_WIN_PANEL_MANAGE | Manager panel. |
| #SYMBOL_WIN_PANEL_MARKER | Marker panel. |
| #SYMBOL_WIN_PANEL_PEN | Pen panel. |

***See also:*** Application; Palette; Symbol

# System

This class provides operating system-specific services.

**Inherits from** Obj

| *Class Requests* | |
| --- | --- |
| BasicEcho (aString, toStdOutBoolean) | Prints a string to standard out or standard error. |
| Beep | Beeps. |
| CanUnloadLibrary (aLibraryReferenceNumber) : aBoolean | Returns true if the specified library can be unloaded. |
| Echo (aString) | Sends a string to standard out. |
| Execute (commandString) | Issues an operating system command. |
| GetArch : SystemArchEnum | Returns the system architecture. |
| GetAvailableMemory : aNumber | Returns bytes of available memory. |
| GetAVProcessID : aNumber | Gets ArcView's process ID. |

| | |
|---|---|
| GetEnvVar (varString) : aString | Gets an environment variable. |
| GetLook : SystemLookEnum | Gets the graphical user interface type. |
| GetOS : SystemOSEnum | Gets the operating system. |
| GetOSVariant : SystemOSVariantEnum | Returns the type of Microsoft Windows. |
| GetUNCFilename (aFileName) : aString | Gets the universal naming convention of the specified filename. |
| IsAltKeyDown : aBoolean | Returns true if the Alt key is held down. |
| IsControlKeyDown : aBoolean | Returns true if the Ctrl key is held down. |
| IsDoubleClick : aBoolean | Returns true when the last mouse click was a double click. |
| IsLibrary (aFileName) : aBoolean | Returns true if the specified filename is a shared library. |
| IsMemoryAvailable : aBoolean | Returns false if no memory is available. |
| IsMetaKeyDown : aBoolean | Returns true if the Meta key is held down. |
| IsShiftKeyDown : aBoolean | Returns true if the Shift key is held down. |
| LoadLibrary (aSharedLibraryName) : aSharedLibraryIdentifier | Loads new objects from a file and returns a handle for that library. |
| PlaySound (soundFileName, playAsynchronouslyBoolean) | Plays a sound file. |
| PrintMemory (aString) | Prints memory information along with the specified string. |
| QueryLocalMachineName (aString, stringSize) : aString | Queries the name of the local machine. |
| RefreshWindows | Allows pending events to process. |
| ReturnScreenSizeInches : aPoint | Returns the width and height of the screen in inches. |
| ReturnScreenSizePixels : aPoint | Returns the width and height of the screen in pixels. |
| SetEnvVar (varString, aString) | Sets the value of the specified environment variable. |
| SetLook (SystemLookEnum) | Sets the GUI type. |
| SetScreenSizeInches (aPoint) | Sets the screen size to inches in the specified point. |
| UnloadLibrary (aSharedLibraryIdentifier) | Unloads a shared library. |

## *Enumerations*

| *SystemLookEnum* | |
|---|---|
| #SYSTEM_LOOK_MOTIF | Motif. |
| #SYSTEM_LOOK_OPENLOOK | OpenLook. |
| #SYSTEM_LOOK_MSW | Microsoft Windows. |
| #SYSTEM_LOOK_PM | Presentation Manager. |
| #SYSTEM_LOOK_PM20 | Presentation Manager Version 2.0. |
| #SYSTEM_LOOK_MAC | Macintosh. |

| *SystemOSEnum* | |
|---|---|
| #SYSTEM_OS_CMS | CMS. |
| #SYSTEM_OS_MAC | Macintosh. |
| #SYSTEM_OS_MSW | Microsoft Windows. |
| #SYSTEM_OS_OS2 | OS2. |
| #SYSTEM_OS_UNIX | UNIX. |
| #SYSTEM_OS_VMS | VMS. |

| *SystemOSVariantEnum* | |
|---|---|
| #SYSTEM_OSVARIANT_MAC68K | The Motorola 68K based version of the MacOS. |
| #SYSTEM_OSVARIANT_MACPPC | The PowerPC based version of the MacOS. |
| #SYSTEM_OSVARIANT_MSW16 | Microsoft Windows 3.1 and 3.11. |
| #SYSTEM_OSVARIANT_MSW95 | Microsoft Windows 95. |
| #SYSTEM_OSVARIANT_MSWNT | Microsoft Windows NT. |
| #SYSTEM_OSVARIANT_UNKNOWN | Unknown. |

| *SystemArchEnum* | |
|---|---|
| #SYSTEM_ARCH_ALPHA | DEC Alpha AXP. |
| #SYSTEM_ARCH_HPPA | HP Precision. |
| #SYSTEM_ARCH_I386 | Intel 80386 family. |
| #SYSTEM_ARCH_IBM370 | IBM 370. |
| #SYSTEM_ARCH_IBMRT | IBM RT. |
| #SYSTEM_ARCH_MC68K | Motorola 68000 family. |
| #SYSTEM_ARCH_MC88K | Motorola 88000 family. |

| | |
|---|---|
| #SYSTEM_ARCH_MIPS | MIPS RISC. |
| #SYSTEM_ARCH_PPC | Motorola PowerPC family. |
| #SYSTEM_ARCH_RS6000 | IBM RS/6000. |
| #SYSTEM_ARCH_SPARC | SUN SPARC. |
| #SYSTEM_ARCH_VAX | DEC VAX. |

***See also:*** Application; DDEClient; DDEServer; RPCClient; RPCServer

***Code example:*** page 281

# Table

This class implements the document that can display tabular data stored in VTab objects.

**Inherits from** Doc

| Class Requests | |
|---|---|
| Make (aVTab) : aTable | Creates a table document to display the specified virtual table. |
| MakeWithGUI (aVTab, aGUIName) : aTable | Creates a table document to display the specified virtual table using the given DocGUI. |

| Instance Requests | |
|---|---|
| BlinkRow (rowNumber) | Blinks the record at the specified row number in a table. |
| BuildQuery | Displays the query builder window for the virtual table in a table document. |
| ConvertRecordToRow (recordNumber) : rowNumber | Returns the row number of a table that holds the specified record number. |
| ConvertRowToRecord (rowNumber) : recordNumber | Returns the record number of the specified row in a table. |
| Copy | Copies the value of current field to the system clipboard. |
| Cut | Cuts the current field to the system clipboard. |
| EditProperties | Displays the property dialog box for a table. |
| EditValues | Allows editing a field in an apply event. |
| Find (aString) : recordNumber | Returns the record number where the specified string is found in a table. |

| | |
|---|---|
| GetActiveField : aField | Gets the active field of a table. |
| GetEditor : anObj | Gets the object through which content of a table is edited. |
| GetEditString : aString | Gets the value of the current field. |
| GetFindString : aString | Gets the string used in the last find request. |
| GetUserField : aField | Gets the selected field in an apply event. |
| GetUserRow : rowNumber | Gets the selected row number in an apply event. |
| GetVTab : aVTab | Gets the virtual table displayed in a table document. |
| MakeField : aField | Creates a new field in a table. |
| Paste | Pastes into the current field from the system clipboard. |
| Print | Sends a table to the printer. |
| PromoteSelection | Moves the selected records to the top of the table. |
| Select | Allows selection of a record in an apply event. |
| SetActiveField (aField) | Sets the active field in a table. |
| ShowCodepage | Displays the codepage associated with a table document. |
| ShowRow (rowNumber) | Scrolls a table in order to show the specified row number. |
| Sort (aField, descendingBoolean) | Sorts a table based on a field. |
| StopEditing | Ends editing after the EditValues request. |

***See also:*** QueryWin; VTab

***Code example:*** page 339

# TabulateAreaDialog (Spatial Analyst)

This class models the dialog box for specifying parameters to compute cross tabulated areas between two grid data sets.

**Inherits from** ModalDialog

| Class Requests | |
|---|---|
| Show (aView) : parameterList | Displays the tabulate area dialog box and returns the parameters specified. |

***See also:*** Grid; View

# Template

This class implements predefined layout documents.

**Inherits from** List

| Class Requests | |
|---|---|
| Make : aTemplate | Creates an empty template. |

| Instance Requests | |
|---|---|
| Edit : aBoolean | Displays the template property dialog box. |
| GetIcon : anIcon | Gets the icon assigned to a template. |
| IsPortrait : aBoolean | Returns true if the template orientation is portrait. |
| SetIcon (anIcon) | Assigns an icon to a template. |
| SetPortrait (aBoolean) | Sets a template's orientation to portrait if the specified Boolean value is true. |

*See also:* Graphic; TGraphic

# TemplateMgr

This class implements the dialog box for managing templates.

**Inherits from** ModalDialog

| Class Requests | |
|---|---|
| Add (aTemplate) | Adds a template to the list of available templates. |
| GetTemplates : templateList | Gets a list of available templates. |
| PutTemplates (templateList) | The specified templates become the available templates list. |
| Show : aTemplate | Displays the template manager dialog box. |

*See also:* Layout; ODB; Project; Template

# TextComposer

This class sets the way text is drawn on a display.

**Inherits from** Obj

| Class Requests | |
|---|---|
| CountLines (aString) : aNumber | Returns the number of lines in the specified string. |
| The : aTextComposer | Gets ArcView's text composer. |

| Instance Requests | |
|---|---|
| GetAngle : aNumber | Gets the angle in a text composer. |
| GetJustification : TextComposerJustEnum | Gets the text justification in a text composer. |
| GetLineSpacing : aNumber | Gets the line spacing in a text composer. |
| GetRightToLeft : aBoolean | Returns true if text is drawn from right to left. |
| SetAngle (aNumber) | Sets the angle of text in a text composer. |
| SetJustification (TextComposerJustEnum) | Sets the text justification in a text composer. |
| SetLineSpacing (aNumber) | Sets the line spacing to the specified value. |
| SetRightToLeft (aBoolean) : aBoolean | Draws text right to left when the specified Boolean value is true. |

## Enumerations

| TextComposerJustEnum | |
|---|---|
| #TEXTCOMPOSER_JUST_CENTER | Centered. |
| #TEXTCOMPOSER_JUST_LEFT | Left justified. |
| #TEXTCOMPOSER_JUST_RIGHT | Right justified. |

***See also:*** Display; GraphicText; String

# *TextFile*

This class implements text files with characters as file elements.

**Inherits from** File

| Class Requests | |
|---|---|
| Make (aFileName, FilePermEnum) : aTextFile | Creates a new file or opens an existing file. |

| Instance Requests | |
|---|---|
| GetCodepage : aNumber | Returns the codepage associated with a text file. |
| Read (aNumber) : aString | Reads the specified number of characters. |

| ReadElt : aString | Reads one character from a text file. |
|---|---|
| SkipLine | Moves the file pointer to the beginning of the next line. |
| Write (aString, aNumber) | Writes the specified string up to the specified number of characters into a text file. |
| WriteElt (aString) | Writes the first character of a string into a text file. |
| WriteQStr (aString) | Adds quotations to a string and writes into a text file. |

## Enumerations

| *FilePermEnum* | |
|---|---|
| #FILE_PERM_READ | Opens an existing file for reading. |
| #FILE_PERM_WRITE | Opens a new file for writing, and overwrites the file if it exists. |
| #FILE_PERM_MODIFY | Opens an existing file for reading and writing. |
| #FILE_PERM_APPEND | Opens a new file appending text, and overwrites the file if it exists. |
| #FILE_PERM_CLEARMODIFY | Opens a new file for reading and writing, and overwrites the file if it exists. |

*See also:* FileName; LineFile

*Code example:* page 362

# TextPositioner

Subclasses of this class implement placement of text in relation to a shape.

**Inherits from** Obj

**Inherited by** PointTextPositioner, PolygonTextPositioner, and PolyLineTextPositioner

| *Class Requests* | |
|---|---|
| Make (aShapeClass) : aTextPositioner | Creates one of the subclasses of text positioner based on the provided class. |

| Instance Requests | |
|---|---|
| Calculate (anchorShape, textExtentPoint, pointMarkerSize, AlternateAnchorPoint) | Computes the position of a text block around the center of the specified shape or alternate anchor point if not nil. The computation is based on the text extent. The third parameter is used when the anchor shape is a point. |
| GetAngle : aNumber | Gets the angle of text in a text positioner. |
| GetHAlign : TextPositionerHAlignEnum | Gets the horizontal alignment in a text positioner. |
| GetOrigin : aPoint | Gets the origin of a text block in a text positioner. |
| GetSpacing : aNumber | Gets the spacing in a text positioner. |
| GetVAlign : TextPositionerVAlignEnum | Gets the vertical alignment in a text positioner. |
| SetHAlign (TextPositionerHAlignEnum) | Sets the horizontal alignment in a text positioner. |
| SetVAlign (TextPositionerVAlignEnum) | Sets the vertical alignment in a text positioner. |

## Enumerations

| TextPositionerHAlignEnum | |
|---|---|
| #TEXTPOSITIONER_HALIGN_AFTER | Position entire text after shape. |
| #TEXTPOSITIONER_HALIGN_BEFORE | Position entire text before shape. |
| #TEXTPOSITIONER_HALIGN_CENTER | Position text centered on shape. |
| #TEXTPOSITIONER_HALIGN_LEFT | Position beginning of text even with left of shape. |
| #TEXTPOSITIONER_HALIGN_RIGHT | Position end of text even with right of shape. |

| TextPositionerVAlignEnum | |
|---|---|
| #TEXTPOSITIONER_VALIGN_ABOVE | Position entire text above shape. |
| #TEXTPOSITIONER_VALIGN_BELOW | Position entire text below shape. |
| #TEXTPOSITIONER_VALIGN_BOTTOM | Position bottom of text even with bottom of shape. |
| #TEXTPOSITIONER_VALIGN_ON | Position centered on shape. |
| #TEXTPOSITIONER_VALIGN_TOP | Position top of text even with top of shape. |

**See also:** FTheme; GraphicText; MultiPoint; Point

# TextSymbol

This class implements characters that are drawn as text.

**Inherits from** Symbol

| Class Requests | |
|---|---|
| Make : aTextSymbol | Creates a text symbol. |

| Instance Requests | |
|---|---|
| = anObj : aBoolean | Returns true if an object is the same as a text symbol. |
| CanSetSize : aBoolean | Returns true if the size of a text symbol can be set. |
| GetFont : aFont | Gets the font of a text symbol. |
| GetSize : aNumber | Gets the size of a text symbol in points. |
| GetType : SymbolEnum | Returns #SYMBOL_TEXT. |
| SetFont (aFont) | Sets the font for a text symbol. |
| SetSize (aNumber) | Sets the size of a text symbol in points. |
| UnHook | Unhooks a text symbol from its scale. |

*See also:* Font

*Code example:* page 369

# TextWin

This class provides the ability to display the contents of a text file.

**Inherits from** Window

| Class Requests | |
|---|---|
| Make (aFileName, titleString) : aTextWin | Displays the contents of the specified text file. |

| Instance Requests | |
|---|---|
| GetFileName : aFileName | Gets the name of the text file displayed in a text window. |
| IsFixedFont : aBoolean | Returns true if the text is displayed in a fixed font. |
| Open | Opens a text window. |
| SetFileName (aFileName) | Sets the file name displayed in a text window. |
| SetFixedFont (aBoolean) | Displays the text in fixed font if the specified Boolean value is true. |

*See also:* FileName; MsgBox

# *TGraphic*

A template graphic object is a component of a template object.

**Inherits from** Obj

| Class Requests | |
|---|---|
| Make (aGraphic) : aTGraphic | Creates a template graphic from a graphic object. |

| Instance Requests | |
|---|---|
| GetPositionedGraphic (aDisplay) : aGraphic | Positions the graphic object associated with the template graphic on a display. |

***See also:*** Display; Layout; Template

# *Theme*

Subclasses of Theme draw images and spatial data in a view document.

**Inherits from** Obj

**Inherited by** DBTheme, FTheme, GTheme, ITheme

| Class Requests | |
|---|---|
| Make (aSrcName) : aTheme | Creates a feature or image theme based on the provided source name. |

| Instance Requests | |
|---|---|
| CanDeleteFromView : aBoolean | Returns true if a theme can be deleted from its view. |
| CanEdit : aBoolean | Returns true if a theme can be edited. |
| CanExportToFTab : aBoolean | Returns true if a theme can be exported to a shapefile. |
| CanFindByPoint : aBoolean | Returns true if a theme's features can be found by point position. |
| CanHotLink : aBoolean | Returns true if a theme can support hot link. |
| CanLabel : aBoolean | Returns true if a theme supports self-labeling. |
| CanProject : aBoolean | Returns true if a theme can be projected. |
| CanReturnClassCounts : aBoolean | Returns true if the number of features in each class of theme's legend can be counted. |

| | |
|---|---|
| CanSelect : aBoolean | Returns true if features in a theme can be selected. |
| ClearSelection | Unselects the selected features in a theme. |
| Clone : aTheme | Clones a theme. |
| EditLegend | Displays the legend editor for a theme. |
| EditProps | Displays the property editor for a theme. |
| ExportToFTab (aFileName) : anFTab | Exports a theme to a shapefile. |
| FindByPoint (aPoint) : aList | Returns a list of identifying keys for the features or cells at the given point. |
| GetComments : aString | Gets the comment property of a theme. |
| GetExtension (anExtensionClass) : aThemeExtension | Gets the associated extension. |
| GetFixedSizeText : aBoolean | Returns true if text is displayed with a fixed size font. |
| GetGraphics : aGraphicSet | Gets the graphics associated with a theme. |
| GetLabelField : aField | Gets the field used in labeling a theme. |
| GetLabelTextSym : aSymbol | Gets the symbol used in labeling a theme. |
| GetLegend : anObj | Gets a legend or an image legend object that is associated with a theme. |
| GetLegendEditorScript : aString | Gets the script that is executed when requesting to edit the legend of a theme. |
| GetName : aString | Gets the name property of a theme. |
| GetObjectTag : anObj | Gets the tag object of a theme. |
| GetPassword : aString | Gets the password property of a theme. |
| GetSrcName : aSrcName | Gets the source name of a theme. |
| GetThreshold : aThreshold | Gets the threshold property of a theme. |
| GetView : aView | Gets the view of a theme. |
| HasAttributes : aBoolean | Returns true when a theme has attribute values. |
| HasTable : aBoolean | Returns true if there is a table document for a theme. |
| Identify (aKeyValue, titleString) | Displays the identify window with the feature specified through its key. |
| Invalidate (invalidateViewBoolean) | Invalidates a theme and redraws the entire view if the specified Boolean value is true. |
| InvalidateLegend | Invalidates the legend of a theme. |
| IsActive : aBoolean | Returns true if a theme is active. |
| IsActiveLocked : aBoolean | Returns true if a theme's active status cannot be changed. |

| | |
|---|---|
| IsLegendVisible : aBoolean | Returns true if the legend of a theme is visible. |
| IsLocked : aBoolean | Returns true if a theme is locked and cannot be edited. |
| IsSuffixUsed : aBoolean | Returns true if the word <169>by<170> and the classification field are added to a theme's name. |
| IsVisible : aBoolean | Returns true if a theme is drawn on the screen. |
| IsVisibleLocked : aBoolean | Returns true if a theme's visible status cannot be changed. |
| RemoveExtension (anExtensionClass) | Removes an associated extension object. |
| ReturnAOI : aRect | Returns the area of interest for a theme. |
| ReturnClassCounts : numberList | Returns the number of features in each class of legend for a theme. |
| ReturnDefaultLegend : aLegend | Returns the single symbol legend for a theme. |
| ReturnExtent : aRect | Returns the extent of a theme. |
| ReturnLabel (aDisplay, aPoint) : aGraphicText | Returns the graphic text label found on the specified display at the given point. |
| ReturnValueString (aField, recordNumber) : aString | Returns the string representation of the value in the specified field at the given record number. |
| SelectByTheme (anotherTheme, FTabRelTypeEnum, aDistance, VTabSelTypeEnum) | Selects the features of a theme if the have the specified relationship with another theme. |
| SetActiveLocked (isLockedBoolean) | If set to true, the active status of a theme cannot be changed. |
| SetDefaultLegend | Sets the legend of a theme's to its default state. |
| SetActive (aBoolean) | Makes a theme active if the specified Boolean value is true. |
| SetAOI (aRect) | Sets the area of interest for a theme. |
| SetComments (aString) | Sets the comment property for a theme. |
| SetExtension (anExtension) | Associates the extension with a theme. |
| SetFixedSizeText (aBoolean) | If set to true then the text is not scaled. |
| SetLabelField (aField) | Sets the specified field as the label field for a theme. |
| SetLabelTextSym (aSymbol) | Sets the symbol for the labels of a theme. |
| SetLegend (anObj) | Associates a theme to a legend or an image legend object. |
| SetLegendEditorScript (aScriptName) | Sets the script to run when user requests to edit the legend of a theme. |
| SetLegendVisible (aBoolean) | Makes a theme's legend visible if the specified Boolean value is true. |

| SetLocked (aBoolean) | Locks a theme if the specified Boolean value is true. |
|---|---|
| SetName (aString) | Sets the name property of a theme. |
| SetObjectTag (anObj) | Sets a theme's object tag. |
| SetPassword (aString) | Sets a theme's password based on the specified string. |
| SetSuffixUsed (aBoolean) | Uses a suffix in the name of a theme if the specified Boolean value is true. |
| SetThreshold (aThreshold) | Sets a theme's threshold. |
| SetVisible (aBoolean) | Draws a theme on the screen if the specified Boolean value is true. |
| SetVisibleLocked (isLockedBoolean) | If set to true, the visibility of a theme cannot be changed. |
| SupportsAOI : aBoolean | Returns true if a theme can have an area of interest. |
| UpdateLegend | Applies the changes in a legend to a theme. |

## Enumerations

| FTabRelTypeEnum | |
|---|---|
| #FTAB_RELTYPE_COMPLETELYCONTAINS | Completely contains. |
| #FTAB_RELTYPE_CONTAINSTHECENTEROF | Contains the center. |
| #FTAB_RELTYPE_HASCENTERWITHIN | Has center within. |
| #FTAB_RELTYPE_INTERSECTS | Intersects. |
| #FTAB_RELTYPE_ISCOMPLETELYWITHIN | Is completely within. |
| #FTAB_RELTYPE_ISWITHINDISTANCEOF | Is within a distance. |

| VTabSelTypeEnum | |
|---|---|
| #VTAB_SELTYPE_AND | Selects from the set of current selection. |
| #VTAB_SELTYPE_NEW | Replaces the existing selection with a new selection. |
| #VTAB_SELTYPE_OR | Adds the selection to the current selections. |
| #VTAB_SELTYPE_XOR | Deselects if already selected, or adds to the selection if not selected. |

**See also:** GraphicSet; ImageLegend; Legend; SrcName; Threshold; View

**Code example:** page 508

# ThemeExtension

This class is used to add ArcView's extension capabilities to theme class.

**Inherits from** ExtensionObject

# ThemeOnThemeDialog

This class implements the theme-on-theme dialog box.

**Inherits from** ModalDialog

| Class Requests | |
|---|---|
| Show (toThemesList, fromThemesList, UnitsLinearEnum) | Displays the theme-on-theme dialog box so that user can select to and from themes, and provides a tolerance distance in the specified measurement units. |

## *Enumerations*

| UnitsLinearEnum | |
|---|---|
| #UNITS_LINEAR_UNKNOWN | Unknown. |
| #UNITS_LINEAR_INCHES | Inches. |
| #UNITS_LINEAR_FEET | Feet. |
| #UNITS_LINEAR_YARDS | Yards. |
| #UNITS_LINEAR_MILES | Miles. |
| #UNITS_LINEAR_MILLIMETERS | Millimeters. |
| #UNITS_LINEAR_CENTIMETERS | Centimeters. |
| #UNITS_LINEAR_METERS | Meters. |
| #UNITS_LINEAR_KILOMETERS | Kilometers. |
| #UNITS_LINEAR_NAUTICALMILES | Nautical miles. |
| #UNITS_LINEAR_DEGREES | Decimal degrees. |

***See also:*** FTheme

# Threshold

This class provides the ability to display a theme only at a certain scale range.

**Inherits from** Obj

| Class Requests | |
| --- | --- |
| Make : aThreshold | Creates an empty threshold object. |

| Instance Requests | |
| --- | --- |
| = anObj : aBoolean | Returns true if the object and a threshold are the same. |
| GetMaximum : aNumber | Gets the high threshold value. |
| GetMinimum : aNumber | Gets the low threshold value. |
| IsMaximumOn : aBoolean | Returns true if the high value is used in thresholding. |
| IsMinimumOn : aBoolean | Returns true if the low value is used in thresholding. |
| IsOff : aBoolean | Returns true if thresholding is not used. |
| IsWithin (aNumber) : aBoolean | Returns true if the specified number is within the threshold range. |
| SetMaximum (aNumber) | Sets the high value of a threshold to the specified scale. |
| SetMaximumOn (aBoolean) | Activates the high value of a threshold if the specified Boolean value is true. |
| SetMinimum (aNumber) | Sets the low value of a threshold to the specified scale. |
| SetMinimumOn (aBoolean) | Activates the low value of a threshold if the specified Boolean value is true. |

*See also:* Theme

# Title

This class implements a chart title.

**Inherits from** ChartPart

| Instance Requests | |
| --- | --- |
| Edit | Displays the title editor. |
| GetLocation : ChartDisplayLocEnum | Gets the location of a title. |
| ReturnRelativeLocation : aPoint | Returns the relative location of a title in percentage of chart size. |
| SetLocation (ChartDisplayLocEnum) | Sets the location of a title. |
| SetName (aString) | Sets the text of a title object. |

| SetRelativeLocation (aPoint) | Sets the location of a title in a percentage of the chart size. |
| SetVisible (aBoolean) | Draws a title if the specified Boolean value is true. |

## Enumerations

| **ChartDisplayLocEnum** | |
| --- | --- |
| #CHARTDISPLAY_LOC_BOTTOM | Below the data area. |
| #CHARTDISPLAY_LOC_LEFT | Left of the data area. |
| #CHARTDISPLAY_LOC_RELATIVE | Within the drawing area using the relative location value. |
| #CHARTDISPLAY_LOC_RIGHT | Right of the data area. |
| #CHARTDISPLAY_LOC_TOP | Above the data area. |

***See also:*** ChartDisplay

# TOC

This class implements the table of contents in a view document.

**Inherits from** Obj

| **Class Requests** | |
| --- | --- |
| GetDefaultSymbol : aTextSymbol | Gets the default text symbol used in displaying text. |
| SetDefaultSymbol (aTextSymbol) | Sets the default text symbol used in displaying text. |

| **Instance Requests** | |
| --- | --- |
| GetSymbol : aTextSymbol | Gets the text symbol used in a table of contents. |
| IsOrderLocked : aBoolean | Returns true if the order of themes in a table of contents cannot be changed. |
| SelectAll | Makes all themes in a table of contents active. |
| SelectFirst | Makes the first theme in a table of contents active. |
| SetOrderLocked (isLockedBoolean) | If set to true, order of themes cannot be changed. |
| SetSymbol (aTextSymbol) | Sets the text symbol in a table of contents. |
| UnselectAll | Deactivates all themes in a table of contents. |

***See also:*** ImageLegend; Legend; Symbol; Theme; View

# Tool

This class implements tool buttons in ArcView's GUI.

**Inherits from** Control

**Inherited by** ToolMenu

| Class Requests | |
|---|---|
| Make : aTool | Creates an empty tool. |

| Instance Requests | |
|---|---|
| Apply | Triggers the apply event. |
| Click | Triggers the click event. |
| GetApply : aString | Gets the name of the apply event script in a tool. |
| GetClick : aString | Gets the name of the click event script in a tool. |
| GetIcon : anIcon | Gets the icon displayed on the face of a tool. |
| HasScript (aScriptName) : aBoolean | Returns true if the specified script is associated with a tool. |
| IsSelected : aBoolean | Returns true if a tool is selected. |
| Select | Selects a tool. |
| SetApply (aString) | Sets the apply event script to the specified script name. |
| SetClick (aString) | Sets the click event script to the specified script name. |
| SetIcon (anIcon) | Sets the icon for a tool. |

*See also:* Button; Choice; ToolBar

*Code example:* page 294

# ToolBar

This class implements the third row of a document GUI.

**Inherits from** ControlSet

| Instance Requests | |
|---|---|
| GetActive : aTool | Returns the selected tool. |
| Empty | Deletes all tools from a tool bar. |

| Remove (aTool) | Removes the specified tool from a tool bar. |
| --- | --- |
| SelectDefault | Selects the default tool. |

***See also:*** ButtonBar; DocGUI; MenuBar; Space; Tool; ToolMenu

***Code example:*** page 346

# ToolMenu

This class implements a pull-down set of tools.

**Inherits from** Tool

| 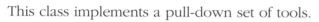Class Requests | |
| --- | --- |
| Make : aToolMenu | Creates an empty tool menu. |

| Instance Requests | |
| --- | --- |
| Add (aTool, positionNumber) | Adds a tool to a tool menu after the specified position. |
| FindByScript (aScriptName) : aTool | Finds a tool associated with the specified script in a tool menu. |
| GetChoice : aTool | Returns the selected tool in a tool menu. |
| GetControls : toolList | Gets the list of tools in a tool menu. |
| HasScript (aScriptName) : aBoolean | Returns true when the specified script is associated with a tool in a tool menu. |
| Remove (aTool) | Removes the specified tool from a tool menu. |
| SetApply (aScriptName) | Sets the script for the apply event of the active tool in a tool menu. |
| SetChoice (aTool) | Selects the tool in a tool menu. |
| SetClick (aScriptName) | Sets the script for the click event of the active tool in a tool menu. |
| SetEnabled (isEnabledBoolean) | If set to true, a tool menu is enabled. |
| SetHelp (aString) | Sets the help property for the active tool in a tool menu. |
| SetHelpTopic (topicString) | Sets the help topic property for the active tool in a tool menu. |
| SetIcon (anIcon) | Sets the icon property for the active tool in a tool menu. |
| SetObjectTag (anObj) | Tags an object to the active tool in a tool menu. |

| SetTag (aString) | Sets the tag property for the active tool in a tool menu. |
|---|---|
| SetUpdate (aScriptName) | Sets the script for the update event of the active tool in a tool menu. |
| Shift (aTool, positionOffset) | Moves the specified tool by the given number within a tool menu. |
| Update | Triggers the update script for all tools in a tool menu. |

*See also:* ToolBar

# TransverseMercator

This class implements the Transverse Mercator projection.

**Inherits from Prj**

| Class Requests | |
|---|---|
| CanDoSpheroid : aBoolean | Always returns true. |
| Make (aRect) : aTrnmerc | Creates a projection bounded by the specified rectangle. |

| Instance Requests | |
|---|---|
| ProjectPt (aPoint) : aBoolean | Projects a point and returns true if successful. |
| Recalculate | Computes derived constants. |
| ReturnCentralMeridian : aNumber | Returns the central meridian in decimal degrees. |
| ReturnFalseEasting : aNumber | Returns the false easting in decimal degrees. |
| ReturnFalseNorthing : aNumber | Returns the false northing in decimal degrees. |
| ReturnReferenceLatitude : aNumber | Returns the reference latitude in decimal degrees. |
| ReturnScale : aNumber | Returns the scale along the central meridian. |
| SetCentralMeridian (aLongitude) | Sets the central meridian to the specified decimal degrees. |
| SetFalseEasting (aNumber) | Sets the value of X at the central meridian. |
| SetFalseNorthing (aNumber) | Sets the value of Y at the central meridian. |
| SetReferenceLatitude (aLatitude) | Sets the reference latitude to the specified decimal degrees. |
| SetScale (aNumber) | Sets the scale along the central meridian. |
| UnProjectPt (aPoint) : aBoolean | Unprojects a point and returns true if successful. |

*See also:* CoordSys; Spheroid

# *Units*

This class implements the conversion of measurement units.

**Inherits from** Obj

| Class Requests | |
|---|---|
| Convert (fromNumber, fromUnitsLinearEnum, toUnitsLinearEnum) : aNumber | Converts the specified length between the specified units. |
| ConvertArea (fromNumber, fromUnitsLinearEnum, toUnitsLinearEnum) : aNumber | Converts the specified area between the specified units. |
| ConvertDecimalDegrees (fromPoint, toPoint, UnitsLinearEnum, toUnitsLinearEnum) : aNumber | Returns the distance between two decimal degree coordinates in the specified units. |
| GetFullUnitString (UnitsLinearEnum) : aString | Gets the name of the unit. |
| GetUnitString (UnitLinearEnum) : aString | Gets the abbreviated name of the unit. |

## *Enumerations*

| UnitsLinearEnum | |
|---|---|
| #UNITS_LINEAR_UNKNOWN | Unknown. |
| #UNITS_LINEAR_INCHES | Inches. |
| #UNITS_LINEAR_FEET | Feet. |
| #UNITS_LINEAR_YARDS | Yards. |
| #UNITS_LINEAR_MILES | Miles. |
| #UNITS_LINEAR_MILLIMETERS | Millimeters. |
| #UNITS_LINEAR_CENTIMETERS | Centimeters. |
| #UNITS_LINEAR_METERS | Meters. |
| #UNITS_LINEAR_KILOMETERS | Kilometers. |
| #UNITS_LINEAR_NAUTICALMILES | Nautical miles. |
| #UNITS_LINEAR_DEGREES | Decimal degrees. |

**See also:** Prj; View

# Value

Subclasses of Value implement literals and enumerations.

**Inherits from** Obj

**Inherited by** Boolean, EnumerationElt, Nil, Number, and String

| Instance Requests | |
|---|---|
| >> (anotherValue) : aBoolean | Returns true if a value is greater than another value. |
| >>= (anotherValue) : aBoolean | Returns true when a value is greater or equal to another value. |
| << (anotherValue) : aBoolean | Returns true if a value is less than another value. |
| <<= (anotherValue) : aBoolean | Returns true when a value is less than or equal to another value. |

*See also:* Collection

# VectorFill

This class implements symbolization of polygons.

**Inherits from** Fill

| Class Requests | |
|---|---|
| Make : aVectorFill | Creates a default vector fill. |

| Instance Requests | |
|---|---|
| = anObj : aBoolean | Returns true when the object is the same as a vector fill. |
| Copy (aSymbol) | Sets the attributes of a vector fill to the specified symbol. |
| GetAngle : aNumber | Gets the angle of a vector fill. |
| GetBgColor : aColor | Gets the background color of a vector fill. |
| GetDotDensity : aNumber | Returns the number of dots in a vector fill. |
| GetDotSymbol : aMarker | Gets the marker that is used to draw dot density fill. |
| GetHeight : aNumber | Gets the vertical height of each line in a vector fill. |
| GetPenSize : aNumber | Gets the pen size used in drawing lines of a vector fill. |

| GetStyle : VectorFillStyleEnum | Gets a vector fill's style. |
|---|---|
| GetWidth : aNumber | Gets the horizontal width of each line segment in a vector fill. |
| GetXOffset : aNumber | Gets the starting horizontal offset in a vector fill. |
| GetXSeparation : aNumber | Gets the horizontal spacing between line segments in a vector fill. |
| GetYOffset : aNumber | Gets the starting vertical offset for a vector fill. |
| GetYSeparation : aNumber | Gets the vertical spacing between line segments in a vector fill. |
| SetAngle (aNumber) | Sets the angle by which a vector fill's lines are drawn to the specified number. |
| SetDotDensity (aNumber) | Sets the number of dots to draw in a vector fill. |
| SetDotSymbol (aMarker) | Sets the marker for drawing dots in a dot density fill. |
| SetHeight (aNumber) | Sets the vertical height of each line in a vector fill. |
| SetPenSize (aNumber) | Sets the pen size by which a vector fill's lines are drawn. |
| SetStyle (VectorFillStyleEnum) | Sets a vector fill's style. |
| SetWidth (aNumber) | Sets the horizontal width of each line segment for a vector fill. |
| SetXOffset (aNumber) | Sets the starting horizontal offset in a vector fill. |
| SetXSeparation (aNumber) | Sets the horizontal spacing between line segments in a vector fill. |
| SetYOffset (aNumber) | Sets the starting vertical offset for a vector fill. |
| SetYSeparation (aNumber) | Sets the vertical spacing between line segments in a vector fill. |

## Enumerations

| VectorFillStyleEnum | |
|---|---|
| #VECTORFILL_STYLE_CROSSHATCH | Cross hatch. |
| #VECTORFILL_STYLE_DOT | Dotted lines. |
| #VECTORFILL_STYLE_DOTDENSITY | Random dots for dot density mapping. |
| #VECTORFILL_STYLE_HATCH | Single hatch. |
| #VECTORFILL_STYLE_RANDDOT | Random dotted lines. |
| #VECTORFILL_STYLE_RECTANGLE | Filled rectangles. |

***See also:*** CompositeFill; RasterFill

# VectorPen

Subclasses of VectorPen implement pen styles.

**Inherits from** Pen

**Inherited by** VectorPenArrow, VectorPenDiamond, VectorPenDot, VectorPenHash, VectorPenHollow, VectorPenScallop, VectorPen-Scrub, VectorPenSlant, and VectorPenZigZag

| Instance Requests | |
|---|---|
| = anObj : aBoolean | Returns true if the specified object is the same as a vector pen. |
| CanSetSize : aBoolean | Returns true if the size can be set. |
| GetAltSize : aNumber | Gets the hollow size of a vector pen in points. |
| GetFlip : aBoolean | Gets the value of flip flag. |
| GetInterval : aNumber | Gets the spacing between line elements in points. |
| GetSize : aNumber | Gets the pen size in a vector pen. |
| SetAltSize (aNumber) | Sets the hollow size of a vector pen in points. |
| SetFlip (aBoolean) | Sets the flip flag of a vector pen to the specified Boolean value. |
| SetInterval (aNumber) | Sets the spacing between line elements in points. |
| SetSize (aNumber) | Sets the pen size in a vector pen to the specified points. |

*See also:* BasicPen; CompositePen

# VectorPenArrow

Objects of this class draw lines of arrows.

**Inherits from** VectorPen

| Class Requests | |
|---|---|
| Make : aVectorPenArrow | Creates a vector pen aroow. |

| Instance Requests | |
|---|---|
| SetArrowLen (aLength) | Sets the length of an arrow. |

# VectorPenDiamond

Objects of this class draw end-to-end diamond symbols.

**Inherits from** VectorPen

| Class Requests | |
|---|---|
| Make : aVectorPenDiamond | Creates a blank vector pen diamond. |

***See also:*** VectorPenDot; VectorPenHollow; VectorPenHash; VectorPenScallop; VectorPenScrub; VectorPenSlant; VectorPenZigZag

# VectorPenDot

Objects of this class draw a line of dots.

**Inherits from** VectorPen

| Class Requests | |
|---|---|
| Make : aVectorPenDot | Creates a blank vector pen dot. |

***See also:*** VectorPenDiamond; VectorPenHollow; VectorPenHash; VectorPenScallop; VectorPenScrub; VectorPenSlant; VectorPenZigZag

# VectorPenHash

Objects of this class draw a hash mark line.

**Inherits from** VectorPen

| Class Requests | |
|---|---|
| Make : aVectorPenHash | Creates a blank vector pen hash. |

***See also:*** VectorPenDiamond; VectorPenDot; VectorPenHollow; VectorPenScallop; VectorPenScrub; VectorPenSlant; VectorPenZigZag

# VectorPenHollow

Objects of this class draw two parallel lines.

**Inherits from** VectorPen

| *Class Requests* | |
|---|---|
| Make : aVectorPenHollow | Creates a blank vector pen hollow. |

*See also:* VectorPenDiamond; VectorPenDot; VectorPenHash; VectorPenScallop; VectorPenScrub; VectorPenSlant; VectorPenZigZag

# VectorPenScallop

Objects of this class draw scalloped lines.

**Inherits from** VectorPen

| *Class Requests* | |
|---|---|
| Make : aVectorPenScallop | Creates a blank vector pen scallop. |

*See also:* VectorPenDiamond; VectorPenDot; VectorPenHollow; VectorPenHash; VectorPenScrub; VectorPenSlant; VectorPenZigZag

# VectorPenScrub

Objects of this class draw lines of scrub symbols.

**Inherits from** VectorPen

| *Class Requests* | |
|---|---|
| Make : aVectorPenScrub | Creates a blank vector pen scrub. |

*See also:* VectorPenDiamond; VectorPenDot; VectorPenHollow; VectorPenHash; VectorPenScallop; VectorPenSlant; VectorPenZigZag

# VectorPenSlant

Objects of this class draw slanted lines.

**Inherits from** VectorPen

| *Class Requests* | |
|---|---|
| Make : aVectorPenSlant | Creates a blank vector pen slant. |

***See also:*** VectorPenDiamond; VectorPenDot; VectorPenHollow; VectorPenHash; VectorPenScallop; VectorPenScrub; VectorPenZigZag

# VectorPenZigZag

Objects of this class draw zigzag lines.

**Inherits from** VectorPen

| Class Requests | |
|---|---|
| Make : aVectorPenZigZag | Creates a blank vector pen zigzag. |

***See also:*** VectorPenDiamond; VectorPenDot; VectorPenHollow; VectorPenHash; VectorPenScallop; VectorPenScrub; VectorPenSlant

# View

This class implements ArcView's view document.

**Inherits from** Doc

| Class Requests | |
|---|---|
| Import (aFileName) : aView | Imports the view from an ArcView version 1.0 project file. |
| Make : aView | Creates a view document. |
| MakeWithGUI (aDocGUIName) : aView | Creates a new view with the specified GUI. |

| Instance Requests | |
|---|---|
| AddTheme (aTheme) | Adds the specified theme to a view document. |
| Clone : aNewView | Clones a view. |
| CopyThemes | Copies the active themes to the application clipboard. |
| CutThemes | Cuts the active themes of a view to the application clipboard. |
| DeleteTheme (aTheme) | Deletes the specified theme from a view document. |
| Draw (aDisplay) | Draws a view on the specified display. |
| Edit | Displays the property dialog box for a view. |
| Export : aFileName | Exports the display of an open view to a graphic file. |

| | |
|---|---|
| ExportToFile (aFileName, formatString, aListOfParameters) : aFileName | Exports the display of an open view to the specified graphic format file. |
| Find (aString) : aBoolean | Returns true when it finds the specified string in the attribute data of a feature. |
| FindTheme (themeNameString) : aTheme | Gets a theme in a view document based on the theme's name. |
| FindThemeByClass (themeName, themeClass) : aTheme | Finds a theme by its name and class. |
| GetActiveThemes : themesList | Gets the list of active themes in a view document. |
| GetAutoLabels (aLabeler, unplacedBoolean) : aBoolean | Gets the label position from the specified labeler and places it into the graphic list of a view. |
| GetDisplay : aDisplay | Gets the display associated with a view. |
| GetEditableTheme : aTheme | Gets the editable theme of a view. |
| GetFindString : aString | Gets the string used in the last find request. |
| GetGraphics : aGraphicList | Gets the graphics list object associated with a view. |
| GetOverlapLabelColor : aColor | Gets the overlap color. |
| GetProjection : aPrj | Gets the projection in a view. |
| GetSelectMode : GraphicsSelectModeEnum | Returns the graphic selection mode in a view. |
| GetThemes : themesList | Gets a list of themes in a view. |
| GetTOC : aTOC | Gets the table of contents object associated with a view. |
| GetTOCWidth : aNumber | Gets the displayed width of table of contents in a view. |
| GetUnits : UnitsLinearEnum | Returns a view's map units. |
| GetVisibleThemes : themesList | Gets a list of visible themes in a view document. |
| Invalidate | Invalidates and redraws a view's display. |
| InvalidateTOC (aTheme) | Invalidates the table of contents for a theme in a view. |
| IsSymWinClient : aBoolean | Returns true if symbol window client. |
| IsTOCUnResizable : aBoolean | Returns true when the table of contents in a view cannot be resized. |
| Label (aPoint) | Places a label at the specified point. |
| LabelThemes (aBoolean) | Places labels on visible and active themes. If the specified Boolean value is true, only the selected features are labeled. |
| MoveUserRect (aRect) : aBoolean | Allows moving of a rectangle in an apply event and returns true if the rectangle is moved. |

| | |
|---|---|
| Paste | Pastes the contents of the application clipboard to a view. |
| Print | Prints a view to the printer. |
| PrintDisplay | Prints a view's display to the printer. |
| PrintTOC | Prints a view's table of contents. |
| ReturnAOI : aRect | Returns the area of interest for the view document. |
| ReturnExtent : aRect | Returns the extent of a view. |
| ReturnScale : aNumber | Returns a scale of a view. |
| ReturnUserCircle : aCircle | Gets a circle from the user in an apply event. |
| ReturnUserLine : aLine | Gets a line from the user in an apply event. |
| ReturnUserPolygon : aPolygon | Gets the polygon from the user in an apply event. |
| ReturnUserPolyLine : aPolyLine | Gets the polyline from the user in an apply event. |
| ReturnUserRect : aRect | Gets a rectangle from the user in an apply event. |
| Select | Allows the user to select or deselect graphics during an apply event. |
| SelectToEdit | Selects a graphics to edit during an apply event. |
| SetAOI (aRect) | Sets the area of interest in a view based on the specified rectangle. |
| SetCoordsVisible (aBoolean) | Displays the coordinates of a view when the specified Boolean value is true. |
| SetEditableTheme (aTheme) : aBoolean | Sets the editable theme in a view to the specified theme and returns true if successful. |
| SetInteractiveSnapping (anFTheme, PointSnapEnum) | Sets the interactive snapping for the specified feature theme of a view. |
| SetInteractiveSnappingPersistent (isPersistentBoolean) | When set to false, the interactive snapping is only applied to the next point. |
| SetOverlapLabelColor (aColor) | Sets the color used for overlapping labels. |
| SetProjection (aPrj) | Sets the projection in a view. |
| SetScaleVisible (aBoolean) | Displays the scale of a view when the specified Boolean value is true. |
| SetSelectMode (GraphicsSelectModeEnum) | Sets the select mode for graphics in a view. |
| SetTOCUnResizable (aBoolean) | Makes the table of contents in a view unresizable if the specified Boolean value is true. |
| SetTOCWidth (aNumber) | Sets the width of the table of contents for a view. |
| SetUnits (UnitsLinearEnum) | Sets the map units of a view. |

## *Enumerations*

| GraphicsSelectModeEnum | |
|---|---|
| #GRAPHICS_SELECT_NORMAL | Drawn with solid handles to be moved, resized, or deleted. |
| #GRAPHICS_SELECT_VERTEX | Drawn with hollow handles to be reshaped by adding, deleting, or moving vertices. |

| PointSnapEnum | |
|---|---|
| #POINT_SNAP_BOUNDARY | Snaps point to closest boundary line. |
| #POINT_SNAP_ENDPOINT | Snaps point to closest endpoint. |
| #POINT_SNAP_INTERSECTION | Snaps point to nearest intersection. |
| #POINT_SNAP_NONE | No point snapping. |
| #POINT_SNAP_VERTEX | Snaps point to closest vertex. |

| UnitsLinearEnum | |
|---|---|
| #UNITS_LINEAR_CENTIMETERS | Centimeters. |
| #UNITS_LINEAR_DEGREES | Decimal degrees. |
| #UNITS_LINEAR_FEET | Feet. |
| #UNITS_LINEAR_INCHES | Inches. |
| #UNITS_LINEAR_KILOMETERS | Kilometers. |
| #UNITS_LINEAR_METERS | Meters. |
| #UNITS_LINEAR_MILES | Miles. |
| #UNITS_LINEAR_MILLIMETERS | Millimeters. |
| #UNITS_LINEAR_NAUTICALMILES | Nautical miles. |
| #UNITS_LINEAR_UNKNOWN | Unknown. |
| #UNITS_LINEAR_YARDS | Yards. |

**See also:** GraphicList; MapDisplay; Theme; TOC

**Code example:** page 482

# ViewFrame

This type of frame draws view documents on a layout.

**Inherits from** Frame

| Class Requests | |
|---|---|
| GetViewName (aView) : aString | Gets the name of the specified view. |
| Make (aRect) : aViewFrame | Creates a blank view frame bounded by the specified rectangle. |

| Instance Requests | |
|---|---|
| CanSimplify : aBoolean | Always returns true. |
| Draw | Draws a view frame on its associated display. |
| Edit (viewFramesList) : aBoolean | Displays the view frame editor. |
| EditSizeAndPos : aBoolean | Displays the size and position editor for a view frame. |
| GetFillObject : anObj | Gets the fill object of a view frame. |
| GetMapDisplay : aDisplay | Gets the map display of a view frame. |
| GetUserScale : aNumber | Gets the user defined scale of a view frame. |
| GetView : aView | Gets the view associated with a view frame. |
| IsExtentPreserved : aBoolean | Returns true when the extent for a view frame is preserved. |
| IsFilled : aBoolean | Returns true if the view frame is filled. |
| IsFilledBy (aClass) : aBoolean | Returns true if the view frame is filled by objects of the specified class. |
| IsLiveLinked : aBoolean | Returns true if the view frame is updated every time the view changes. |
| IsScalePreseved : aBoolean | Returns true if a view frame is drawn at the same scale as the view. |
| Offset (aPoint) | Moves a view frame by the specified amounts. |
| ReturnScale : aNumber | Returns the scale at which a view frame is drawn. |
| SetBounds (aRect) | Sets the location and size of a view frame. |
| SetDisplay (aDisplay) | Associates the display with a view frame. |
| SetExtentPreserved (aBoolean) | If set to true, the extent of a view frame is preserved. |
| SetFillObject (anObj) | Fills a view frame with the specified object. |
| SetScalePreserved (aBoolean) | If set to false, a view frame draws the entire view display. |
| SetUserScale (aScale) | Sets the user-defined scale of a view frame. |
| SetView (aView, liveLinkBoolean) | Associates the view with a view frame. |

***See also:*** GraphicList; Display; Template; View

# VTab

The virtual table class handles tabular data.

**Inherits from** Obj

**Inherited by** FTab

| Class Requests | |
|---|---|
| CanMake (aFileName) : aBoolean | Returns true if a virtual table can be created from a disk file. |
| Make (aFileName, editableBoolean, skipFirstRecordBoolean) : aVTab | Creates a virtual table from a disk file. |
| MakeNew (aFileName, aClass) : aVTab | Creates a new virtual table and a new disk file. The file is defined by its class (dBASE, INFO or DText). |
| MakeSQL (anSQLCon, queryString) : aVTab | Creates a virtual table based on a query to an SQL database. |

| Instance Requests | |
|---|---|
| Activate | Makes a virtual table available after deactivation. |
| AddFields (fieldsList) | Adds the fields in the provided list to a virtual table. |
| AddRecord : aNumber | Adds a blank record to a virtual table and returns its record number. |
| BeginTransaction | Establishes the beginning of transaction in an edit session. |
| Calculate (aString, aField) : aBoolean | Calculates the result of a string and places it in the specified field for every selected record or for all records if records are not selected. The string may contain field names and constants. |
| CanAddFields : aBoolean | Returns true if fields can be added to a virtual table. |
| CanAddRecords : aBoolean | Returns true if records can be added to a virtual table. |
| CanEdit : aBoolean | Returns true if a virtual table can be edited. |
| CanModifyIndex (aField) : aBoolean | Returns true if the index of the specified field can be modified. |
| CanRedo : aBoolean | Returns true if there are undo operations that can be redone. |
| CanRemoveFields : aBoolean | Returns true if fields can be removed from a virtual table. |

| | |
|---|---|
| CanRemoveRecords : aBoolean | Returns true if records can be deleted from a virtual table. |
| CanUndo : aBoolean | Returns true if there is a transaction session that can be undone. |
| CreateIndex (aField) | Creates an index for the specified field of a virtual table. |
| DeActivate | Makes a virtual table inaccessible. |
| EndTransaction | Ends a transaction in an editing session. |
| Export (aFileName, aClass, selectedRecordsOnlyBoolean) : aVTab | Creates a new virtual table and file name from an existing virtual table. The new file type is based on its class (dBASE, INFO, or DText). |
| FindField (fieldNameString) : aField | Gets the field specified by its name in a virtual table. |
| Flush | Writes the memory buffer to the disk file. |
| GetBaseTableClass : aClass | Returns the class of base table. |
| GetBaseTableFileName : aFileName | Returns the filename of the base table. |
| GetDefBitmap : aBitMap | Gets the bitmap that indicates the records which meet the table definition. |
| GetDefinition : aString | Gets the string that establishes which records of a file are considered part of the virtual table. |
| GetFields : fieldsList | Gets the list of fields in a virtual table. |
| GetLastSelection : aBitMap | Gets the bitmap saved by the RememberSelection request. |
| GetLockError : LockErrorEnum | Returns the lock error. |
| GetNumRecords : aNumber | Gets the number of records in a virtual table. |
| GetNumSelRecords : aNumber | Gets the number of selected records in a virtual table. |
| GetResultBitmap : aBitMap | Gets the work bitmap used by the query request. |
| GetResultField : aField | Gets the result field while processing the calculate request. |
| GetSelection : aBitMap | Gets the bitmap indicating which records in a virtual table are selected. |
| HasError : aBoolean | Returns true if the create process encountered errors. |
| HasLockError : aBoolean | Returns true if getting a lock was unsuccessful. |
| Identify (recordNumber, titleString) | Shows the specified record in an identify window. |
| IsBase : aBoolean | Returns true if a virtual table is from a single file source. |
| IsBeingEditedWithRecovery : aBoolean | Returns true if a transaction is open. |

| | |
|---|---|
| IsEditable : aBoolean | Returns true if a virtual table can be edited. |
| IsFieldIndexed (aField) : aBoolean | Returns true if the specified field has an index. |
| IsJoinedField (aField) : aBoolean | Returns true if the specified field was joined to a VTab. |
| IsLinked : aBoolean | Returns true if a virtual table has links. |
| Join (toField, fromVTab, fromField) | Joins two virtual tables. |
| Link (fromField, toVTab, toField) | Links two virtual tables. |
| Query (queryString, aBitmap, VTabSelTypeEnum) : aBoolean | Evaluates the query and places the results in the provided bitmap. |
| Redo | Performs the redo operation. |
| Refresh | Updates a virtual table from its file source. |
| RememberSelection | Saves the current selection bitmap. |
| RemoveFields (fieldsList) | Removes the fields in the specified list from a virtual table. |
| RemoveIndex (aField) | Removes the index that belongs to the specified field of a virtual table. |
| RemoveRecord (recordNumber) | Removes the specified record from a virtual table. |
| RemoveRecords (aBitmap) | Removes the records that are set in the specified bitmap from a virtual table. |
| ReturnValue (aField, recordNumber) : anObj | Returns the value of the field at the specified record number. |
| ReturnValueNumber (aField, recordNumber) : aNumber | Returns the numerical equivalence of a field at the specified record number. |
| ReturnValueString (aField, recordNumber) : aString | Returns the string equivalence of a field at the specified record number. |
| SaveEditsAs (aFileName) : newVTab | Saves the edits of an editing session to a file instead of the current VTab. |
| SetDefBitmap (aBitmap) | Includes the records set in the bitmap as part of the virtual table. |
| SetDefinition (queryString) : aBoolean | Applies the query to a virtual table as its definition. |
| SetEditable (aBoolean) | Makes a virtual table editable if the specified Boolean value is true. |
| SetResultBitmap (aBitmap) | Sets the working bitmap for processing queries. |
| SetResultField (aField) | Sets the result field for processing the calculate request. |
| SetSelection (aBitmap) | Applies the bitmap to a virtual table as its selected records. |
| SetValue (aField, recordNumber, anObj) | Sets the value of the field at the specified record number to the specified object. |

| | |
|---|---|
| SetValueNumber (aField, recordNumber, aNumber) | Sets the value of the field at the specified record number to the specified number. |
| SetValueString (aField, aRecordNumber, aString) | Sets the value of the field at the specified record number to the specified string. |
| StartEditingWithRecovery : aBoolean | Starts an editing session with recovery functions enabled. |
| StopEditingWithRecovery (saveEditsBoolean) : aBoolean | Stops an editting session and saves the edits if the boolean value is set to true. |
| Summarize (aFile name, aClass, groupingField, fieldsList, VTabSummaryEnumList) : aVTab | Creates a new virtual table and file name for the summary operation on the fields in the list. Records are grouped by the specified field. |
| UnjoinAll | Removes all joins to a virtual table. |
| Undo | Peforms the undo operation. |
| UnlinkAll | Removes all links to a virtual table. |
| UpdateDefBitmap | Matches the definition records to the associated definition bitmap. |
| UpdateSelection | Matches the selected records to the associated selection bitmap. |

## Enumerations

| LockErrorEnum | |
|---|---|
| #LOCK_ERROR_EDIT | Error in getting an edit Lock. |
| #LOCK_ERROR_NONE | No lock error. |
| #LOCK_ERROR_READ | Error in getting a read Lock. |
| #LOCK_ERROR_WRITE | Error in getting a write Lock |

| VTabSelTypeEnum | |
|---|---|
| #VTAB_SELTYPE_NEW | Creates a new set. |
| #VTAB_SELTYPE_AND | Selects only from the existing selection. |
| #VTAB_SELTYPE_OR | Adds to the selection set. |
| #VTAB_SELTYPE_XOR | Adds to the selection if not selected, and removes from the selection if already selected. |

| VTabSummaryEnum | |
|---|---|
| #VTAB_SUMMARY_AVG | Adds up the values and divides by the number of values in the group. |
| #VTAB_SUMMARY_COUNT | Count of non-null values. |

| #VTAB_SUMMARY_FIRST | First value. |
|---|---|
| #VTAB_SUMMARY_LAST | Last value. |
| #VTAB_SUMMARY_MAX | Maximum. |
| #VTAB_SUMMARY_MIN | Minimum. |
| #VTAB_SUMMARY_STDEV | Standard deviation. |
| #VTAB_SUMMARY_SUM | Summation. |
| #VTAB_SUMMARY_VAR | Variance. |

*See also:* BitMap; Field; IdentifyWin; SQLWin; SummaryDialog; Table

*Code example:* page 464

# Window

Subclasses of Window implement ArcView windows.

**Inherits from** Obj

**Inherited by** DBTQueryWin, IdentifyWin, ImageWin, NetworkWin, QueryWin, SQLWin, and TextWin

| *Class Requests* | |
|---|---|
| Make (resourceLibraryFile name, resourceLibraryNameString, resourceModuleNameString, windowsResourceNameString, reloadBoolean) : aWindow | Creates a window by loading the specified resource from the resource module of the specified resource library of Neuron Data's Open Interface. |

| *Instance Requests* | |
|---|---|
| Activate | Opens a window and makes it active. |
| Close | Unloads a window. |
| IsOpen : aBoolean | Returns true when a window is open. |
| IsSymWinClient : aBoolean | Returns true when a window is a symbol window client. |
| Maximize | Maximizes a window. |
| Minimize | Minimizes a window. |
| Move (xNumber, yNumber) | Moves the window by the specified amount in pixels. |
| MoveTo (xNumber, yNumber) | Moves a window to a new location indicated by the specified number of pixels. |

| Open | Loads and displays a window. |
|---|---|
| Resize (widthNumber, heightNumber) | Resizes the window to the specified pixels. |
| Restore | Restores from minimized or maximized state. |
| ReturnExtent : aRect | Returns the extent of a window in pixels. |
| SetExtent (aRect) | Sets the extent of a window to the specified pixels. |
| SetName (aString) | Sets the name for a window. |

***See also:*** ModalDialog; MsgBox

# XAxis

This class implements the X axis on a chart document.

## Inherits from Axis

| *Instance Requests* | |
|---|---|
| IsBottom : aBoolean | Returns true if an X axis is shown on the bottom of its associated chart. |
| IsValueAxis : aBoolean | Returns true for a value axis versus a group axis. |
| SetAxisVisible (aBoolean) | Makes an X axis visible if the Boolean value is true. |
| SetBottom (aBottom) | Displays the axis on the bottom if the Boolean value is true. |
| SetBoundsUsed (aBoolean) | The established bounds are used when the Boolean value is true. |
| SetColor (aColor) | Sets the color of an X axis. |
| SetCrossValueUsed (aBoolean) | The established cross value is used if the Boolean value is true. |
| SetLabelVisible (aBoolean) | Displays the label of an X axis if the Boolean value is true. |
| SetLog (aBoolean) | Makes an X axis logarithmic when the Boolean value is true. |
| SetMajorGridVisible (aBoolean) | Displays the major grid in an X axis when the Boolean value is true. |
| SetMinorGridVisible (aBoolean) | Displays the minor grid in an X axis when the Boolean value is true. |
| SetName (aString) | Sets the name of an X axis object. |
| SetTickLabelsVisible (aBoolean) | Displays the tick mark labels on an X axis when the Boolean value is true. |

***See also:*** ChartDisplay; ChartPart, YAxis

# XYName

This class implements sources for XY themes.

**Inherits from** SrcName

| Class Requests | |
| --- | --- |
| Make (aVTab, anXField, aYField) : anXYName | Creates an XYName object from a virtual table with X and Y fields. |

| Instance Requests | |
| --- | --- |
| GetVTab : aVTab | Gets the underlying virtual table. |
| GetXField : aField | Gets the X field for an XYName object. |
| GetYField : aField | Gets the Y field for an XYName object. |
| SetXField (aField) | Sets the X field in an XYName object. |
| SetYField (aField) | Sets the Y field in an XYName object. |

*See also:* DynName; GeoName; Theme

*Code example:* page 496

# YAxis

This class implements the Y axis on a chart document.

**Inherits from** Axis

| Instance Requests | |
| --- | --- |
| IsLeft : aBoolean | Returns true if a Y axis is shown on the left side of its associated chart. |
| IsValueAxis : aBoolean | Returns true for a value axis versus a group axis. |
| SetAxisVisible (aBoolean) | Makes a Y axis visible if the Boolean value is true. |
| SetBoundsUsed (aBoolean) | The established bounds are used when the Boolean value is true. |
| SetColor (aColor) | Sets the color of a Y axis. |
| SetCrossValueUsed (aBoolean) | The established cross value is used if the Boolean value is true. |
| SetLabelVisible (aBoolean) | Displays a Y axis if the specified Boolean value is true. |

| | |
|---|---|
| SetLeft (aBoolean) | Displays a Y axis on the left side of chart when the Boolean value is true. |
| SetLog (aBoolean) | Makes a Y axis logarithmic when the Boolean value is true. |
| SetMajorGridVisible (aBoolean) | Displays the major grid in a Y axis when the Boolean value is true. |
| SetMinorGridVisible (aBoolean) | Displays the minor grid in a Y axis when the Boolean value is true. |
| SetName (aString) | Sets the name of a Y axis object. |
| SetTickLabelsVisible (aBoolean) | Displays the tick mark labels on a Y axis when the Boolean value is true. |

***See also:*** ChartDisplay; ChartPart; XAxis

# 100+ Scripts

# Applications

## ◆ *CheckEnvironmentVariable*

Checks for the existence of a DOS environment variable and returns its value. If used as a standalone program, the DOS environment variable *AVEXT* is assigned to the *v* variable. If called from another program, the name of the DOS environment variable passed in SELF is returned.

**Topics**: Application, Environment Variable

**Search Keys**: System.GetEnvVar

```
' Variable Initialization
t = "CheckEnvironmentVariable"
if (SELF = nil) then

   DOSEnvVar = "avext"

else

   DOSEnvVar = SELF

end


' Get the DOS environment variable (DOSEnvVar).

v = System.GetEnvVar(DOSEnvVar)

if (v = nil) then

  MsgBox.Error("DOS Environment Variable" ++ DOSEnvVar ++ "not set"
+ nl + "Please Set Environment Variable in DOS and Restart
Windows","DOS Environment Variable")

   exit

end

Return v
```

# ◆ *SetEnvironmentVariable*

Sets the value of a DOS environment variable. If used as a standalone program, the DOS environment variable *AVEXT* is assigned to the *v* variable. If called from another program, the name of the DOS environment variable passed in SELF is returned.

**Topics**: Application, Environment Variable

**Search Keys**: System.GetEnvVar

```
' Variable Initialization
t = "SetEnvironmentVariable"
if (SELF = nil) then
   DOSEnvVar = "avext"
else
   DOSEnvVar = SELF
end


' Set the DOS environment variable (DOSEnvVar).
l = "C:\ESRI\AV_GIS\arcview\ext"
System.SetEnvVar(DOSEnvVar,l)
v = System.GetEnvVar(DOSEnvVar)
if (v = nil) then
   MsgBox.Error("DOS Environment Variable" ++ DOSEnvVar ++ "not set"
+ nl +
   "Please Set Environment Variable in DOS and Restart Windows","DOS
Environment Variable")
   exit
else
   MsgBox.Info("DOS Environment Variable" ++ DOSEnvVar ++ "set to"
+ NL +
```

```
    l,"DOS Environment Variable")
end
Return v
```

## ◆  *UserExtVariable*

*USEREXT* is a new environment variable used by ESRI to point at the extensions created by a user. ArcView will look in this location in addition to the *Ext32* sub-directory of your ArcView installation.

**Topics**: Extensions

```
' Variable Initialization
t = "UserExtVariable"
System.SetEnvVar("USEREXT","c:\")
MsgBox.Info(System.GetEnvVar("USEREXT"),t)
```

# DocGUIs

## ◆  *ButtonAdd*

Adds a button to the chosen GUI button bar.

**Topics**: Button, ControlSet, ButtonBar

**Search Keys**: Button, Add, Remove

**Requirements**: Script will not run properly without the compiled script named *IconGet* called by av.Run.

```
' Variable Initialization
```

```
t = "ButtonAdd"
'***Critical Resource Test***
if ((av.GetProject.FindDoc("IconGet").is(SEd)) = false) then
  MsgBox.Error("Critical Resource NOT Available" + NL +
                "Program Cannot Continue" + NL +
                "Press OK to EXIT",t)
  exit
end
'***      ***       ***       ***
p = av.GetProject
guis = p.GetGUIs
parm = SELF
if (parm.Is(ButtonBar).Not) then
    if (((parm = nil) or parm.Is(Choice)).Not) then
      MsgBox.Warning("No ButtonBar Chosen: Exiting",t)
      return nil
    else
' Dialog to select a GUI and get button bar for that GUI.
      gui = MsgBox.List(guis,"Pick a GUI: ",t)
      if (gui = nil) then
        MsgBox.Warning("No GUI Picked: Exiting",t)
        exit
      end
      parm = gui.GetButtonBar
    end
end
index = parm.GetControls.Count
itemAttributes = MsgBox.MultiInput("Enter item
```

```
attributes:",t,{"Click","Help","Icon"},
                 {"Project.HelpAvenue","Type Status Bar Message
Here","H"})
if (itemAttributes.Count = 0) then
    MsgBox.Warning("No Attributes Entered: Exiting",t)
    return nil
end
' Making and setting attributes for the new button to add to the
' button bar.
b = Button.Make
b.SetClick(itemAttributes.Get(0))
b.SetHelp(itemAttributes.Get(1))
ikahn = av.Run("IconGet",itemAttributes.Get(2))
b.SetIcon(ikahn)
parm.Add(b,index)
return true
```

## ◆ *ButtonBar*

Cycles through the button bar elements for the current document type and displays a short report for each element.

**Topics**: DocGUI, ButtonBar, Button

**Search Keys**: av.GetActiveGUI, ActiveGUI.GetButtonBar, Control.GetClick

**Requirements**: Script will not run properly without the compiled *ButtonReport* script called by av.Run.

```
' Variable Initialization
t = "ButtonBar"
'***Critical Resource Test***
```

```
if ((av.GetProject.FindDoc("ButtonReport").is(SEd)) = false) then
  MsgBox.Error("Critical Resource NOT Available" + NL +
               "Program Cannot Continue" + NL +
               "Press OK to EXIT",t)
  exit
end
' ***      ***      ***      ***
buttonDict = Dictionary.Make(25)
activeGUI = av.GetActiveGUI
bbar = activeGUI.GetButtonBar
for each b in bbar
    if (b.Is(Space).Not) then
      buttonDict.Add(b.GetClick,b)
    end
end
while (true)
    butt = MsgBox.ListasString(buttonDict.ReturnKeys,"Pick a
Button:",av.GetActiveGui.GetName + " Button")
    if (butt = nil) then exit end
    MsgBox.Report(av.Run("ButtonReport",
{buttonDict.Get(butt)}),av.GetActiveGui.GetName + " Button")
end
```

## ◆ *ButtonCopy*

Copies a button to a button bar in another GUI.

**Topics**: DocGUI, Menu, ButtonBar, Button

**Search Keys**: Menu, ButtonBar, Button

```
' Variable Initialization
t = "ButtonCopy"
p = av.GetProject
guis = p.GetGUIs
' Pick a from DocGUI.
fromGUI = MsgBox.List(guis,"Select GUI to copy FROM: ",t)
if (fromGUI = nil) then
    MsgBox.Warning("No GUI Picked: Exiting",t)
    exit
end
' Information to get the from button.
fromDict = Dictionary.Make(10)
for each b in fromGUI.GetButtonBar
    if (b.Is(Space).Not) then
       fromDict.Add(b.GetClick,b)
    end
end
fromButton = MsgBox.ListasString(fromDict.ReturnKeys,"Pick a button
to copy",t)
if (fromButton = nil) then
    MsgBox.Warning("No button picked: Exiting",t)
    exit
end
' Pick a to DocGUI.
toGUI = MsgBox.List(guis,"Select GUI to copy TO: ",t)
if (toGUI = nil) then
    MsgBox.Warning("No GUI Picked: Exiting",t)
    exit
```

```
end
' Add from button to toGUI.
bb = toGUI.GetButtonBar
index = bb.GetControls.Count
bb.Add(fromDict.Get(fromButton).Clone,index)
```

## ◆ *ButtonDeleteEdit*

Deletes or edits the attributes of a button from the chosen GUI.

**Topics**: ButtonBar, Button, ControlSet

**Search Keys**: ButtonBar, Button, Remove, SetIcon, SetClick, SetHelp

```
' Variable Initialization
t = "ButtonDeleteEdit"
p = av.GetProject
guis = p.GetGUIs
' Pick a from DocGUI.
gui = MsgBox.List(guis,"Select GUI: ",t)
if (gui = nil) then
    MsgBox.Warning("No GUI Picked: Exiting",t)
    exit
end
while (true)
    bb = gui.GetButtonBar
    bDict = Dictionary.Make(20)
' Initialize dictionary with values for this button bar.
    for each b in bb
      if (b.Is(Space).Not) then
          bDict.Add(b.GetClick,b)
```

```
        end

    end
' Select a button.
    bLabel = MsgBox.ListasString(bDict.ReturnKeys,"Pick a
Button:",t)

    if (bLabel = nil) then exit end

    bButton = bDict.Get(bLabel)

    ch = MsgBox.ChoiceasString({"delete","edit"},"Pick option to
change active ButtonBar:",t)

    if (ch = "delete") then
' Verify desire to delete.
        yn = MsgBox.YesNo("Do you want to delete" + nl +
"       Button:" ++ bButton.GetClick,t,true)

        if (yn = true) then

            bb.Remove(bButton)

        end

    elseif (ch = "edit") then
' Get current values and place them in an input box for editing.
        currentAttributes = {bButton.GetClick,bButton.GetHelp,
bButton.GetIcon.GetName}

        itemAttributes = MsgBox.MultiInput("Enter item
attributes:",t,{"Click","Help","Icon"},currentAttributes)

        if (itemAttributes.Count = 0) then

            MsgBox.Warning("No attributes entered: Exiting",t)

            return nil

        end
' Making and setting attributes for the new button to add to the
' current button bar.
        bButton.SetClick(itemAttributes.Get(0))

        bButton.SetHelp(itemAttributes.Get(1))
```

```
        ikahn = av.Run("xIconGet",itemAttributes.Get(2))

        bButton.SetIcon(ikahn)
' index = bb.ReturnIndex(bButton)
' bb.Remove(bButton)
' bb.Add(bButton,index)
      else
        exit
      end
end
```

## ◆ *ChoiceAdd*

Adds a menu item to the menu passed as a parameter. All additions are inserted at the bottom of the menu. A major upgrade to the script would be an improved selection for the index value (the position of the menu item). There is no provision for adding separators.

**Topics**: DocGUI, Menu, ControlSet

**Search Keys**: Menu, ControlSet, Add, Remove

**Requirements**: Script must be called by another program and the menu passed in the first parameter. If SELF is nil, the user is prompted to select a menu.

```
' Variable Initialization
t = "ChoiceAdd"
p = av.GetProject
guiList = p.GetGUIs
parm = SELF
if (parm.Is(Menu).Not) then
    if (parm = nil) then
      gui = MsgBox.List(guiList,"Pick a GUI:",t)
```

```
        if (gui = nil) then exit end
        mb = gui.GetMenuBar
        menuList = Dictionary.Make(2)
        for each x in mb
            if (x.Is(Menu)) then
                menuList.Add(x.GetLabel,x)
            end
        end
      ch = MsgBox.ListasString(menuList.ReturnKeys,"Pick a Menu:",t)
        if (ch = nil) then exit end
        parm = menuList.Get(ch)
    else
        MsgBox.Warning("This script must be passed the Menu as the
First Parameter",t)
        return nil
    end
end
index = parm.GetControls.Count
itemAttributes = MsgBox.MultiInput("Enter item attributes:",t,
{"Label","Click","Help"},{"","",""})
if (itemAttributes.Count = 0) then
    MsgBox.Warning("No attributes entered: Exiting",t)
    return nil
end
' Making and setting attributes for the new choice to add to this
' menu.
ch = Choice.Make
ch.SetLabel(itemAttributes.Get(0))
ch.SetClick(itemAttributes.Get(1))
```

```
ch.SetHelp(itemAttributes.Get(2))
parm.Add(ch,index)
return true
```

# ◆ *ChoiceDeleteEdit*

Allows editing or deleting of a choice in a menu on the active GUI.

**Topics**: DocGUI, Menu, Choice, ControlSet

**Search Keys**: DocGUI, Menu, Choice, ControlSet, GetActiveGUI, Remove

```
' Variable Initialization
t = "ChoiceDeleteEdit"
menuDict = Dictionary.Make(20)
choiceDict = Dictionary.Make(20)
mDict     = Dictionary.Make(5)
' Initialize dictionary with values for this menu bar.
for each m in av.GetActiveGUI.GetMenuBar
    mDict.Add(m.GetLabel,m)
    mc = m.GetControls
    mList = List.Make
    for each x in mc
      if (x.Is(Space).Not) then
         mList.Add(x.GetLabel)
         choiceDict.Add(m.GetLabel + ":" + x.GetLabel,x)
      end
    end
    menuDict.Add(m.GetLabel,mList)
end
' Pick a menu and a choice. Verify desire to delete.
```

```
while (true)

    mLabel = MsgBox.ListasString(menuDict.ReturnKeys,"Pick a
Menu:",t)

    if (mLabel = nil) then exit end

    mChoice = menuDict.Get(mLabel)

    mc = MsgBox.ListasString(mChoice,"Pick a choice:",t)

    if (mc = nil) then exit end

    ch = choiceDict.Get(mLabel+":"+mc)

    if (ch = nil) then exit end

    ed = MsgBox.ChoiceasString({"delete","edit"},"Pick option to
change active ButtonBar:",t)

    if (ed = "delete") then
' Verify desire to delete.

    yn = MsgBox.YesNo("Do you want to delete" + nl + "    Choice:"
++ mc + nl + "on" + nl + "       menu:" ++ mLabel,t,true)

      if (yn = true) then

        mDict.Get(mLabel).Remove(ch)

      end

    elseif (ed = "edit") then
' Get current values and place them in an input box for editing.

        currentAttributes = {ch.GetLabel,ch.GetClick,ch.GetHelp}

        itemAttributes = MsgBox.MultiInput("Enter item
attributes:",t,{"Label","Click","Help"},currentAttributes)

        if (itemAttributes.Count = 0) then

          MsgBox.Warning("No attributes entered: Exiting",t)

          return nil

        end
' Making and setting attributes for the new button to add to the
' button bar.
```

```
      ch.SetLabel(itemAttributes.Get(0))
      ch.SetClick(itemAttributes.Get(1))
      ch.SetHelp(itemAttributes.Get(2))
' index = bb.ReturnIndex(bButton)
' bb.Remove(bButton)
' bb.Add(bButton,index)
   else
      exit
   end
end
```

## ◆ InstallaTool

Installs a tool on the Table tool bar and assigns an icon, an apply event, and a help string to the tool. Use this script to install the Examine tool used in the *TableChange* script.

**Topics**: Icon, IconMgr, DocGUI

```
' Variable Initialization
t = "InstallaTool"
' Reference to the project object.
p = av.GetProject
' Reference to the table DocGUI.
gui = p.FindGUI("Table")
' Reference to the tb on the table DocGUI.
tb  = gui.GetToolbar
' All on one line.
' tb = av.GetProject.FindGUI("Table").GetToolbar
' Make a tool.
newTool = Tool.Make
```

```
' Pick an icon.
iconList = IconMgr.GetIcons
newIcon = NIL
for each x in iconList
  if (x.GetName = "Examine") then
    newIcon = x
    break
  end
end
if (newIcon = NIL) then
  MsgBox.Warning("Icon ""examine"" not found.",t)
  exit
end
' Set the apply event and icon, and install.
newTool.SetApply("TableChange")
newTool.SetHelp("Edit field values.")
newTool.SetIcon(newIcon)
tb.Add(newTool,-1)
```

# ◆ *Menu*

Allows the user to browse the controls in a menu bar and menu. For the current DocGUI, a menu selection followed by a choice selection results in the display of the action associated with the choice.

**Topics**: DocGUI, MenuBar, Menu, Choice, Control

**Search Keys**: MenuBar, DocGUI, Menu, Choice, Control

**Requirements**: Script will not run properly without a compiled script named *ChoiceReport* called by av.Run.

```
Variable Initialization
t = "Menu"
' ***Critical Resource Test***
if ((av.GetProject.FindDoc("ChoiceReport").is(SEd)) = false) then
   MsgBox.Error("Critical Resource NOT Available" + NL +
                "Program Cannot Continue" + NL +
                "Press OK to EXIT",t)
   exit
end
' ***     ***     ***     ***
menuDict = Dictionary.Make(25)
choiceDict = Dictionary.Make(25)
for each m in av.GetActiveGUI.GetMenuBar
    mc = m.GetControls
    mList = List.Make
    for each x in mc
      if (x.Is(Space).Not) then
         mList.Add(x.GetLabel)
         choiceDict.Add(m.GetLabel + ":" + x.GetLabel,x)
      end
    end
    menuDict.Add(m.GetLabel,mList)
end
while (true)
    k = MsgBox.ListasString(menuDict.ReturnKeys,"Pick a Menu:",t)
    if (k = nil) then exit end
    mlist = menuDict.Get(k)
    mx = MsgBox.ListasString(menuDict.Get(k),"Pick a Choice:",t)
```

```
   if (mx = nil) then exit end
     ch = choiceDict.Get(k+":"+mx)
     if (ch = nil) then exit end
     MsgBox.Report(av.Run("ChoiceReport",ch),t)
end
```

## ◆ *MenuAdd*

Adds a new menu or a choice to an existing menu and updates the chosen GUI.

**Related to**: ButtonAdd, ToolAdd

**Topics**: Menu, Choice, DocGUI

**Search Keys**: Menu, Choice, DocGUI

**Requirements**: Script will not run properly without a compiled script named *ChoiceAdd* called by av.Run.

```
' Variable Initialization
t = "MenuAdd"
' ***Critical Resource Test***
if ((av.GetProject.FindDoc("ChoiceAdd").is(SEd)) = false) then
  MsgBox.Error("Critical Resource NOT Available" + NL +
               "Program Cannot Continue" + NL +
               "Press OK to EXIT",t)
  exit
end
' ***     ***     ***     ***
p = av.GetProject
guis = p.GetGUIs
' Pick a DocGUI from the current project.
gui = MsgBox.List(guis,"Pick a GUI: ",t)
```

```
if (gui = nil) then

    MsgBox.Warning("No GUI Picked: Exiting",t)

    exit

end

' Information about current menus.

menuDict = Dictionary.Make(10)

for each m in gui.GetMenuBar

    menuDict.Add(m.GetLabel,m)

end

index = menuDict.Count

' To examine the elements of the chosen GUI, uncomment the next line.

' av.Run("GUIReport",gui)

ch = MsgBox.YesNo("Do you want to add a new menu to" ++
gui.GetName,t,true)

if (ch = true) then

    menuLabel = MsgBox.Input("Enter a label for the new
Menu:",t,"Menu Label")

    if (menuLabel = nil) then

      MsgBox.Warning("No Label Entered: Exiting",t)

      exit

    end

    newMenu = Menu.Make

    newMenu.SetLabel(menuLabel)

    gui.GetMenuBar.Add(newMenu,index)

    while (true)

      if (av.Run("ChoiceAdd",newMenu) = nil) then

        break

      end

    end
```

```
else
    m = MsgBox.ListasString(menuDict.ReturnKeys,"Pick a Menu:",t)
    if (m = nil) then
      MsgBox.Warning("No Menu Picked: Exiting",t)
      exit
    end
    while (true)
      if (av.Run("ChoiceAdd",menuDict.Get(m)) = nil) then
        break
      end
    end

end
```

## ◆ *MenuCopy*

Dialog that copies a menu from one menu bar to another.

**Topics**: MenuBar, Menu, ControlSet, Control

**Search Keys**: MenuBar, Menu, ControlSet, Control, Add

```
' Variable Initialization
t = "MenuCopy"
p = av.GetProject
guis = p.GetGUIs
' Pick a from DocGUI.
fromGUI = MsgBox.List(guis,"Select GUI to copy FROM: ",t)
if (fromGUI = nil) then
    MsgBox.Warning("No GUI Picked: Exiting",t)
    exit
```

```
end
' Information to get the from menu.
fromDict = Dictionary.Make(10)
for each m in fromGUI.GetMenuBar
    fromDict.Add(m.GetLabel,m)
end
fromMenu = MsgBox.ListasString(fromDict.ReturnKeys,"Pick a Menu to
Copy",t)
if (fromMenu = nil) then
    MsgBox.Warning("No Menu Picked: Exiting",t)
    exit
end
' Pick a to DocGUI.
toGUI = MsgBox.List(guis,"Select GUI to copy TO: ",t)
if (toGUI = nil) then
    MsgBox.Warning("No GUI Picked: Exiting",t)
    exit
end
' Add fromMenu to toGUI.
mb = toGUI.GetMenuBar
index = mb.GetControls.Count
mb.Add(fromDict.Get(fromMenu).Clone,index)
```

## ◆ *MenuDelete*

Deletes a menu or a choice on an existing menu and updates the chosen GUI.

**Topics**: Menu, Choice, DocGUI, ControlSet

**Search Keys**: Menu, Choice, DocGUI, ControlSet, Remove

```
' Variable Initialization
t = "MenuDelete"
p = av.GetProject
guis = p.GetGUIs
' Pick a DocGUI from the current project.
gui = MsgBox.List(guis,"Pick a GUI: ",t)
if (gui = nil) then
    MsgBox.Warning("No GUI Picked: Exiting",t)
    exit
end
' Information about current menus.
menuDict = Dictionary.Make(10)
for each m in gui.GetMenuBar
    menuDict.Add(m.GetLabel,m)
end
index = menuDict.Count
' To examine the elements of the chosen GUI, uncomment the next line.
'av.Run("xGUIReport",gui)
ch = MsgBox.YesNo("Do you want to delete a menu to" ++ gui.
GetName,t,true)
if (ch = true) then
    m = MsgBox.ListasString(menuDict.ReturnKeys,"Pick a Menu: ",t)
    if (m = nil) then
      MsgBox.Warning("No label entered: Exiting",t)
        exit
    end
    if (MsgBox.YesNo("Are You Sure You Want to Delete: " +
m,t,true).Not) then
        MsgBox.Warning("Not sure? No action taken: Exiting",t)
```

```
      exit

   end

   gui.GetMenuBar.Remove(menuDict.Get(m))

else

   MsgBox.Warning("No Menu Picked: Exiting",t)

   exit

end
```

## ◆ *ToolAdd*

Adds a tool to the chosen GUI's tool bar.

**Related to**: AddButton, AddChoice

**Topics**: Tool, ToolBar, Control, ControlSets, Icon

**Search Keys**: Tool, ToolBar, Control, ControlSets, Icon

**Requirements**: This script will not run properly without a compiled script named *IconGet* called by av.Run.

```
' Variable Initialization

t = "ToolAdd"

' ***Critical Resource Test***

if ((av.GetProject.FindDoc("IconGet").is(SEd)) = false) then

   MsgBox.Error("Critical Resource NOT Available" + NL +

                "Program Cannot Continue" + NL +

                "Press OK to EXIT",t)

   exit

end

' ***      ***      ***       ***

p = av.GetProject

guis = p.GetGUIs
```

```
parm = SELF
if (parm.Is(ToolBar).Not) then
    if (((parm = nil) or parm.Is(Choice)).Not) then
      MsgBox.Warning("No Button Bar Chosen: Exiting",t)
      return nil
    else
' Dialog to select a GUI and get button bar for the GUI.
      gui = MsgBox.List(guis,"Pick a GUI: ",t)
      if (gui = nil) then
         MsgBox.Warning("No GUI Picked: Exiting",t)
         exit
      end
      parm = gui.GetToolBar
    end
end
index = parm.GetControls.Count
itemAttributes = MsgBox.MultiInput("Enter item
attributes:",t,{"Click","Apply","Help","Icon"},{"","","",""})
if (itemAttributes.Count = 0) then
    MsgBox.Warning("No attributes entered: Exiting",t)
    return nil
end
' Making and setting attributes for the new button to add to the
' button bar.
h = Tool.Make
h.SetClick(itemAttributes.Get(0))
h.SetApply(itemAttributes.Get(1))
h.SetHelp(itemAttributes.Get(2))
ikahn = av.Run("IconGet",itemAttributes.Get(3))
```

```
h.SetIcon(ikahn)
parm.Add(h,index)
return true
```

## ◆ ToolDeleteEdit

Deletes or edits tools on the tool bar of the active GUI.

**Topics**: Tool, ToolBar, Control, ControlSets, Icon

**Search Keys**: Tool, ToolBar, Control, ControlSets, Icon, Remove, Add

**Requirements**: Script will not run properly without a compiled script named *IconGet* called by av.Run.

```
' Variable Initialization
t = "ToolDeleteEdit"
' ***Critical Resource Test***
if ((av.GetProject.FindDoc("IconGet").is(SEd)) = false) then
   MsgBox.Error("Critical Resource NOT Available" + NL +
               "Program Cannot Continue" + NL +
               "Press OK to EXIT",t)
   exit
end
' ***      ***      ***      ***
while (true)
    tb = av.GetActiveGUI.GetToolBar
    tDict = Dictionary.Make(20)
' Initialize dictionary with values for this button bar.
    for each h in tb
      if (h.Is(Space).Not) then
        tDict.Add(h.GetApply,h)
```

```
        end

      end

 ' Select a tool.

      tLabel = MsgBox.ListasString(tDict.ReturnKeys,"Pick a tool:",t)

      if (tLabel = nil) then exit end

      hTool = tDict.Get(tLabel)

      ch = MsgBox.ChoiceasString({"delete","edit"},"Pick option to
 change active tool bar:",t)

      if (ch = "delete") then
 ' Verify desire to delete.

        yn = MsgBox.YesNo("Do you want to delete" + nl + "     Tool:"
 ++ hTool.GetApply,t,true)

        if (yn = true) then

          tb.Remove(hTool)

        end

      elseif (ch = "edit") then
 ' Get current values and put them in an input box for editing.

        currentAttributes = {hTool.GetClick,hTool.GetApply,
 hTool.GetHelp,hTool.GetIcon.GetName}

        itemAttributes = MsgBox.MultiInput("Enter item attributes:",
 t, {"Click","Apply","Help","Icon"}, currentAttributes)

        if (itemAttributes.Count = 0) then

          MsgBox.Warning("No attributes entered: Exiting",t)

          return nil

        end

 ' Making and setting attributes for the new
 ' button to add to the button bar.

        hTool.SetClick(itemAttributes.Get(0))

        hTool.SetApply(itemAttributes.Get(1))
```

```
    hTool.SetHelp(itemAttributes.Get(2))
    ikahn = av.Run("IconGet",itemAttributes.Get(3))
    hTool.SetIcon(ikahn)
  else
    exit
  end
end
```

# Documentation

## ◆ ButtonBarDocumentation

Marks up a button bar documentation report as an HTML document. The script can be called by another program if it is passed a single object (ButtonBar), and can also be run as a standalone. In the standalone mode, the user is led through a sequence of choices to pick a button bar in a GUI to document.

**Topics**: Documentation, DocGUI, ButtonBar, Button

**Search Keys**: DocGUI, ButtonBar, Button, av.Run, HTML

**Requirements**: Script will not run properly without the script named *ButtonDocumentation* called by av.Run.

```
' Variable Initialization
t = "ButtonBarDocumentation"
' ***Critical Resource Test***
if ((av.GetProject.FindDoc("ButtonDocumentation").is(SEd)) =
false) then
  MsgBox.Error("Critical Resource NOT Available" + NL +
               "Program Cannot Continue" + NL +
```

```
                    "Press OK to EXIT",t)
   exit
 end
 ' ***      ***      ***      ***
 ' Make this script callable or standalone.
 parm = SELF
 if (parm.Is(ButtonBar).Not) then
    p = av.GetProject
    guis = p.GetGUIs
    gui = MsgBox.List(guis,"Pick a GUI:",t)
    if (gui = nil) then exit end
    parm = gui.GetButtonBar
 end
 rpt = "<h2>ButtonBar</h2>" + nl
 buttonList = parm.GetControls
 if (buttonList.Count = 0) then
    rpt = rpt + "<p>No Buttons for this DocGUI." + nl
 else
    rpt = rpt + "<ul>" + nl
    for each b in buttonList
      if (b.Is(Space)) then continue end
      s = "<li>" + nl
      rpt = rpt + s + av.Run("ButtonDocumentation",b)
    end
    rpt = rpt + "</ul>" + nl
 end
 if (SELF.Is(ButtonBar)) then
    return(rpt)
```

```
else
  MsgBox.Report(rpt,t)
end
```

# ◆ *ButtonDocumentation*

Displays the attributes and characteristics of a button.

**Topics**: Documentation, Button, Icon

**Search Keys**: GUI.GetButtonBar.GetControls, Control.GetClick, Control.Get-Icon, Control.GetHelp, Control.GetUpdate

**Comments**: Script is callable if it is passed a button object. If it is not passed a button object, the script will prompt the user to select a GUI and then a button from the GUI's button bar.

```
' Variable Initialization
t = "ButtonDocumentation"
parm = SELF
if (parm.Is(Button).Not) then
    p = av.GetProject
    guiList = p.GetGUIs
    gui = MsgBox.List(guiList,"Pick a GUI:",t)
    if (gui = nil) then exit end
    bb = gui.GetButtonBar.GetControls
    if (bb.Count = 0) then
      MsgBox.Warning("No Buttons in this GUI: Exiting",t)
      exit
    end
    bbDict = Dictionary.Make(1)
    for each x in bb
      if (x.Is(Space).Not) then
```

```
        bbDict.Add(x.GetClick,x)
      end
    end
    ch = MsgBox.ListasString(bbDict.ReturnKeys,"Pick a button:",t)
    if (ch = nil) then exit end
    parm = bbDict.Get(ch)
end


rpt = "<dl>" + nl
rpt = rpt + "<dt>Icon" + nl + "<dd>" + parm.GetIcon.GetName + nl
rpt = rpt + "<dt>Click" + nl + "<dd>" + parm.GetClick + nl
rpt = rpt + "<dt>Help"  + nl + "<dd>" + parm.GetHelp + nl
rpt = rpt + "<dt>Update" + nl + "<dd>" + parm.GetUpdate + nl
rpt = rpt + "</dl>" + nl
if (SELF.is(Button).Not) then
    MsgBox.Report(rpt,t)
else
    return rpt
end
```

# ◆ *ButtonReport*

Displays the attributes and characteristics of a button.

**Topics**: Documentation, Button, Icon

**Search Keys**: GUI.GetButtonBar.GetControls, Control.GetClick, Control.Get-Icon, Control.GetHelp, Control.GetUpdate

**Comments:** Script is callable if it is passed a button object. If it is not passed a button object, the script will prompt the user to select a GUI and then a button from the GUI's button bar.

```
' Variable Initialization
t = "ButtonReport"
if (SELF = nil ) then
  parm = SELF
else
  parm = SELF.Get(0)
end
if (parm.Is(Button).Not) then
    p = av.GetProject
    guiList = p.GetGUIs
    gui = MsgBox.List(guiList,"Pick a GUI:",t)
    if (gui = nil) then exit end
    bb = gui.GetButtonBar.GetControls
    if (bb.Count = 0) then
      MsgBox.Warning("No buttons in this GUI: Exiting",t)
      exit
    end
    bbDict = Dictionary.Make(1)
    for each x in bb
      if (x.Is(Space).Not) then
        bbDict.Add(x.GetClick,x)
      end
    end
    ch = MsgBox.ListasString(bbDict.ReturnKeys,"Pick a button:",t)
    if (ch = nil) then exit end
    parm = bbDict.Get(ch)
  end
```

```
rpt = " " + nl
rpt = rpt + "Icon: " + parm.GetIcon.GetName + nl
rpt = rpt + "Click: " + parm.GetClick + nl
rpt = rpt + "Help: " + parm.GetHelp + nl
rpt = rpt + "Update: " + parm.GetUpdate + nl
rpt = rpt + nl
if (SELF = nil ) then
    MsgBox.Report(rpt,t)
else
    return rpt
end
```

# ◆ *ChartDocumentation*

Shell program for chart documentation. Revise the script to include the metadata about charts for your projects.

**Topics**: Metadata, Chart

**Search Keys**: Chart

```
' Variable Initialization
t = "ChartDocumentation"
parm = SELF
p = av.GetProject
cList = List.Make
for each x in p.GetDocs
    if (x.Is(Chart)) then
      cList.Add(x)
    end
end
```

```
if (parm.Is(Chart).Not) then
    if (cList.Count > 0) then
      parm = MsgBox.List(cList,"Pick a Chart",t)
      if (parm = nil) then exit end
    else
      MsgBox.Warning("No Charts in this Project",t)
      exit
    end
end
' Add your chart documentation here.
cRpt = "<h1>Charts</h1>" + nl
cRpt = cRpt + "<ul>" + nl
for each x in cList
    cRpt = cRpt + "<li>" + x.GetName + nl
end
cRpt = cRpt + "</ul>" + nl
return cRpt
```

## ◆ *ChoiceDocumentation*

Gets the attributes and characteristics of a choice. Choice is the class of a selection item on a drop-down menu.

**Topics**: DocGUI, Choice, Documentation

**Search Keys**: GUI.GetMenuBar.GetControls, Control.GetLabel

**Requirements**: Script may be called from another script and must be passed a single parameter that is an object of the choice type. The script can also be run as a standalone.

```
' Variable Initialization
t = "ChoiceDocumentation"
```

```
' Make this script callable or standalone.
parm = SELF
if (parm.Is(Choice).Not) then
   p = av.GetProject
   guis = p.GetGUIs
   gui = MsgBox.List(guis,"Pick a GUI:",t)
   if (gui = nil) then exit end
   mbar = gui.GetMenuBar.GetControls
   if (mbar.Count = 0) then exit end
   mbarDict = Dictionary.Make(2)
   for each m in mbar
     if (m.Is(Menu)) then
       mbarDict.Add(m.GetLabel,m)
     end
   end
   ch = MsgBox.ListasString(mbarDict.ReturnKeys,"Pick a Menu:",t)
   if (ch = nil) then exit end
   mbarChoice = mbarDict.Get(ch).GetControls
   if (mbarChoice.Count = 0) then exit end
   chDict = Dictionary.Make(1)
   for each ch in mbarChoice
     if (ch.Is(Choice)) then
       chDict.Add(ch.GetLabel,ch)
     end
   end
   chLabel = MsgBox.ListAsString(chDict.ReturnKeys,"Pick a Menu
item:",t)
   if (chLabel = nil) then exit end
```

```
        parm = chDict.Get(chLabel)
end
rpt = "<dl>" + nl
rpt = rpt + "<dt>Label" + nl + "<dd>" + parm.GetLabel + nl
rpt = rpt + "<dt>Control Set Member" + nl + "<dd>" +
parm.GetControlSet.GetLabel + nl
rpt = rpt + "<dt>Click" + nl + "<dd>" + parm.GetClick + nl
rpt = rpt + "<dt>Help"  + nl + "<dd>" + parm.GetHelp + nl
rpt = rpt + "<dt>Update" + nl + "<dd>" + parm.GetUpdate + nl
rpt = rpt + "</dl>" + nl
if (SELF.Is(Choice)) then
    return rpt
else
    MsgBox.Report(rpt,t)
end
```

## ◆ *ChoiceReport*

Gets the attributes and characteristics of a choice. Choice is the class of a selection item on a drop-down menu.

**Topics**: Choice, Control

**Search Keys**: Choice, GetLabel, GetScript, GetUpdate

**Requirements**: Script must be called from another script and must be passed a single parameter that is an object of the choice type.

```
' Variable Initialization
t = "ChoiceReport"
parm = SELF
if (parm.Is(Choice).Not) then
```

```
    MsgBox.Error("This script must be passed a choice.",t)
    exit
end
nltab = nl + "      "
rpt = ""
rpt = "Label:" + nltab + parm.GetLabel + nl
rpt = rpt + "Control Set Member:" + nltab +
parm.GetControlSet.GetLabel + nl
rpt = rpt + "Click:" + nltab + parm.GetClick + nl
rpt = rpt + "Help:"  + nltab + parm.GetHelp + nl
rpt = rpt + "Update:" + nltab + parm.GetUpdate + nl
return rpt
```

## ◆ *EmbeddedScriptDocumentation*

Runs as a standalone or is called by another script. The script formats as an HTML document containing basic information about an embedded script.

**Topics**: Script, HTML, Metadata

**Search Keys**: Script, HTML

```
' Variable Initialization
t = "EmbeddedScriptDocumentation"
' Make this script callable or standalone.
parm = SELF
if (parm.Is(Script).Not) then
    p = av.GetProject
    scriptList = p.GetScripts
    if (scriptList.Count = 0) then
        MsgBox.Warning("No Embedded Scripts in this Project:
Exiting",t)
```

```
      exit
  end
  ch = MsgBox.ListasString(scriptList.ReturnKeys,"Pick a
Script:",t)
  if (ch = nil) then
    MsgBox.Warning("No Embedded Script Picked: Exiting",t)
    exit
  end
  parm = scriptList.Get(ch)
end


s = "<h2>" + parm.GetName + "</h2>" + nl
source = parm.AsString
s = s + "<pre>" + nl + source + nl + "</pre>" + nl
if (SELF.Is(SEd).Not) then
  MsgBox.Report(s,t)
else
  return s
end
```

## ◆ *FThemeDocumentation*

Driver program for documenting a feature theme. The implementation details are left to the reader.

**Topics**: Documentation, View, FTheme

**Search Keys**: Theme.GetComments, Theme.GetExtract, Theme.GetSrcName, Theme.GetHotField, Theme.GetHotScriptname, MsgBox.Report

```
' Variable Initialization
t = "FThemeDocumentation"
```

```
' Make this script callable or standalone.
parm = SELF
if (parm.Is(FTheme).Not) then
  docList = av.GetProject.GetDocs
  vList = List.Make
  for each d in docList
    if (d.Is(View)) then vList.Add(d) end
  end
  if (vList.Count = 0) then
    MsgBox.Warning("No Views in This Document: Exiting",t)
    exit
  end
  vparm = MsgBox.List(vList,"Pick a View:",t)
  if (vparm = nil) then
    MsgBox.Warning("No View Picked: exiting",t)
    exit
  end
  ftList = List.Make
  for each t in vparm.GetThemes
    if (t.Is(FTheme)) then
      ftList.Add(t)
    end
  end
  if (ftList.Count = 0) then
    MsgBox.Warning("No Feature Themes in this View: Exiting",t)
    exit
  end
  parm = MsgBox.List(ftList,"Pick an FTheme",t)
```

```
  if (parm = nil) then
    MsgBox.Warning("No Feature Theme picked: Exiting",t)
    exit
  end
end
ftRpt = "<h3>Feature Theme:" ++ parm.GetName + "</h3>" + nl
ftRpt = ftRpt + "<dl>" + nl
ftRpt = ftRpt + "<dt>Comments" + nl + "<dd>" + parm.GetComments + nl
ftRpt = ftRpt + "<dt>Extent" + nl + "<dd>" +
parm.ReturnExtent.asString + nl
ftRpt = ftRpt + "<dt>SrcName" + nl + "<dd>" +
parm.GetSrcName.GetFileName.asString + nl
ftRpt = ftRpt + "<dt>HotField" + nl + "<dd>" +
parm.GetHotField.GetName + nl
ftRpt = ftRpt + "<dt>HotScript" + nl + "<dd>" +
parm.GetHotScriptName + nl
if (SELF.Is(FTheme).Not) then
    MsgBox.Report(ftRpt,t)
else
    return ftRpt
end
```

## ◆ *GUIReport*

Creates a report marked up using HTML and passes it back to the calling program. The report for each GUI component is called to create the text message. If this script is called by itself, it displays a report in a MsgBox.Report dialog about each element of the GUI, that is, Menubar/Menu/Choice, Buttonbar/Button, and Toolbar/Tool.

**Topics**: Menu, Choice, Control, Button, Tool

**Search Keys**: Menu, Choice, Control, Button, Tool, av.Run

**Requirements**: Script may be called or standalone, or it may be passed a DocGUI in the active project. This script will not run properly without the following critical resources: compiled script named *MenuBarDocumentation* called by av.Run; compiled script named *MenuDocumentation* called by av.Run in *MenuBarDocumentation* script; compiled script named *ChoiceDocumentation* called by av.Run in *MenuDocumentation* script; compiled script named *ButtonBarDocumentation* called by av.Run; compiled script named *ButtonDocumentation* called by av.Run in *ButtonBarDocumentation* script; compiled script named *ToolBarDocumentation* called by av.Run; and compiled script named *ToolDocumentation* called by av.Run in *ToolBarDocumentation* script.

```
' Variable Initialization

t = "GUIReport"

' ***Critical Resource Test***

if ((av.GetProject.FindDoc("MenuBarDocumentation").is(SEd)) =
false) then

  MsgBox.Error("Critical Resource 1 NOT Available" + NL +

              "Program Cannot Continue" + NL +

              "Press OK to EXIT",t)

  exit

end

if ((av.GetProject.FindDoc("MenuDocumentation").is(SEd)) = false)
then

  MsgBox.Error("Critical Resource 2 NOT Available" + NL +

              "Program Cannot Continue" + NL +

              "Press OK to EXIT",t)

  exit

end

if ((av.GetProject.FindDoc("ChoiceDocumentation").is(SEd)) =
false) then

  MsgBox.Error("Critical Resource 3 NOT Available" + NL +

              "Program Cannot Continue" + NL +
```

```
                  "Press OK to EXIT",t)
  exit
end
if ((av.GetProject.FindDoc("ButtonBarDocumentation").is(SEd)) =
false) then
  MsgBox.Error("Critical Resource 4 NOT Available" + NL +
              "Program Cannot Continue" + NL +
              "Press OK to EXIT",t)
  exit
end
if ((av.GetProject.FindDoc("ButtonDocumentation").is(SEd)) =
false) then
  MsgBox.Error("Critical Resource 5 NOT Available" + NL +
              "Program Cannot Continue" + NL +
              "Press OK to EXIT",t)
  exit
end
if ((av.GetProject.FindDoc("ToolBarDocumentation").is(SEd)) =
false) then
  MsgBox.Error("Critical Resource 6 NOT Available" + NL +
              "Program Cannot Continue" + NL +
              "Press OK to EXIT",t)
  exit
end
if ((av.GetProject.FindDoc("ToolDocumentation").is(SEd)) = false)
then
  MsgBox.Error("Critical Resource 7 NOT Available" + NL +
              "Program Cannot Continue" + NL +
              "Press OK to EXIT",t)
```

```
    exit
  end
  '***      ***       ***       ***
  ' Make the script callable or standalone.
  parm = SELF
  if (parm.Is(DocGUI).Not) then
     p = av.GetProject
     guis = p.GetGUIs
  ' Pick a DocGUI from the current project.
     parm = MsgBox.List(guis,"Pick a GUI: ",t)
     if (parm = nil) then
       MsgBox.Warning("No GUI Picked: Exiting",t)
       exit
     end
  end
  ' For the selected GUI, loop through the elements of the interface
  ' and call the appropriate report function.
  rpt = "<h2>DocGUI:" ++ parm.GetName + "</h2>" + nl
  if (parm.GetType <> nil) then
     rpt = rpt + "<p>Type: " ++ parm.GetType + "<br>" + nl
  end
  rpt = rpt + "<p>Date:" ++ Date.Now.AsString + "<br>" + nl
  ' Menu bar
  mbar = parm.GetMenubar
  rpt = rpt + av.Run("MenuBarDocumentation",mbar)
  ' Button bar
  bbar = parm.GetButtonBar
  rpt = rpt + av.Run("ButtonBarDocumentation",bbar)
```

```
' Tool bar
tbar = parm.GetToolBar
rpt = rpt + av.Run("ToolBarDocumentation",tbar)


' Show the report.
if (SELF.Is(DocGUI)) then
    return rpt
else
    MsgBox.Report(rpt,"DocGUI: " + parm.GetName)
' Write the report to an external file.
    f = MsgBox.Input("Enter a File Name:",t,
"c:\arcview\projects\foo.doc")
    if (f = nil) then exit end
    lf = LineFile.Make(f.asFileName,#FILE_PERM_WRITE)
    for each ln in rpt.asTokens(nl)
      lf.WriteElt(ln)
    end
    lf.Close
end
' Name: Header
' Example Header SEd.
```

# ◆ *Indent*

Splits a string passed as the first parameter into tokens along the newline character, nl, and indents each line by an amount specified in the second parameter.

**Topics**: av.Run, String

**Search Keys**: av.Run, String, asTokens

**Requirements**: Script must be called from another script, and passed the text block in the first parameter and the amount of the indent in the second.

```
' Variable Initialization
t = "Indent"
parm = SELF
if ((parm.Is(List).Not) or (parm.Count <> 2)) then exit end
str = parm.Get(0).asTokens(nl)
out = ""
indent = parm.Get(1)
for each ln in str
   out = out + indent + ln + nl
end
return out
```

# ◆ *IThemeDocumentation*

Documentation shell for image themes. Details are left to the user.

**Topics**: ITheme, Documentation, Metadata, HTML

**Search Keys**: ITheme, Documentation, Metadata

```
' Variable Initialization
t = "IThemeDocumentation"
' Make this script callable or standalone.
```

```
parm = SELF
if (parm.Is(ITheme).Not) then
   docList = av.GetProject.GetDocs
   vList = List.Make
   for each d in docList
     if (d.Is(View)) then vList.Add(d) end
   end
   if (vList.Count = 0) then
     MsgBox.Warning("No Views in This Document:Exiting",t)
     exit
   end
   vparm = MsgBox.List(vList,"Pick a View:",t)
   if (vparm = nil) then
     MsgBox.Warning("No View Picked: exiting",t)
     exit
   end
   itList = List.Make
   for each t in vparm.GetThemes
     if (t.Is(ITheme)) then
       itList.Add(t)
     end
   end
   if (itList.Count = 0) then
     MsgBox.Warning("No Image Themes in this View: Exiting",t)
     exit
   end
   parm = MsgBox.List(itList,"Pick an ITheme",t)
   if (parm = nil) then
```

```
      MsgBox.Warning("No Image Theme picked: Exiting",t)

      exit

   end

end

itRpt = "<h3>Image Theme:" ++ parm.GetName + "</h3>" + nl

itRpt = itRpt + "<dl>" + nl

itRpt = itRpt + "<dt>Extent" + nl

itRpt = itRpt + "<dd>" + parm.ReturnExtent.AsString + nl

itRpt = itRpt + "<dt>Source" + nl

itRpt = itRpt + "<dd>" + parm.GetSrcName.GetFileName.AsString + nl

itRpt = itRpt + "<dt>Name" + nl

itRpt = itRpt + "<dd>" + parm.GetName + nl

itRpt = itRpt + "</dl>" + nl

if (SELF.Is(ITheme).Not) then

   MsgBox.Report(itRpt,t)

else

   return itRpt

end
```

## ◆ *LayoutDocumentation*

Documentation shell for layouts. Implementation details are left to the user.

**Topics**: Layout, Documentation

**Search Keys**: Is(Layout), MsgBox.Report

```
' Variable Initialization

t = "LayoutDocumentation"

parm = SELF

p = av.GetProject
```

```
d = p.GetDocs
layList = List.Make
if (SELF.Is(Layout).Not) then
   for each x in d
      if (x.Is(Layout)) then
         layList.Add(x)
      end
   end
   if (layList.Count = 0) then
      MsgBox.Warning("No Layouts in This Project",t)
      exit
   end
   parm = MsgBox.List(layList,"Pick a layout:",t)
   if (parm = nil) then exit end
end
layRpt = "<h1>Layout:" ++ parm.GetName + "</h1>" + nl
if (SELF.Is(Layout).Not) then
   MsgBox.Report(layRpt,t)
else
   return layRpt
end
```

## ◆ *MenuBarDocumentation*

Marks up a menu bar documentation report as an HTML document. It can be called by another program if passed a single object (MenuBar). The script can also be run standalone. In the standalone mode, the user is led through a sequence of choices to pick a menu bar in a GUI to document.

**Topics**: DocGUI, MenuBar, Menu, Choice

**Search Keys**: DocGUI, MenuBar, Menu, Choice, av.Run, HTML

**Requirements**: Script will not run properly without a compiled script named *MenuDocumentation* called by av.Run, and a compiled script named *ChoiceDocumentation* called by av.Run in the *MenuDocumentation* script.

```
' Variable Initialization
t = "MenuBarDocumentation"
' ***Critical Resource Test***
if ((av.GetProject.FindDoc("MenuDocumentation").is(SEd)) = false)
then
  MsgBox.Error("Critical Resource NOT Available" + NL +
                "Program Cannot Continue" + NL +
                "Press OK to EXIT",t)
  exit
end
if ((av.GetProject.FindDoc("ChoiceDocumentation").is(SEd)) =
false) then
  MsgBox.Error("Critical Resource NOT Available" + NL +
                "Program Cannot Continue" + NL +
                "Press OK to EXIT",t)
  exit
end
' Make this script callable or standalone.
parm = SELF
if (parm.Is(MenuBar).Not) then
  p = av.GetProject
  guis = p.GetGUIs
  gui = MsgBox.List(guis,"Pick a GUI:",t)
  if (gui = nil) then exit end
  parm = gui.GetMenuBar
```

```
end
rpt = "<h2>MenuBar</h2>" + nl
menuList = parm.GetControls
if (menuList.Count = 0) then
    rpt = rpt + "<p>No Menus for This DocGUI." + nl
else
    rpt = rpt + "<ul>" + nl
    for each m in menuList
      if (m.Is(Space)) then continue end
      s = "<li>" + nl
      rpt = rpt + s + av.Run("MenuDocumentation",m)
    end
    rpt = rpt + "</ul>" + nl
end
if (SELF.Is(MenuBar)) then
    return(rpt)
else
    MsgBox.Report(rpt,t)
end
```

## ◆ *MenuDocumentation*

Script can be called from another program and passed a single object (Menu), or it can be run standalone. In the standalone case, the user is presented with a series of dialog boxes to select a menu. The script then calls the *ChoiceDocumentation* script to document each choice on the menu.

**Topics**: Menu, Choice

**Search Keys**: Menu, Choice, av.Run

**Requirements**: Script will not run properly without a compiled script named *ChoiceDocumentation* called by av.Run.

```
' Variable Initialization
t = "MenuDocumentation"
' ***Critical Resource Test***
if ((av.GetProject.FindDoc("ChoiceDocumentation").is(SEd)) =
false) then
  MsgBox.Error("Critical Resource NOT Available" + NL +
               "Program Cannot Continue" + NL +
               "Press OK to EXIT",t)
  exit
end
'  ***     ***      ***       ***
' Make this script callable or standalone.
parm = SELF
if (parm.Is(Menu).Not) then
  p = av.GetProject
  guis = p.GetGUIs
  gui = MsgBox.List(guis,"Pick a GUI:",t)
  if (gui = nil) then exit end
  mbar = gui.GetMenuBar.GetControls
  if (mbar.Count = 0) then exit end
  mbarDict = Dictionary.Make(2)
  for each m in mbar
    if (m.Is(Menu)) then
      mbarDict.Add(m.GetLabel,m)
    end
  end
  ch = MsgBox.ListasString(mbarDict.ReturnKeys,"Pick a Menu:",t)
  if (ch = nil) then exit end
  parm = mbarDict.Get(ch)
```

```
end
rpt = "<h3>Menu:" ++ parm.GetLabel + "</h3>" + nl
chList = parm.GetControls
if (chList.Count = 0) then
   rpt = rpt + "<p>No choices on this Menu." + nl
else
   rpt = rpt + "<ul>" + nl
   for each ch in chList
     if (ch.Is(Choice)) then
       s = "<li>" + nl
       rpt = rpt + s + av.Run("ChoiceDocumentation",ch)
     end
   end
   rpt = rpt + "</ul>" + nl
end
if (SELF.Is(Menu)) then
   return rpt
else
   MsgBox.Report(rpt,t)
end
```

## ◆ *ProjectDocumentation*

Driver script for the current project documentation or metadata. If you have a large script collection in the project you may have trouble with virtual memory. The documentation is stored temporarily as a string, and the string will include all sources for SEds and embedded scripts. Results are shown in a MsgBox.Report, and the option to save to a file is provided at the end.

**Topics**: Project, HTML

**Search Keys**: Project, Documentation, HTML

**Requirements**: This script will not run properly without the following critical resources: compiled script named *ViewDocumentation* called by av.Run; compiled script named *TableDocumentation* called by av.Run; compiled script named *ChartDocumentation* called by av.Run; compiled script named *LayoutDocumentation* called by av.Run; compiled script named *SEdDocumentation* called by av.Run; compiled script named *GUIReport* called by av.Run; and compiled script named *EmbeddedScriptDocumentation* called by av.Run.

```
' Variable Initialization
t = "ProjectDocumentation"
' ***Critical Resource Test***
if ((av.GetProject.FindDoc("ViewDocumentation").is(SEd)) = false)
then

  MsgBox.Error("Critical Resource 1 NOT Available" + NL +
               "Program Cannot Continue" + NL +
               "Press OK to EXIT",t)

  exit

end

if ((av.GetProject.FindDoc("TableDocumentation").is(SEd)) = false)
then

  MsgBox.Error("Critical Resource 2 NOT Available" + NL +
               "Program Cannot Continue" + NL +
               "Press OK to EXIT",t)

  exit

end

if ((av.GetProject.FindDoc("ChartDocumentation").is(SEd)) = false)
then

  MsgBox.Error("Critical Resource 3 NOT Available" + NL +
               "Program Cannot Continue" + NL +
               "Press OK to EXIT",t)

  exit
```

```
end
if ((av.GetProject.FindDoc("LayoutDocumentation").is(SEd)) =
false) then
  MsgBox.Error("Critical Resource 4 NOT Available" + NL +
                "Program Cannot Continue" + NL +
                "Press OK to EXIT",t)
  exit
end
if ((av.GetProject.FindDoc("SEdDocumentation").is(SEd)) = false)
then
  MsgBox.Error("Critical Resource 5 NOT Available" + NL +
                "Program Cannot Continue" + NL +
                "Press OK to EXIT",t)
  exit
end
if ((av.GetProject.FindDoc("GUIReport").is(SEd)) = false) then
  MsgBox.Error("Critical Resource 6 NOT Available" + NL +
                "Program Cannot Continue" + NL +
                "Press OK to EXIT",t)
  exit
end
if ((av.GetProject.FindDoc("EmbeddedScriptDocumentation").is(SEd))
= false) then
  MsgBox.Error("Critical Resource 7 NOT Available" + NL +
                "Program Cannot Continue" + NL +
                "Press OK to EXIT",t)
  exit
end
html = "c:\".asFileName
```

```
p = av.GetProject

' HTML header

header = "<html>" + nl

header = header + "<head>" + nl

header = header + "<title>Project:" ++ p.GetFileName.asString ++ "</
title>" + nl

header = header + "</head>" + nl

header = header + "<body>" + nl

document = header

' Project documentation

s = "<h1>Project:" ++ p.GetFileName.asString + "</h1>" + nl

attr = "<dl>" + nl

attr = attr + "<dt>Selection Color" + nl + "<dd>" + p.GetSelCol-
or.asString + nl

attr = attr + "<dt>Serial Number" + nl + "<dd>" + p.GetSerialNumber
+ nl

attr = attr + "<dt>Startup Script" + nl + "<dd>" + p.GetStartup + nl

attr = attr + "<dt>Shutdown Script" + nl + "<dd>" + p.GetShutdown +
nl

attr = attr + "</dl>" + nl

document = document + s + attr

' Document lists

vList = List.Make

tList = List.Make

lList = List.Make

cList = List.Make

sList = List.Make

for each d in p.GetDocs
    if (d.Is(View)) then
```

```
      vList.Add(d)
   elseif (d.Is(Table)) then
      tList.Add(d)
   elseif (d.Is(Layout)) then
      lList.Add(d)
   elseif (d.Is(Chart)) then
      cList.Add(d)
   elseif (d.Is(SEd)) then
      sList.Add(d)
   else
      MsgBox.Warning(d.GetName ++ "has an invalid Doc class.",t)
      exit
   end
end
' Call for view documentation.
vdoc = ""
s = "<h2>View Documents</h2>" + nl
if (vList.Count = 0) then
   vdoc = "<p>No Views in this Project." + nl
else
   for each v in vList
      vdoc = vdoc + av.Run("ViewDocumentation",v)
   end
end
document = document + s + vdoc
' Call for table documentation.
tdoc = ""
s = "<h2>Table Documents</h2>" + nl
```

```
if (tList.Count = 0) then
    tdoc = "<p>No Tables in this Project." + nl
else
    for each t in tList
       tdoc = tdoc + av.Run("TableDocumentation",t)
    end
end
document = document + s + tdoc
' Call for chart documentation.
cdoc = ""
s = "<h2>Chart Documents</h2>" + nl
if (cList.Count = 0) then
    cdoc = "<p>No Charts in this Project." + nl
else
    for each c in cList
       cdoc = cdoc + av.Run("ChartDocumentation",c)
    end
end
document = document + s + cdoc
' Call for layout documentation.
ldoc = ""
s = "<h2>Layout Documents</h2>" + nl
if (lList.Count = 0) then
    ldoc = "<p>No Layouts in this Project." + nl
else
    for each l in lList
       ldoc = ldoc + av.Run("LayoutDocumentation",l)
    end
```

```
end
document = document + s + ldoc
' Call for SEd documentation.
sdoc = ""
s = "<h2>SEd Documents</h2>" + nl
if (sList.Count = 0) then
    sdoc = "No SEds in this Project." + nl
else
    sdoc = "<ul>" + nl
    for each s1 in sList
      sdoc = sdoc + "<li>" + nl + av.Run("SEdDocumentation",s1)
    end
    sdoc = sdoc + "</ul>" + nl
end
document = document + s + sdoc
' Call for DocGUI documentation.
guiDoc = ""
s = "<h2>DocGUI Documentation</h2>" + nl
guis = p.GetGUIs
if (guis.Count = 0) then
    guiDoc = "<p>No GUIs for this Project." + nl
else
    guiDoc = "<ul>" + nl
    for each gui in guis
      guiDoc = guiDoc + "<li>" + nl + av.Run("GUIReport",gui)
    end
    guiDoc = guiDoc + "</ul>" + nl
end
```

```
document = document + s + guiDoc
' Call for script documentation.
scriptDoc = ""
s = "<h2>Embedded Script Documentation</h2>" + nl
embeddedScripts = p.GetScripts
if (embeddedScripts.Count = 0) then
    guiDoc = "<p>No embedded Scripts for this Project." + nl
else
    scriptDoc = "<ul>" + nl
    for each emb in embeddedScripts
      scriptDoc = scriptDoc + "<li>" + nl +
av.Run("wtpc.EmbeddedScriptDocumentation",emb)
    end
    scriptDoc = scriptDoc + "</ul>" + nl
end
document = document + s + scriptDoc
' HTML footer
document = document + "</body>" + nl + "</html>" + nl
' Now write the result to text file.
html.SetCWD
tf = FileDialog.Put("projdoc.htm".asFileName,"*.htm",t)
if (tf = nil) then
    MsgBox.Warning("No output file defined: exiting",t)
    exit
end
htmlFile = LineFile.Make(tf,#FILE_PERM_WRITE)
if (htmlFile = nil) then
    MsgBox.Warning("Output file" ++ tf.asString ++ "is invalid:
exiting",t)
```

```
      exit
end
numLines = document.AsTokens(nl).Count
htmlFile.Write(document.AsTokens(nl),numLines)
```

# ◆ *SEdDocumentation*

A shell script for SEd documentation. It returns the SEd source as part of the documentation and the text is marked up with HTML.

**Topics**: SEd, Documentation, HTML

**Search Keys**: SEd, Documentation, HTML

```
' Variable Initialization
t = "SEdDocumentation"
' Make this script callable or standalone.
parm = SELF
if (parm.Is(SEd).Not) then
   p = av.GetProject
   docList = p.GetDocs
   sedList = List.Make
   for each d in docList
      if (d.Is(SEd)) then sedList.Add(d) end
   end
   if (sedList.Count = 0) then
      MsgBox.Warning("No SEds in this Project: Exiting",t)
      exit
   end
   parm = MsgBox.List(sedList,"Pick an SEd:",t)
   if (parm = nil) then
```

```
      MsgBox.Warning("No SEd Picked: Exiting",t)
      exit
    end
  end


s = "<h2>" + parm.GetName + "</h2>" + nl
source = parm.GetSource
s = s + "<pre>" + nl + source + nl + "</pre>" + nl
if (SELF.Is(SEd).Not) then
   MsgBox.Report(s,t)
else
   return s
end
```

## ◆ *TableDocumentation*

Template script for table documentation. Currently, it returns the name of the table. Implementation details are left to the user.

**Topics**: Table, VTab

**Search Keys**: Table, Documentation

**Requirements**: Script will not run properly without a compiled script named *TableReportFile* called by av.Run.

```
' Variable Initialization
t = "TableDocumentation"
' ***Critical Resource Test***
if ((av.GetProject.FindDoc("TableReportFile").is(SEd)) = false)
then
  MsgBox.Error("Critical Resource NOT Available" + NL +
```

```
                    "Program Cannot Continue" + NL +
                    "Press OK to EXIT",t)
  exit
end
rpt = ""
' Make this script callable or standalone.
parm = SELF
if (parm.Is(Table).Not) then
   p = av.GetProject
   docList = p.GetDocs
   tblList = List.Make
   for each d in docList
      if (d.Is(Table)) then tblList.Add(d) end
   end
   if (tblList.Count = 0) then
     MsgBox.Warning("No Tables in this Project: Exiting",t)
     exit
   end
   parm = MsgBox.List(tblList,"Pick a Table:",t)
   if (parm = nil) then
     MsgBox.Warning("No Table Picked: Exiting",t)
     exit
   end
end
rpt = av.Run("TableReportFile",parm)
if (SELF.Is(Table).Not) then
   MsgBox.Report(rpt,t)
   out = FileDialog.Put("d:\temp\out.tbl".asFileName,"*.tbl",t)
```

```
    if (out = nil) then exit end
    outFile = LineFile.Make(out,#FILE_PERM_WRITE)
    if (outFile = nil) then exit end
    outFile.Write(rpt.asTokens(nl),rpt.asTokens(nl).Count)
    outFile.Flush
    outFile.Close
else
    return rpt
end
```

# ◆ *TableReport*

Utility to select a table available on the table list in a project, and to obtain a report about the table and all fields in the table or a single field from the table. The report is organized in a MsgBox.Report format.

**Topics**: Project, Documentation, Table

**Search Keys**: Table.GetVTab, VTab.GetFields, Table.GetObjectTag, Field.IsTypeNumber/String/Shape, Field.IsBase, Field.IsVisible

```
' Variable Initialization
t = "TableReport"
' Get the project.
theProj = av.GetProject
' Select the table from table list.
theDocList = theProj.GetDocs
theDict = Dictionary.Make(theDocList.Count)
for each i in theDocList
    if (i.is(Table)) then
        theDict.Add(i.GetName,i)
```

```
    end
end
theList = theDict.ReturnKeys
theTable = theDict.Get(MsgBox.ListAsString(theList,"Choose a
Table","TABLES"))
if (theTable = NIL) then exit end
' Get the VTab.
theVTab = theTable.GetVTab
' Get the fields in the VTab.
theFieldList = theVTab.GetFields
msgbox.report("TABLE NAME:" ++ theTable.GetName + NL +
            "Creator:" ++ theTable.GetCreator + NL +
            "Creation Date:" ++ theTable.GetCreationDate + NL +
            "Comments:" ++ theTable.GetComments + NL + NL +
            "Object Tag:" ++ theTable.GetObjectTag.asstring + NL +
            "DocGUI:" ++ theTable.GetGui + NL +
            "Table Joined:" ++ theVTab.IsBase.NOT.asstring + NL +
            "Table Linked:" ++ theVTab.IsLinked.asstring + NL +
            "Number of Fields:" ++ theFieldList.Count.AsString,
            "TABLE REPORT")
theDict = Dictionary.Make(theFieldList.Count + 1)
theDict.Add("ALL FIELDS","ALL FIELDS")
for each i in theFieldList
    theDict.Add(i.GetName,i)
end
theFLDList = theDict.ReturnKeys


theChoice = msgBox.ListAsString(theFLDList,"Choose a Single Field
or ALL FIELDS","FIELDS")
```

```
if (theChoice = NIL) then exit end
if (theChoice = "ALL FIELDS") then
    for each i in theFieldList
        theField = i
        msgbox.report("FIELD NAME:" ++ theField.GetName + NL +
                "Field Is Number Field:" ++
        theField.IsTypeNumber.AsString + NL +
                "Field Is Character Field:" ++
        theField.IsTypeString.AsString + NL +
                "Field Is Shape Field:" ++
        theField.IsTypeShape.asString + NL +
                "Field Type:" ++ theField.GetType.asString + NL +
                "Field Width:" ++ theField.GetWidth.asString + NL +
                "Field Precision:" ++ theField.GetPrecision.asString
        + NL + NL +
                "Alias:" ++ theField.GetAlias + NL +
                "Field Is Visible:" ++ theField.IsVisible.asstring
        + NL +
                "Table Pixel Width:" ++
        theField.GetPixelWidth.asString,
                theField.GetName ++ "FIELD REPORT")
    end
else
    thefield = theVTab.FindField(theChoice)
    msgbox.report("FIELD NAME:" ++ theField.GetName + NL +
                "Field Is Number Field:" ++
        theField.IsTypeNumber.AsString + NL +
                "Field Is Character Field:" ++
        theField.IsTypeString.AsString + NL +
                "Field Is Shape Field:" ++
        theField.IsTypeShape.asString + NL +
```

```
                      "Field Type:" ++ theField.GetType.asString + NL +
                    "Field Width:" ++ theField.GetWidth.asString + NL +
                  "Field Precision:" ++ theField.GetPrecision.asString
+ NL + NL +

                   "Alias:" ++ theField.GetAlias + NL +
                  "Field Is Visible:" ++ theField.IsVisible.asstring
+ NL +

                  "Table Pixel Width:" ++
theField.GetPixelWidth.asString,

                    theField.GetName ++ "FIELD REPORT")
end
```

## ◆ *TableReportFile*

Utility to select a table available on the table list in a project, and to obtain a report about the table and all fields in the table or a single field from the table. The report is saved to a temporary ASCII text file.

**Topics**: Project, Documentation, Table

**Search Keys**: Table.GetVTab, VTab.GetFields, Table.GetObjectTag, Field.IsType-Number/ String/Shape, Field.IsBase, Field.IsVisible

**Requirements**: Script must be called by another script which supplies the table name to be documented, and then displays the output of the report, such as the *TableDocumentation* script.

```
' Variable Initialization
t = "TableReportFile"
rpt = ""
theTable = SELF
' Get the project.
theProj = av.GetProject
' Get the VTab.
```

```
theVTab = theTable.GetVTab
' Get the fields in the VTtab.
theFieldList = theVTab.GetFields
theDict = Dictionary.Make(theFieldList.Count + 1)
theDict.Add("ALL FIELDS","ALL FIELDS")
for each i in theFieldList
  theDict.Add(i.GetName,i)
end
theFLDList = theDict.ReturnKeys
rpt = rpt + "<h1>TABLE NAME:" ++ theTable.GetName + "</h1>" + nl
rpt = rpt + "<pre>"+ nl
rpt = rpt + "Creator:" ++ theTable.GetCreator + nl
rpt = rpt + "Creation Date:" ++ theTable.GetCreationDate + nl
rpt = rpt + "Comments:" ++ theTable.GetComments + nl
rpt = rpt + "Object Tag:" ++ theTable.GetObjectTag.asstring + nl
rpt = rpt + "DocGUI:" ++ theTable.GetGui + nl
rpt = rpt + "Table Joined:" ++ theVTab.IsBase.NOT.asstring + nl
rpt = rpt + "Table Linked:" ++ theVTab.IsLinked.asstring + nl
rpt = rpt + "Number of Fields:" ++ theFieldList.Count.AsString + nl
rpt = rpt + "</pre>" + nl
' The report for each field in the VTab.
  rpt = rpt + "<dl>" + nl
  for each i in theFieldList
      theField = i
    rpt = rpt + "<dt><h2>FIELD NAME:" ++ theField.GetName+"</h2>" + nl
      rpt = rpt + "<dd><pre>" + nl
      rpt = rpt + "     Field Is Number Field:" ++
theField.IsTypeNumber.AsString + nl
```

```
        rpt = rpt + "     Field Is Character Field:" ++
theField.IsTypeString.AsString + nl

        rpt = rpt + "     Field Is Shape Field:" ++
theField.IsTypeShape.asString + nl

    rpt = rpt + "   Field Type:" ++ theField.GetType.asString + nl

      rpt = rpt + "    Field Width:" ++ theField.GetWidth.asString
+ nl

        rpt = rpt + "     Field Precision:" ++
theField.GetPrecision.asString + nl

      rpt = rpt + nl

      rpt = rpt + "     Alias:" ++ theField.GetAlias + nl

        rpt = rpt + "     Field Is Visible:" ++
theField.IsVisible.asstring + nl

        rpt = rpt + "     Table Pixel Width:" ++
theField.GetPixelWidth.asString + nl

      rpt = rpt + "</pre>" + nl

   end

   rpt = rpt + "</dl>" + nl

return rpt
```

## ◆ *ToolBarDocumentation*

Marks up a tool bar documentation report as an HTML document. This script can be called by another program if it is passed a single object (ToolBar). It can also run as a standalone. In the standalone mode, the user is led through a sequence of choices to select a tool bar in a GUI to be documented.

**Related to**: All other documentation scripts.

**Topics**: DocGUI, ToolBar, Tool

**Search Keys**: DocGUI, ToolBar, Tool, av.Run, HTML

**Requirements**: Script will not run properly without a compiled script named *ToolDocumentation* called by av.Run.

```
' Variable Initialization
t = "ToolBarDocumentation"
' ***Critical Resource Test***
if ((av.GetProject.FindDoc("ToolDocumentation").is(SEd)) = false)
then
   MsgBox.Error("Critical Resource NOT Available" + NL +
                "Program Cannot Continue" + NL +
                "Press OK to EXIT",t)
   exit
end
'  ***      ***      ***      ***
' Make this script callable or standalone.
parm = SELF
if (parm.Is(ToolBar).Not) then
   p = av.GetProject
   guis = p.GetGUIs
   gui = MsgBox.List(guis,"Pick a GUI:",t)
   if (gui = nil) then exit end
   parm = gui.GetToolBar
end
rpt = "<h2>Toolbar</h2>" + nl
toolList = parm.GetControls
if (toolList.Count = 0) then
   rpt = rpt + "<p>No Tools for this DocGUI." + nl
else
   rpt = rpt + "<ul>" + nl
   for each t in toolList
     if (t.Is(Space)) then continue end
```

```
        s = "<li>" + nl
        rpt = rpt + s + av.Run("ToolDocumentation",t)
    end
    rpt = rpt + "</ul>" + nl
end
if (SELF.Is(ToolBar)) then
    return(rpt)
else
    MsgBox.Report(rpt,t)
end
```

## ◆ *ToolDocumentation*

Displays a report as an HTML document about the properties of a tool. The script can be called from another script if it is passed a single object (Tool). Otherwise, the user is taken through a sequence of choices to pick a tool to document.

**Topics**: Control, ToolBar, DocGUI, Tool

**Search Keys**: DocGUI, ToolBar, Tool, av.Run, HTML, Click Event, Apply Event, Update Event

```
' Variable Initialization
t = "ToolDocumentation"
' Make this script callable or standalone.
parm = SELF
if (parm.Is(Tool).Not) then
    p = av.GetProject
    guiList = p.GetGUIs
    gui = MsgBox.List(guiList,"Pick a GUI:",t)
    if (gui = nil) then exit end
    tb = gui.GetToolBar.GetControls
```

```
      if (tb.Count = 0) then
        MsgBox.Warning("No Tools in this GUI: Exiting",t)
        exit
      end
      tbDict = Dictionary.Make(1)
      for each x in tb
        if (x.Is(Space).Not) then
          tbDict.Add(x.GetApply,x)
        end
      end
      ch = MsgBox.ListasString(tbDict.ReturnKeys,"Pick a Tool:",t)
      if (ch = nil) then exit end
      parm = tbDict.Get(ch)
    end
rpt = "<dl>" + nl
rpt = rpt + "<dt>Apply" + nl + "<dd>" + parm.GetApply + nl
rpt = rpt + "<dt>Icon" + nl + "<dd>" + parm.GetIcon.GetName + nl
rpt = rpt + "<dt>Click" + nl + "<dd>" + parm.GetClick + nl
rpt = rpt + "<dt>Help"  + nl + "<dd>" + parm.GetHelp + nl
rpt = rpt + "<dt>Update" + nl + "<dd>" + parm.GetUpdate + nl
rpt = rpt + "</dl>" + nl
if (SELF.Is(Tool)) then
    return rpt
else
    MsgBox.Report(rpt,t)
end
```

# ◆ *ToolReport*

Displays a report about the properties of a tool.

**Topics**: DocGUI, ToolBar, Tool

**Search Keys**: Is(Tool), Tool.GetApply, Tool.GetIcon.GetName, Tool.GetClick, Tool.GetHelp, Tool.GetUpdate

**Requirements**: Script must be called from another script, and must be passed two parameters. The first parameter is the name of the GUI containing the tool, and the second is the tool as an object.

```
' Variable Initialization
t = "ToolReport"
parmList = SELF
if (parmList.Is(List).Not) then exit end
if (parmList.Count <> 2) then exit end
guiName = ParmList.Get(0)
if (guiName.Is(String).Not) then
   MsgBox.Error("First parameter must be the name of a GUI",t)
   exit
end
tl = parmList.Get(1)
if (parmList.Get(1).Is(Tool).Not) then
   MsgBox.Error("This script must be passed a tool.",t)
   exit
end
rpt = "<h3>Tool Report</h3>" + nl
rpt = rpt + "<dl>" + nl
rpt = rpt + "<dt>Apply" + nl + "<dd>" + tl.GetApply + nl
rpt = rpt + "<dt>Icon" + nl + "<dd>" + tl.GetIcon.GetName + nl
rpt = rpt + "<dt>Control Set Member" + nl + "<dd>" + guiName + "/
```

```
ToolBar" + nl

rpt = rpt + "<dt>Click" + nl + "<dd>" + tl.GetClick + nl

rpt = rpt + "<dt>Help"  + nl + "<dd>" + tl.GetHelp + nl

rpt = rpt + "<dt>Update" + nl + "<dd>" + tl.GetUpdate + nl

rpt = rpt + "</dl>" + nl

return rpt
```

## ◆ *ViewDocumentation*

Driver for the documentation of a view. It calls other scripts to get the documentation for FThemes and IThemes.

**Topics**: View, Documentation

**Search Keys**: Doc.Is(View), View.GetThemes, av.Run

**Requirements**: Script will not run properly without a compiled script named *FThemeDocumentation* called by av.Run and a compiled script named *ITheme-Documentation* called by av.Run.

```
' Variable Initialization

t = "ViewDocumentation"

' ***Critical Resource Test***

if ((av.GetProject.FindDoc("FThemeDocumentation").is(SEd)) =
false) then

  MsgBox.Error("Critical Resource 1 NOT Available" + NL +

               "Program Cannot Continue" + NL +

               "Press OK to EXIT",t)

  exit

end

if ((av.GetProject.FindDoc("IThemeDocumentation").is(SEd)) =
false) then

  MsgBox.Error("Critical Resource 2 NOT Available" + NL +

               "Program Cannot Continue" + NL +
```

```
                    "Press OK to EXIT",t)
  exit
end
' ****      *****      *****
' Make this script callable or standalone.
parm = SELF
if (parm.Is(View).Not) then
    p = av.GetProject
    vList = List.Make
    for each d in p.GetDocs
      if (d.Is(View)) then vList.Add(d) end
    end
    if (vList.Count = 0) then
      MsgBox.Warning("No Views in this Project: Exiting",t)
      exit
    end
    parm = MsgBox.List(vList,"Pick a View:",t)
    if (parm = nil) then
      MsgBox.Warning("No View Picked: Exiting",t)
      exit
    end
end
' Write the view documentation.
document = "<h2>" + parm.GetName + "</h2>" + nl
' Get feature themes.
tDoc = ""
tList = parm.GetThemes
s = "<h3>Feature Themes</h3>" + nl
```

```
s = s + "<ul>" + nl
for each t in tList
    if (t.Is(FTheme)) then
      tDoc = tDoc + "<li>" + av.Run("FThemeDocumentation",t)
    end
end
document = document + s + tDoc + "</ul>" + nl
' Get image themes.
tDoc = ""
s = "<h3>Image Themes</h3>" + nl
s = s + "<ul>" + nl
for each t in tList
    if (t.Is(ITheme)) then
      tDoc = tDoc + "<li>" + av.Run("IThemeDocumentation",t)
    end
end
document = document + s + tDoc + "</ul>" + nl
if (SELF.Is(View).Not) then
    MsgBox.Report(document,t)
else
  return document
end
```

# ◆ *ViewReport*

Utility to select a view available on the view list in a project, and to obtain a report about the view, themes, and so forth.

**Topics**: Project, Documentation, View

**Search Keys**: View.GetGUI, View.GetDisplay.GetUnits, View.GetDisplay.GetReportUnits, View.ReturnExtent, View.ReturnScale, TextWin.Make

```
' Variable Initialization
t = "ViewReport"
aFileName = "fooview.txt".asFileName
' Get the project.
theProj = av.GetProject
' Select the view from view list.
theDocList = theProj.GetDocs
theDict = Dictionary.Make(theDocList.Count)
for each i in theDocList
    if (i.is(View)) then
       theDict.Add(i.GetName,i)
    end
end
theList = theDict.ReturnKeys
theView = theDict.Get(MsgBox.ListAsString(theList,"Choose a
View","VIEWS"))
if (theView = nil) then exit end
' Open a line file in write mode.
lf = LineFile.Make(afilename,#FILE_PERM_WRITE)
' Write view information to line file.
lf.WriteElt("VIEW NAME:" ++ theView.GetName)
lf.WriteElt("Creator:" ++ theView.GetCreator)
lf.WriteElt("Creation Date:" ++ theView.GetCreationDate)
lf.WriteElt("Comments:" ++ theView.GetComments)
lf.WriteElt("  Object Tag:" ++ theView.GetObjectTag.asstring)
lf.WriteElt("  DocGUI:" ++ theView.GetGui)
```

```
lf.WriteElt(" Map Units:" ++ theView.GetDisplay.GetUnits.asstring)

lf.WriteElt(" Distance Units:" ++
theView.GetDisplay.GetDistanceUnits.asstring)

lf.WriteElt(" View Extent:" ++ theView.ReturnExtent.asstring)

lf.WriteElt(" View Scale:" ++ theView.ReturnScale.asstring)

lf.WriteElt(" Table Of Contents Width:" ++
theView.GetTOCWidth.asstring ++ "pixels")

lf.WriteElt(" Table Of Contents Is Resizeable:" ++
theView.IsTOCUnResizable.not.asstring)

lf.WriteElt("")

lf.WriteElt("THEMES:")

' Get themes.

theThemes = theView.GetThemes

for each i in theThemes

    if (i.Is(FTheme)) then

      lf.WriteElt("      FEATURE THEME:" ++ i.GetName)

    else

      if (i.Is(ITheme)) then

        lf.WriteElt("      IMAGE THEME:" ++ i.GetName)

      else

        lf.WriteElt("      NOT Feature or Image Theme??" ++ i.GetName)

      end

    end

    lf.WriteElt("      Theme Extent:" ++ i.ReturnExtent.asstring)

    if (i.GetThreshold.IsOff.Not) then

      lf.WriteElt("      Theme Threshold:" ++ "maximum-" ++
i.GetThreshold.GetMaximum.asstring ++ "minimum-" ++
i.GetThreshold.GetMinimum.asstring)

    end

    lf.WriteElt("      Theme Source Name:" ++
```

```
i.GetSrcName.getfilename.getfullname)
    if (i.IsActive) then
      lf.WriteElt("       Theme Is Active")
    end
    if (i.IsVisible) then
      lf.WriteElt("       Theme Is Visible")
    end
    if (i.IsLocked) then
      lf.WriteElt("       Theme Is Locked")
    end
    if (i.IsLegendVisible) then
      lf.WriteElt("       Theme Legend Is Visible")
    else
      lf.WriteElt("       Theme Legend Is NOT Visible")
    end
    if (i.Is(FTheme)) then
      theFtab = i.GetFtab
      theFieldList = theFTab.GetFields
     lf.WriteElt("    Theme Table: " ++ theFieldList.Count.AsString
++ " Fields")
      lf.WriteElt("         Field Names:")
      for each j in theFieldList
          theString = ""
          theString = "              " ++ j.GetName
          if (j.IsTypeShape) then
          theString = theString ++ "    TYPE: SHAPE" ++ "     WIDTH:
" + j.GetWidth.AsString
          else
              if (j.IsTypeString) then
```

```
              theString = theString ++ "      TYPE: CHARACTER" ++ "
WIDTH:    " + j.GetWidth.AsString ++ "        ALIAS:" ++ j.GetAlias
              else
                 if (j.IsTypeNumber) then
                     theString = theString ++
                     "       TYPE:" ++ j.GetType.AsString ++
          "      WIDTH:    " + j.GetWidth.AsString ++
          "      PRECISION:" ++ j.GetPrecision.asstring ++
          "      ALIAS:" ++ j.GetAlias
                end
             end
           end
        theString = theString ++ "     VISIBLE:" ++ j.isVisible.asstring
          lf.WriteElt(theString)
        end
     end
   lf.WriteElt("")
   lf.WriteElt("")
end
lf.Flush
lf.Close
thetextwin = textwin.make(afilename,theView.asstring ++
"Report").setfixedfont(true)
```

# Files

## ◆ CurrentWorkingDirectoryInitialize

Initializes a dictionary as a global variable. The dictionary is used by system scripts to set the current working directory when a project, script, view, or similar file oriented dialog box is invoked.

**Topics**: File, Dictionary

**Search Keys**: Dictionary, Add, Global Variable

**Requirements**: Should be run in the open script for a project if you intend to use it. Rewrites for system scripts such as *Project.Open* are included in this collection of scripts.

```
' Variable Initialization
t = "CurrentWorkingDirectoryInitialize"
_dirDict = Dictionary.Make(5)
_dirDict.Add("Feature Themes","c:\projects\data".asFileName)
_dirDict.Add("Image Themes","c:\projects\images\tif".asFileName)
_dirDict.Add("Projects","d:\projects\hmp".asFileName)
_dirDict.Add("Scripts","d:\projects\hmp\scripts".asFileName)
_dirDict.Add("Tables","d:\projects\hmp".asFileName)
```

## ◆ FileExists

Checks for the existence of a file.

**Topic**: File

**Search Keys**: File, Exists

```
' Variable Initialization
t = "FileExists"
if (SELF.Is(FileName)) then
   parm = SELF
else
   parm = MsgBox.Input("Filename:",t,"")
   if (parm = nil) then exit end
   if (File.Exists(parm.AsFileName).Not) then
     MsgBox.Error("File" ++ parm ++ "Does Not Exist: Exiting",t)
     exit
   end
   MsgBox.Info(parm ++ "exists",t)
   parm = parm.asFilename
end
av.ShowMsg("Returning:" ++ parm.GetFullName)
return parm
```

## ◆ *FileExistsInSearchPath*

Looks for files in the paths that match a given pattern.

**Topic**: File

**Search Keys**: File, IsDir, asPattern, Filter

```
' Variable Initialization
t = "FileExistsInSearchPath"
aFileNameList = List.Make
if (SELF.Is(FileName)) then
   if (SELF.IsDir) then
     parm = SELF
```

```
      end
else
   parm = nil
   while (true)
      parm = MsgBox.Input("Enter a Directory: ",t,"c:\arcview")
      if (parm = nil) then
        MsgBox.Warning("No Directory Entered: Exiting",t)
        exit
      end
      if (parm.asFileName.IsDir.Not) then
        MsgBox.Warning("The directory entered:" ++ parm ++ "is not
valid. Please try again.",t)
      else
        break
      end
   end
end
root = parm.AsFileName
newdirlst = {}
dirthislevel = {root}
while (dirthislevel.Count <> 0)
   newdirlst = newdirlst + dirthislevel.clone
   old = dirthislevel.clone
   dirthislevel = {}
   for each x in old
     for each y in x.Read("*".asPattern)
       if (y.IsDir) then
         dirthislevel.add(y)
       end
```

```
         end
      end
   end
   dirrpt = ""
   ' newdirlst.Sort(true)
   for each x in newdirlst
      dirrpt = dirrpt + x.asString + nl
   end
   msgbox.Report(dirrpt,t)
   FileName.SetSearchPaths(newdirlst)
   f = MsgBox.Input("Enter a FileName:",t,"")
   if (f = nil) then exit end
   if (FileName.ExistsInPaths(f)) then
      MsgBox.Info(f ++ "is in this subdirectory.",t)
      if (FileName.UniqueInPaths(f)) then
         MsgBox.Info(FileName.MakeExisting(f).asString,t)
      else
         fileReport = ""
         for each path in newdirlst
            if (File.Exists(FileName.Merge(path.asString,f))) then
               fileReport = fileReport +
   FileName.Merge(path.asString,f).asString + nl
            end
         end
         MsgBox.Report(fileReport,"Files in this subdirectory tree")
      end
   else
      MsgBox.Info(f ++ "not found!",t)
   end
```

# ◆ *PLasXYEvent*

Reads a PL94-171 file and, by using the centroid for each census block, creates a comma-separated variable file with the census block ID, longitude of the centroid, and latitude of the centroid.

**Topics**: File, TextFile, LineFile, XYEvent, View, Theme

**Search Keys**: TextFile, LineFile, XYEvent, View, Theme, Read, WriteElt, Make

**Comments**: Should be extended to include population attributes with the centroids, but this can be done with a join because the centroid carries the census block ID.

**Requirements**: A PL94-171 file for a county. Be careful with the size of the file because script execution can be very slow. This script will not run properly without a PL9-4171 file for a county referenced by the *plfile* variable.

```
' Variable Initialization
t = "PLasXYEvent"
plfile = "c:\arcview\data\pl48491"
' ***Critical Resource Test***
pl = TextFile.Make(plfile.asFileName,#FILE_PERM_READ)
if (pl = nil) then
  MsgBox.Error("Critical Resource NOT Available" + NL +
               "Program Cannot Continue" + NL +
               "Press OK to EXIT",t)
  exit
end
' ***     ***      ***       ***
pl = TextFile.Make(plfile.asFileName,#FILE_PERM_READ)
flds = {
        {"sumlev",10,3},
        {"ctbna",51,6},
```

```
           {"block",46,4},
           {"intptlng",277,10},
           {"intptlat",268,9}
         }
out = LineFile.Make((plfile + ".txt").asFileName,#FILE_PERM_WRITE)
if (out = nil) then exit end
reclen   = 516
numRec   = pl.GetSize/reclen
n        = 0
implied = 1000000
while (pl.IsAtEnd.Not)
   rec = pl.Read(recLen)
   n = n + 1
' Update the thermometer.
   proceed = av.SetStatus((n/numRec) * 100)
   if ( proceed.Not ) then
     av.ClearStatus
     av.ShowMsg( "Stopped" )
     exit
   end
   sumlev = rec.Middle(10,3)
' You are interested only in block records, summary level 750.
   if (sumlev <> "750") then
     continue
   end
   ctbna    = rec.Middle(51,6)+rec.Middle(46,4)
   long     = rec.Middle(277,10).asNumber/implied
   lat      = rec.Middle(268,9).asNumber/implied
```

```
    outline = ctbna + "," + long.asString + "," + lat.asString
' Preview
' MsgBox.Info(x,"out")
' if (MsgBox.MiniYesNo("Continue?",true).Not) then exit end
    out.WriteElt(outline)
end
out.Close
pl.Close


v = View.Make
v.SetName("PL94-171")
```

## ◆ *SetDirectories*

Sets the current working directory. The current working directory is the starting point for dialog boxes such as Add Theme and Load Text File.

**Topics**: File, FileName

**Search Keys**: File, FileName, SetCWD, IsDir

```
' Variable Initialization
t = "SetDirectories"
dir = MsgBox.Input("Enter a directory name:",t,"c:\arcview")
if (dir = nil) then exit end
if (dir.asFileName.IsDir.Not) then
    MsgBox.Warning(dir ++ "is not a directory: Exiting",t)
    exit
end
dir.asFileName.SetCWD
MsgBox.Info(dir.asString,t)
```

```
FileDialog.Show("*.apr","Projects *.apr","Pick a File")
dir = MsgBox.Input("Enter a directory name:",t,"c:\arcview")
if (dir = nil) then exit end
if (dir.asFileName.IsDir.Not) then
   MsgBox.Warning(dir ++ "is not a directory: Exiting",t)
   exit
end
dir.asFileName.SetCWD
dir = FileName.GetCWD
MsgBox.Info(dir.asString,t)
FileDialog.Show("*.apr","Projects *.apr","Pick a File")
```

## ◆ *Tiger*

Avenue has a TextFile class that supports reading and writing streams of ASCII characters. The TIGER files are good examples of fixed length record and fixed length field files, with no record or field separators. This script reads the record numbers from long/lat and to long/lat from the first TIGER file, and writes them to a comma-separated LineFile variable. ArcView will attempt to read any TextFile into memory before it begins processing the file. For large files, this can be time-consuming, so be careful.

**Topics**: LineFile, TextFile, PolyLine, Shape, FTab

**Search Keys**: LineFile, TextFile, PolyLine, Shape, Make, Read, Write

**Requirements**: Script will not run properly without the following critical resources: TIGER file for a county referenced by *tigerfile* variable, and a valid file name and path for output file referenced by *tigerfileout* variable.

```
' Variable Initialization
t = "Tiger"
implied = 1000000
```

```
tigerfile = "c:\arcview\data\tgr48491.f41"

fn1 = tigerfile.asFileName

tf  = TextFile.Make(fn1,#FILE_PERM_READ)

if (tf = nil) then

   MsgBox.Warning("File" ++ fn1.GetName ++ "Does not exist:
Exiting",t)

   exit

end

' Output file

tigerfileout = "c:\arcview\data\tgr48491.av1"

fn1Out = tigerfileout.asFileName

lf = LineFile.Make(fn1Out,#FILE_PERM_WRITE)

if (lf = nil) then

   MsgBox.Warning("File" ++ fn1Out.GetName ++ "Does not exist:
Exiting",t)

   exit

end

' List containing field names, offsets, and lengths.

fldList = {{"recnum",5,10},

           {"fromlong",190,10},

           {"fromlat",200,9},

           {"fromlong",209,10},

           {"fromlong",219,9}}

' Record length

recLen  = 228

' In the event the process is slow, provide an out after a fixed

' number of records.

n = 0

size = tf.GetSize/recLen
```

```
while (tf.IsAtEnd.Not)
   rec = tf.Read(recLen)
   n = n + 1
' Update the thermometer.
   proceed = av.SetStatus((n/size) * 100)
   if ( proceed.Not ) then
     av.ClearStatus
     av.ShowMsg( "Stopped" )
     exit
   end


   lineOut = ""
   comma   = ""
   for each x in fldList
     fldVal = rec.Middle(x.Get(1),x.Get(2))
     if (x.Get(0) <> "recnum") then
       fldVal = ((fldVal.asNumber)/implied).asString
     else
       fldVal = fldVal + ",  0"
     end
     lineOut = lineOut+comma+fldVal
     comma = ","
   end
   lf.WriteElt(lineOut)
end
lf.Close
tf.Close
```

# Graphics

## ◆ DrawSomething

Draws a set of lines in several different colors and adds the lines to the graphics list for the current view. The colors are stored in a dictionary.

**Topics**: Color, ColorMap, Dictionary, Graphic

**Search Keys**: Color, ColorMap, Dictionary, Graphic, MakeColorWheel, Graphic-Shape, Symbol

**Requirement**: This script will not run properly unless the active document is a view.

```
' Variable Initialization
t = "DrawSomething"
' ***Critical Resource Test***
if (av.GetActiveDoc.Is(View).Not) then
  MsgBox.Error("Critical Resource NOT Available" + NL +
               "Program Cannot Continue" + NL +
               "Press OK to EXIT",t)
  exit
end
'***     ***      ***       ***
rainbow = Dictionary.Make(5)
rainbow.Add(1,Color.GetRed)
rainbow.Add(2,Color.GetCyan)
rainbow.Add(3,Color.GetGreen)
rainbow.Add(0,Color.GetBlue)
colorWheel = ColorMap.MakeColorWheel(256)
d = av.GetActiveDoc
```

```
if (d.Is(View).Not) then exit end
av.ClearStatus
av.ShowStopButton
for each i in (1 .. 255)
   progress = ((i/255)*100)
   status = av.SetStatus(progress)
   if (status = False) then exit end

   y = 0@i
   x = (255 - i + 1)@0
   xLine = Line.Make(x,y)
   clr = colorWheel.Get(i)
   xShape = GraphicShape.Make(xLine)
   xShape.GetSymbol.SetSize(2)
   xShape.GetSymbol.SetColor(clr)
   d.GetDisplay.BeginClip
   d.GetGraphics.Add(xShape)
   d.GetDisplay.EndClip
end
av.ClearStatus
d.GetDisplay.Invalidate(true)
```

## ◆ *DrawText*

Draws a text graphic at a given location on the display in a given font.

**Topics**: Display, Graphic, Font, GraphicText

**Search Keys**: Display, Graphic, Font, GraphicText, SetFont, SetSize, TextSymbol

**Requirements**: This script will not run properly unless the active document is a view.

```
' Variable Initialization
t = "DrawText"
' ***Critical Resource Test***
if (av.GetActiveDoc.Is(View).Not) then
  MsgBox.Error("Critical Resource NOT Available" + NL +
               "Program Cannot Continue" + NL +
               "Press OK to EXIT",t)
  exit
end
'***      ***      ***      ***
v = av.GetActiveDoc
if (v.Is(View).Not) then exit end
dis = v.GetDisplay
gl = v.GetGraphics
gl.UnselectAll
symba = TextSymbol.Make
timesBI = Font.Make("Times New Roman","Bold Italic")
symba.SetFont(timesBI)
symba.SetSize(256)
pt = 0@0
t = GraphicText.Make(t,pt)
t.SetSymbols({symba})
t.SetDisplay(dis)
dis.BeginClip
gl.Add(t)
dis.EndClip
dis.ZoomToRect(Rect.Make(0@0,4000@1000))
dis.Invalidate(true)
```

# ◆ *GraphicsButtons*

Reads the data section at the end of the script and draws a raised button as a graphic in the current view.

**Topics**: Graphic, Color, Polygon

**Search Keys**: Graphic, Color, Polygon, GetGraphics, ColorMap, Make, Graphic-Shape

**Requirements**: Script will not run properly unless the script name is referenced by the variable *t*. Next, a view must exist in the project with the view name referenced by the *vname* variable.

```
' Variable Initialization
t = "GraphicsButtons"
vName = "Buttons"
'***Critical Resource Test***
if ((av.GetProject.FindDoc(t).is(SEd)) = false) then
  MsgBox.Error("Critical Resource 1 NOT Available" + NL +
               "Program Cannot Continue" + NL +
               "Press OK to EXIT",t)
  exit
end
if ((av.GetProject.FindDoc(vname).is(View)) = false) then
  MsgBox.Error("Critical Resource 2 NOT Available" + NL +
               "Program Cannot Continue" + NL +
               "Press OK to EXIT",t)
  exit
end
'***     ***     ***     ***
start = "'Start: Data"
```

```
fin   = "'End: Data"
flag = false
size = 0
 p = av.GetProject
 d = p.FindDoc(t)
if ((d.Is(SEd).Not) or (d = nil)) then exit end
d.GetWin.Open
 v = p.FindDoc(vName)
if ((v.Is(View).Not) or (v = nil)) then exit end
 colorWheel = ColorMap.MakeRandom(255)
grayWheel  = ColorMap.MakeGrayScale(255)
buttonSides = ColorMap.MakeGrayScale(32)
 s = d.GetSource.asTokens(nl)
for each i in (0 .. (s.Count - 1))
  x = s.Get(i)
  if (x = start) then
    flag = true
    continue
  end
  if (x = fin) then
    flag = false
    continue
  end
  if (flag) then
    data = x.Right(x.Count - 1).asTokens(",")
    if (data.Get(0) = "rect") then
      x = data.Get(1).asNumber
      y = data.Get(2).asNumber
```

```
    h = data.Get(3).asNumber
    w = data.Get(4).asNumber
    shp = Rect.Make(x@y,h@w)
  elseif (data.Get(0) = "point") then
    x = data.Get(1).asNumber
    y = data.Get(2).asNumber
    shp = x@y
  elseif ((data.Get(0) = "polyline") or (data.Get(0) = "polygon"))
then
    listOfLists = List.Make
    n = data.Get(1).asNumber
    pos = 2
    for each i in (1 .. n)
      m = data.Get(pos).asNumber
      tmpList = List.Make
      for each j in (1 .. (2*m)) by 2
        x = data.Get(pos+j).asNumber
        y = data.Get(pos+j+1).asNumber
        tmpList.Add(x@y)
      end
      listOfLists.Add(tmpList)
      pos = pos + (2*m + 1)
    end
    if (data.Get(0) = "polyline") then
      shp = PolyLine.Make(listOfLists)
    elseif (data.Get(0) = "polygon") then
      shp = Polygon.Make(listOfLists)
    end
```

```
    end
    g = GraphicShape.Make(shp)
    g.SetSelected(true)
    symb = g.GetSymbol
    symb.SetOutlined(TRUE)
    symb.SetOlWidth(2)
    symb.SetStyle(#RASTERFILL_STYLE_OPAQUESTIPPLE)
    v.GetGraphics.Add(g)
  end
end
v.GetGraphics.GroupSelected
gl = v.GetGraphics.get(0)
gl.SetSelected(false)
 symba = gl.GetGraphics.Get(0).GetSymbol
symba.SetOlColor(grayWheel.Get(Number.MakeRandom(0,255)))
symba.SetBgColor(colorWheel.Get(Number.MakeRandom(0,255)))
for each i in (1 .. 4)
  symba = gl.GetGraphics.Get(i).GetSymbol
  symba.SetOlColor(buttonSides.Get(4*i -1 + 8))
  symba.SetBgColor(buttonSides.Get(4*i -1 + 8))
end
v.GetDisplay.Invalidate(true)
v.GetDisplay.Flush
 'Start: Data
'rect,0,0,1,9
'polygon,1,4,0,0,.1,.1,.1,8.9,0,9
'polygon,1,4,0,0,.1,.1,.9,.1,1,0
'polygon,1,4,1,0,.9,.1,.9,8.9,1,9
```

```
'polygon,1,4,1,9,.9,8.9,.1,8.9,0,9
'End: Data
```

## ◆ *PrettyButtons*

Draws an array of raised buttons in a quilt-like pattern. The description of the buttons is in the data section of this script. After running the script, go to the Buttons view and zoom to the extent to see the buttons.

**Topics**: Graphic, ColorMap

**Search Keys**: Graphic, ColorMap, GraphicShape, Symbol

**Requirements**: Script will not run properly unless the script name is referenced by the *t* variable. Next, the view must exist in the project with the view name referenced by the *vname* variable.

```
' Variable Initialization
t = "PrettyButtons"
vName = "Buttons"
' ***Critical Resource Test***
if ((av.GetProject.FindDoc(t).is(SEd)) = false) then
  MsgBox.Error("Critical Resource 1 NOT Available" + NL +
              "Program Cannot Continue" + NL +
              "Press OK to EXIT",t)
  exit
end
if ((av.GetProject.FindDoc(vname).is(View)) = false) then
  MsgBox.Error("Critical Resource 2 NOT Available" + NL +
              "Program Cannot Continue" + NL +
              "Press OK to EXIT",t)
  exit
```

```
end
' ***      ***      ***        ***
start = "'Start: Data"
fin   = "'End: Data"
flag = false

ch = MsgBox.MultiInput("Parameters:",t,{"Size:",
"Highlighting:"},{"0","0.05"})
if (ch = nil) then
  MsgBox.Warning("No values entered: Exiting",t)
  exit
end
size = ch.Get(0).asNumber
butt = ch.Get(1).asNumber
p = av.GetProject
d = p.FindDoc(t)
if ((d.Is(SEd).Not) or (d = nil)) then exit end
d.GetWin.Open
v = p.FindDoc(vName)
if ((v.Is(View).Not) or (v = nil)) then exit end
colorWheel = ColorMap.MakeRandom(255)
grayWheel = ColorMap.MakeGrayScale(255)
buttonSides = ColorMap.MakeGrayScale(32)
s = d.GetSource.asTokens(nl)
for each i in (0 .. (s.Count - 1))
  x = s.Get(i)
  if (x = start) then
    flag = true
    continue
```

```
  end
  if (x = fin) then
    flag = false
    continue
  end
  if (flag) then
    data = x.Right(x.Count - 1).asTokens(",")
    if (data.Get(0) = "rect") then
      x = data.Get(1).asNumber
      y = data.Get(2).asNumber
      h = data.Get(3).asNumber
      w = data.Get(4).asNumber
      shp = Rect.Make(x@y,h@w)
    elseif (data.Get(0) = "point") then
      x = data.Get(1).asNumber
      y = data.Get(2).asNumber
      shp = x@y
    elseif ((data.Get(0) = "polyline") or (data.Get(0) = "polygon"))
then
        listOfLists = List.Make
        n = data.Get(1).asNumber
        pos = 2
        for each i in (1 .. n)
          m = data.Get(pos).asNumber
          tmpList = List.Make
          for each j in (1 .. (2*m)) by 2
            x = data.Get(pos+j).asNumber
            y = data.Get(pos+j+1).asNumber
```

```
          tmpList.Add(x@y)
        end
        listOfLists.Add(tmpList)
        pos = pos + (2*m + 1)
      end
      if (data.Get(0) = "polyline") then
        shp = PolyLine.Make(listOfLists)
      elseif (data.Get(0) = "polygon") then
        shp = Polygon.Make(listOfLists)
      end
    end
    g = GraphicShape.Make(shp)
    g.SetSelected(true)
    symb = g.GetSymbol
    symb.SetOutlined(true)
    symb.SetOlWidth(2)
    symb.SetStyle(#RASTERFILL_STYLE_OPAQUESTIPPLE)
    v.GetDisplay.BeginClip
    v.GetGraphics.Add(g)
    v.GetDisplay.EndClip
  end
end
v.GetGraphics.GroupSelected
g1 = v.GetGraphics.get(0)
g1.SetSelected(false)
bg = Rect.Make((-1)@(-1),((size+1)*10 + 1)@((size+1)*10 + 1))
g = GraphicShape.Make(bg)
symb = g.GetSymbol
```

```
symb.SetOutlined(TRUE)
symb.SetOlWidth(2)
symb.SetStyle(#RASTERFILL_STYLE_OPAQUESTIPPLE)
symb.SetOlColor(Color.GetBlack)
symb.SetBgColor(Color.GetGray)
v.GetGraphics.Add(g)
for each i in (0 .. size)
  for each j in (0 .. size)
    glnew = gl.clone
    glnew.SetOrigin((10*i)@(10*j))
    gLst = glnew.GetGraphics
    for each k in (0 .. (gLst.Count - 1))
      g = gLst.Get(k)
      symba = g.GetSymbol
      symba.SetOlColor(grayWheel.Get(Number.MakeRandom(0,255)))
      symba.SetBgColor(colorWheel.Get(Number.MakeRandom(0,255)))
      r = g.GetBounds
      o = r.ReturnOrigin
      h = r.GetHeight
      w = r.GetWidth
' Bottom
      p1 = o
      p2 = p1 + ((butt) @ (butt))
      p3 = p2 + (0@(h-(2*butt)))
      p4 = p3 + (-(butt) @ (butt))
      pList = {{p1,p2,p3,p4}}
      p = Polygon.Make(pList)
      gs = GraphicShape.Make(p)
```

```
glnew.Add(gs)

symba = gs.GetSymbol

symba.SetOutlined(true)

symba.SetStyle(#RASTERFILL_STYLE_OPAQUESTIPPLE)

symba.SetOlColor(Color.GetBlack)

symba.SetBgColor(buttonSides.Get(10))
```

' Right

```
p1 = o + (w@0)

p2 = p1 + ((-(butt)) @ (butt))

p3 = p2 + (((-w)+(2*butt))@ 0)

p4 = p3 + (-(butt) @ (-(butt)))

pList = {{p1,p2,p3,p4}}

p = Polygon.Make(pList)

gs = GraphicShape.Make(p)

glnew.Add(gs)

symba = gs.GetSymbol

symba.SetStyle(#RASTERFILL_STYLE_OPAQUESTIPPLE)

symba.SetOutlined(true)

symba.SetOlColor(Color.GetBlack)

symba.SetBgColor(buttonSides.Get(14))
```

' Top

```
p1 = o + (w@h)

p2 = p1 + ((-(butt)) @ (-(butt)))

p3 = p2 + (0@((-h)+(2*butt)))

p4 = p3 + ((butt) @ (-(butt)))

pList = {{p1,p2,p3,p4}}

p = Polygon.Make(pList)

gs = GraphicShape.Make(p)
```

```
        glnew.Add(gs)
        symba = gs.GetSymbol
        symba.SetStyle(#RASTERFILL_STYLE_OPAQUESTIPPLE)
        symba.SetOutlined(true)
        symba.SetOlColor(Color.GetBlack)
        symba.SetBgColor(buttonSides.Get(18))
' Left
        p1 = o + (0@h)
        p2 = p1 + ((butt) @ (-(butt)))
        p3 = p2 + (((w-(2*butt)))@0)
        p4 = p3 + ((butt) @ (butt))
        pList = {{p1,p2,p3,p4}}
        p = Polygon.Make(pList)
        gs = GraphicShape.Make(p)
        glnew.Add(gs)
        symba = gs.GetSymbol
        symba.SetStyle(#RASTERFILL_STYLE_OPAQUESTIPPLE)
        symba.SetOutlined(true)
        symba.SetOlColor(Color.GetBlack)
        symba.SetBgColor(buttonSides.Get(22))
    end
    v.GetDisplay.BeginClip
    v.GetGraphics.Add(glnew)
    v.GetDisplay.EndClip
  end
end
v.GetGraphics.Remove(0)
v.GetDisplay.Invalidate(true)
```

```
' Start: Data
' rect,0,0,1,9
' rect,1,1,1,7
' rect,2,2,1,5
' rect,3,3,1,3
' rect,4,3,1,3
' rect,5,3,1,3
' rect,6,2,1,5
' rect,7,1,1,7
' rect,8,0,1,9
' rect,1,0,7,1
' rect,2,1,5,1
' rect,3,2,3,1
' rect,3,6,3,1
' rect,2,7,5,1
' rect,1,8,7,1
' End: Data
```

# Message Boxes

## ◆ *MessageBoxFlowControl*

The YesNo, YesNoCancel, and MiniYesNo dialogs make it possible for the user to control the flow of a process. In general, pressing Yes returns true, pressing No returns false, and pressing Cancel returns nil to the dialog. Nothing is less exciting than putting a MsgBox.Info statement into a loop, and then realizing the loop has 1,000 iterations and you have no way to stop it, except to turn off your machine. The MiniYesNo dialog can be used in a simple and efficient way to break out of loops.

**Topics**: MsgBox

**Search Keys**: MsgBox, YesNo, YesNoCancel, MiniYesNo

```
' Variable Initialization
t = "MessageBoxFlowControl"
' Again loop forever.
while (true)
    ch = MsgBox.YesNo("Do you want to continue?",t,true)
    if (ch = true) then
      MsgBox.Info("You pressed yes",t)
    else
      MsgBox.Info("You pressed no",t)
    end
    ch = MsgBox.YesNoCancel("Do you want to continue?",t,true)
    if (ch = true) then
      MsgBox.Info("You pressed yes",t)
    elseif (ch = false) then
```

```
    MsgBox.Warning("You pressed no",t)
  elseif (ch = nil) then
    MsgBox.Error("You pressed cancel",t)
  end
  ch = MsgBox.MiniYesNo("Do you want to continue?",true)
  if (ch = true) then
    MsgBox.Info("You pressed yes",t)
  else
    MsgBox.Info("You pressed no",t)
  end
  if (MsgBox.MiniYesNo("Do you want to quit?",true)) then
    break
  end
end
```

## ◆ *MessageBoxInput*

Exercises all the dialogs for using MsgBox to get data or attribute values from a user. MsgBox.Input returns a single value—either a string or nil—and Msg-Box.MultiInput always returns a list of strings, but it may be empty. Testing for nil or empty lists determines whether the dialog was canceled.

**Topics**: MsgBox

**Search Keys**: MsgBox, Input, MultiInput

```
' Variable Initialization
t = "MessageBoxInput"
' Loop forever.
while (true)
    ch = MsgBox.Input("Enter your name:",t,"John")
```

```
    if (ch = nil) then

       MsgBox.Warning("Cancel Pressed: No Data Entered","Canceling
the Dialog")

       exit

    else

       MsgBox.Info("Your Name Is" ++ ch,t)

    end

    ch = MsgBox.MultiInput("Enter your Name/Phone/
Email:",t,{"Name","Phone","Email"},{"Bill","415-555-1212",
"president@whitehouse.gov"})

    if (ch.Count = 0) then

       MsgBox.Warning("Cancel Pressed: No Data Entered","Canceling
the Dialog")

       exit

    else

       MsgBox.Report("Your Name is" ++ ch.Get(0) + nl + "Your Phone
is" ++ ch.Get(1) + nl + "and" + nl + "Your Email is" ++ ch.Get(2),t)

    end

end
```

## ◆ *MessageBoxReport*

MsgBox.Report is not a frequently used dialog, but it is very helpful for examining text representations of graphics objects that do not fit in a smaller dialog, such as polylines. With some work this script can be used to present formatted reports, but the fonts may cause problems.

**Topic**: MsgBox

**Search Keys**: MsgBox, Report

```
' Variable Initialization

t = "MessageBoxReport"
```

```
' First, a point and a rectangle in a report.
pt = 1@1
r  = Rect.Make(0@0,2@2)
MsgBox.Report(pt.asString + nl + r.asString,t)
' Now a polyline. A polyline is made from a list of points.
' First, you make a list of points. Note that as a string, this
' polyline is reported to have one segment of seven parts.
first = {0@0, 1@(-1), 2@0, 3@1, 4@0, 5@(-1), 6@0}
pline1 = Polyline.Make({first})
MsgBox.Report(pline1.asString,t)
' As a string, pline2 is reported to have two segments, each with
' seven points.
second = {0@1, 1@0, 2@(-1), 3@0, 4@1, 5@0, 6@(-1)}
pline2 = PolyLine.Make({first,second})
MsgBox.Report(pline2.asString,t)
' Sources for SEds or scripts can also be displayed in
' a MsgBox.Report. Highlight some text and the next
' section will display it in a MsgBox.Report.
aSEd = av.GetActiveDoc
if (aSEd.is(SEd).not) then
    MsgBox.Warning("The active document" ++ aSEd.GetName ++ "is not
an SEd",t)
    exit
end
MsgBox.Report(aSEd.GetSelected,t)
' MsgBox.Report can be used for very simple text reports. There are
' no tabs or other facilities to line up text
' and the font is usually proportionally spaced.
```

```
' Consequently, reasonable tables are a challenge.

oneList = {1,2,3,4}

twoList = {"John","Susan","Helen","Mary"}

threeList = {1.00, 2.25, 5.58, 7.23}

rpt = ""

newline = ""

for each i in 0 .. 3

    rpt = rpt + newline + oneList.Get(i).asString + tab +
twoList.Get(i) + tab + threeList.Get(i).asString

    newline = nl

end

MsgBox.Report(rpt,t)
```

## ◆ *MessageBoxSelection*

Selection Message Boxes come in six flavors: MsgBox.List, MsgBox.ListasString, MsgBox.Choice, MsgBox.ChoiceasString, MsgBox.MultiList, and MsgBox.Multi-ListAsString. The difference between list and choice is in the way the selection is formatted. List has a scrolling box whereas choice has a drop-down box. MultiList is like list except that multiple choices can be made from the list and a list is the returned object type. For list, multilist, and choice, the lists are created by sending the GetName request to every object in the list. This does not work for strings and numbers, which return an empty string for their names. The ListasString, Multi-ListAsString, and ChoiceasString send the asString request to every object in the list. The return value from these selections is not usually a string (unless the class of the item in the list is a string), but rather the object itself. Thus, the selection from any of these dialogs can and will be views, themes, tables, and so forth.

**Topic**: MsgBox

**Search Keys**: MsgBox, List, ListasString, Choice, ChoiceasString, MultiList, Multi-ListAsString

```
' Variable Initialization
```

```
t = "MessageBoxSelection"
testList = {"ESRI", "Avenue", 5, 6@7, av.GetActiveDoc,
av.GetProject, {1, "one", 0@0}}
```

' This is an eclectic list of two strings, and a number, point, Doc,
' project, and another list. Note that when a string is sent the
' GetName request, its position in the list of choices is blank. Note
' also that the final element of testList, a short list of its own,
' never appears. It answers both GetName and asString with an empty
' string. Note that it can still be selected by choosing the blank
' area where it could be displayed.

```
while (true)
```

' Loop forever, or until you cancel and invoke the break.

' MsgBox.List

```
    ch = MsgBox.List(testList,"List:Select an item:",t)
    if (MsgBox.YesNo(ch.GetClass.GetClassName,"Do you want to
continue?",True).Not) then
      break
    end
    ch = MsgBox.ListasString(testList,"ListasString:Select an
item:",t)
    if (MsgBox.YesNo(ch.GetClass.GetClassName,"Do you want to
continue?",True).Not) then
      break
    end
    ch = MsgBox.Choice(testList,"Choice:Select an item:",t)
    if (MsgBox.YesNo(ch.GetClass.GetClassName,"Do you want to
continue?",True).Not) then
      break
    end
    ch = MsgBox.ChoiceasString(testList,"ChoiceasString:Select an
```

```
item:",t)

    if (MsgBox.YesNo(ch.GetClass.GetClassName,"Do you want to
continue?",True).Not) then

        break

    end

    ch = MsgBox.MultiList(testList,"MultiList:Select Multiple
Items:",t)

    if (MsgBox.YesNo(ch.GetClass.GetClassName,"Do you want to
continue?",True).Not) then

        break

    end

    ch = MsgBox.MultiListasString(testList,
"MultiListAsString:Select Multiple Items:",t)

    if (MsgBox.YesNo(ch.GetClass.GetClassName,"Do you want to
continue?",True).Not) then

        break

    end


end
```

# Miscellaneous

## ◆ *Clone*

Demonstrates how and when to use the clone request when working with objects. Variables in Avenue are pointers to objects, such as places on a city map. The variable is the map, and the actual data objects are real places in a city. When you make a copy of the variable, it is very important to understand if you copied the map or the city. These are certainly two very different operations.

**Topics**: List, Clone, DeepClone

**Search Keys**: List, Clone, DeepClone

**Comments**: This is a simple script, but it is important to understand. When you make copies of objects, you must know what you copied—the real object or a reference to the real object. If you want to copy GUI elements, you must be careful to clone the items before you install them on the new GUI. Otherwise, you may have installed a new reference to the same object, and a deletion or change may have unintended ramifications. Consult Help for more information about Clones and DeepClones.

```
' Variable Initialization
t = "Clone"
five = 5
six = 6
seven = 7
john = "john"
helen = "helen"
mary = "mary"
susan = "susan"
family = {john, helen, mary, susan}
list1 = {av.GetActiveDoc, {1, 2}, five, family}
list2 = list1
list3 = list1.Clone
list4 = list1.DeepClone
' How many elements are in the fourth item in this list?
MsgBox.Info(list1.Get(3).Count.asString,"list1")
MsgBox.Info(list2.Get(3).Count.asString,"list2")
MsgBox.Info(list3.Get(3).Count.asString,"list3")
```

```
MsgBox.Info(list4.Get(3).Count.asString,"list4")
' Add one more person to the family.
family.Add("Ruth")
' Now how many items are in the fourth position in each list?
MsgBox.Info(list1.Get(3).Count.asString,"list1")
MsgBox.Info(list2.Get(3).Count.asString,"list2")
MsgBox.Info(list3.Get(3).Count.asString,"list3")
MsgBox.Info(list4.Get(3).Count.asString,"list4")
' Delete the first item in list1.
list1.Remove(0)
' What is the class type of each item in all of the lists?
MsgBox.Info(list1.Get(0).GetClass.GetClassName,"list1")
MsgBox.Info(list2.Get(0).GetClass.GetClassName,"list2")
MsgBox.Info(list3.Get(0).GetClass.GetClassName,"list3")
MsgBox.Info(list4.Get(0).GetClass.GetClassName,"list4")
' After the previous manipulations, is the second
' item of list1 the same as the first item of list3 or list4?
if (list1.Get(0) = list3.Get(1)) then
   MsgBox.Info("Items are the same!","List1/List3")
   if (list1 = list3) then
  IST =  MsgBox.Info("The Lists are the same!!!","List1/List3")
   else
 IST =  MsgBox.Info("These Lists are different
Objects!!?!!","List1/List3")
   end
else
   MsgBox.Warning("Not the same!","List1/List3")
end
```

```
if (list1.Get(0) = list4.Get(1)) then
   MsgBox.Info("Items are the same!","List1/List4")
else
   MsgBox.Warning("Not the same!","List1/List4")
end
if (list3.Get(1) = list4.Get(1)) then
   MsgBox.Info("Items are the same!","List3/List4")
else
   MsgBox.Warning("Not the same!","List3/List4")
end
' Finally, delete something from list3 and note the impact on list1.
list3.Remove(1)
MsgBox.Info(list1.Count.asString,"Size of List1")
for each x in list1
   MsgBox.Info("List1:" ++ x.GetClass.GetClassName,"List1")
end
MsgBox.Info(list3.Count.asString,"Size of List3")
for each x in list3
   MsgBox.Info("List3:" ++ x.GetClass.GetClassName,"List3")
end
```

## ◆ *ColorDict*

Builds a color dictionary with the nine basic ArcView colors. The user selects a color from the list, and a color object is returned. If SELF is nil or does not contain a list of two elements, the title and message for the message box will be *COLOR* and *Pick a Color*, respectively. If SELF is a list of two elements, the first element of the list will be used as the message, and the second element of the list will be used as the title for the message box.

**Topics**: Dictionary, Colors

**Search Keys**: Dictionary.Make, Dictionary.Add, Dictionary.ReturnKeys, Dictionary.Get, Color(GetBlack)

```
' Variable Initialization
t = "ColorDict"
if (SELF = nil) then
    theMsg = "Pick a Color"
    theTitle = "COLOR"
else
  if (SELF.Count <> 2) then
    theMsg = "Pick a Color"
    theTitle = "COLOR"
  else
    theMsg = SELF.Get(0)
    theTitle = SELF.Get(1)
  end
end
' Set up dictionary.
colorDict = Dictionary.Make(9)
colorDict.Add("Black",color.getBlack)
colorDict.Add("Blue",color.getBlue)
colorDict.Add("Cyan",color.getCyan)
colorDict.Add("Gray",color.getGray)
colorDict.Add("Green",color.getGreen)
colorDict.Add("Magenta",color.getMagenta)
colorDict.Add("Red",color.getRed)
colorDict.Add("White",color.getWhite)
colorDict.Add("Yellow",color.getYellow)
```

```
' Get list of dictionary keys.
theColorList = colorDict.ReturnKeys
' Prompt user for selection.
theChoice = MsgBox.ChoiceAsString(theColorList,theMsg,theTitle)
if (theChoice = NIL) then
  theColor = nil
else
  theColor = colorDict.Get(theChoice)
end
return theColor
```

## ◆ *Dictionary*

Shows the functionality of a name dictionary and a dictionary. The script shows creating the dictionaries, populating the dictionaries, and accessing entries in the dictionaries.

**Topics**: Dictionary

**Search Keys**: Dictionary.Add, Dictionary.Get

```
' Variable Initialization
t = "Dictionary"
theChoice = MsgBox.ListAsString(
      {"Dictionary","Name Dictionary"},
      "Select Type of Dictionary",
       "Dictionary Sample")
if (theChoice = nil) then exit end
if (theChoice = "Name Dictionary") then
' Get list of document objects in this project.
   theList = av.GetProject.GetDocs
```

```
' Create a name dictionary with the number of document objects in
' this project.
   theNameDict = NameDictionary.Make(theList.Count)
' Populate the name dictionary with the document objects.
   for each i in theList
      theNameDict.Add(i)
   end
' Obtain a list of the names of the document objects.
      theNewList = theNameDict.ReturnKeys
   ' Let user select a document object from the list by name.
      theKey = msgbox.ListAsString(theNewList,
                  "Select Document Name",
                  "Name Dictionary")
   ' Report back to the user the name
   ' of the object and its class name.
      if (theKey = nil) then exit end
      msgbox.info(theKey ++ "is a" ++
                  theNameDict.Get(theKey).getclass.getclassname,
                  "Name Dictionary: Key and Object Class")
   ' **********
   ' Another way of executing this process is to take advantage of the
   ' List request for a MsgBox object as follows:
   ' theList = av.GetProject.GetDocs
   ' theKey = MsgBox.List(theList,
   '                        "SelectDocumentName",
   '                        "Name Dictionary")
   ' Report back to the user the name of the object and its class
   ' name.
```

```
' if (theKey = nil) then exit end
' msgbox.info(theKey ++ "is a" ++
'              theKey.getclass.getclassname,
'              "Name Dictionary: Key and Object Class")
' **********
else
' Create dictionary.
   SampleDict = Dictionary.Make(10)
' Add entries to dictionary.
   SampleDict.Add("String",
              "This is a Text String")
   SampleDict.Add("Number",
              100)
   SampleDict.Add(1000,
              "1000")
   SampleDict.Add("Boolean-True",
              True)
   SampleDict.Add(False,
              "Boolean-False")
   SampleDict.Add("List of Strings",
              {"String 1","String 2","String 3"})
   SampleDict.Add("List of Numbers",
              {10,20,30})
   SampleDict.Add("Mixed List",
              {"String 1",10,"String 2",20})
   SampleDict.Add("List of Objects",
              {av.GetProject.GetDocs})
   theList = SampleDict.ReturnKeys
```

```
        theChoice = MsgBox.ListAsString(theList,
                        "Select Key Value",
                        "Dictionary")
    if (theChoice = nil) then exit end
    if (SampleDict.Get(theChoice).getClass.getClassname <> "List")
then
  ' Report back to the user the key and the value from the
  ' dictionary.
        MsgBox.Info("KEY:" ++ theChoice.asstring ++
                "CLASS:" ++ theChoice.GetClass.GetClassName + nl +
                "VALUE:" ++ SampleDict.Get(theChoice).asstring ++
                 "CLASS:" ++ SampleDict.Get(theChoice)
.GetClass.GetClassName,
                    "Dictionary: Key and Value")
    else
      if (theChoice <> "List of Objects") then
        MsgBox.ListAsString(SampleDict.Get(theChoice),
              "KEY:" ++ theChoice.asstring,
              "Value is List")
      else
  ' Create a string of the names of the objects in the list.
        for each i in SampleDict.Get(theChoice)
          MsgBox.ListAsString(i,
              "KEY:" ++ theChoice.asstring,
              "Value is List of Document Objects in the Project")
        end
      end
    end
  end
```

## ◆ *IconGet*

Gets an icon from the IconMgr.

**Topics**: Icon

**Search Keys**: Icon

**Requirements**: Script is called from another script, and returns the name of the icon as it appears in the IconMgr.GetIcons list. For examples of a calling program, see *ButtonAdd*.

```
' Variable Initialization
t = "IconGet"
parm = SELF
iconDict = Dictionary.Make(1)
for each i in IconMgr.GetIcons
  iconDict.Add(i.GetName,i)
end
if (parm = nil) then
  ikahn = MsgBox.ListasString(iconDict.ReturnKeys,"Choose an
Icon:",t)
  if (ikahn = nil) then
    MsgBox.Warning("No Icon Selected: Setting to Empty",t)
    return iconDict.Get("Empty")
  end
else
  ikahn = iconDict.Get(parm)
  if (ikahn = nil) then
   MsgBox.Warning("No Icon available with name" ++ parm + ": Setting
to Empty",t)
    return iconDict.Get("Icon9")
  end
```

```
end

return   ikahn
```

# ◆ *LabelButtons*

Shows the current values for LabelButtons. The block of code commented out at the end sets new values for the click and update events. New values have not been set in this script. Using the values in the commented block merely sets the values to the defaults used by ArcView.

**Topics**: LabelButtons

**Search Keys**: LabelButtons, Update Events

```
' Variable Initialization

t = "LabelButtons"

p = av.GetProject

if (p = nil) then exit end

bList = p.GetButtons

for each b in bList

    MsgBox.Info("Label:" ++ b.GetLabel + NL +

                "Click:" ++ b.GetClick + NL +

                "Update:" ++   b.GetUpdate,"Project" ++ t)

end

' lButton = LabelButton.Make

' lButton.SetLabel("&Open")

' lButton.SetUpdate("Project.Button2Update")

' lButton.SetClick("Project.Button2")

' bList.Set(1,lButton)
```

# ◆ *ReportCoordinates*

Returns the coordinates of points in the view where the mouse is clicked.

**Topics**: View, Tool

**Search Keys**: View, Tool, DocGUI, Install, Apply

**Requirements**: Must be installed as a tool.

```
' ***The following comments make up the install
' script for this tool. Specify the DocGUI in the first line.***
' parm = av.FindGUI("View").GetToolBar
' index = parm.GetControls.Count
' MsgBox.Info(index.asString,"")
' Making and setting attributes for the new button
' to add to the button bar.
' h = Tool.Make
' h.SetApply("ReportCoordinates")
' h.SetHelp("Report Mouse Coordinates")
' ikahn = IconMgr.Show(0)
' h.SetIcon(ikahn)
' parm.Add(h,index)
' End install script for this tool.


' Variable Initialization
t = "ReportCoordinates"
v = av.GetActiveDoc
if (v.Is(View).Not) then
   MsgBox.Warning("Active document," ++ v.GetName ++ "is not a
View",t)
   exit
```

```
end
pt = v.GetDisplay.ReturnUserPoint
if (pt = nil) then exit end
MsgBox.Report(pt.AsString,t)
```

## ◆ *StackTest*

Exercises the stack class and its requests. Stacks are useful data structures for storing last-in-first-out (LIFO) sequences of elements. Beginning with ArcView 2.1, it became possible to use a stack to create a navigation tool that remembers the path a user took through a set of projects.

**Topics**: Stack

**Search Keys**: Stack, Push, Pop

```
' Variable Initialization
t = "StackTest"
st = Stack.Make
st.Push(1)
st.Push("John")
st.Push("Susan")
st.Push(4)
st.Push("Mary")
st.Push("Helen")
stackDict = Dictionary.Make(st.Depth/4)
for each i in (0 .. (st.Depth - 1))
   stackDict.Add(st.Peek(i),i)
end
ch = MsgBox.ListasString(stackDict.ReturnKeys,"Pick one",t)
if (ch = nil) then
```

```
    MsgBox.Warning("No stack element picked",t)
    exit
end
i = stackDict.Get(ch)
for each j in (0 .. (i-1))
    st.Pop
end
stackDict = Dictionary.Make(st.Depth/4)
for each i in (0 .. (st.Depth - 1))
    stackDict.Add(st.Peek(i),i)
end
ch = MsgBox.ListasString(stackDict.ReturnKeys,"Pick one","")
while (st.Depth > 0)
    MsgBox.Info(st.Pop.asString,"")
end
MsgBox.Info(st.Depth.asString,t)
```

# Projects

## ◆ *ExtractScript*

Browses a selected project file and prompts the user with a list of the scripts in the project. The user can then select a single script that will be imported into the current project, compiled, and named.

**Topics**: Linefile.Make

**Search Keys**: Linefile.Make

```
' Variable Initialization
t = "ExtractScript"
n = ""
' Escape sequences.
escape = Dictionary.Make(2)
escape.Add("\\","\")
escape.Add("\"","""")
escape.Add("\n",nl)
' Pick a project file.
fname = FileDialog.Show ("*.apr","Project *.apr", "Pick a Project")
if (fname.Is(FileName).Not) then
   MsgBox.Error("No Project File Picked: Extract Stopped",t)
   exit
end
f = LineFile.Make(fname,#FILE_PERM_READ)
nameList = List.Make
srcDict  = Dictionary.Make(20)
flag = 0
while (f.IsAtEnd.Not)
   ln = f.ReadElt
' ***Read the source for the SEd.***
   if ((flag = 2) and (ln.Contains("Source:") or
ln.Contains("SourceCode:"))) then
      flag = 0
      j = ln.IndexOf("""")
      if (j = -1) then
        continue
      end
```

```
      source = ln.Right(ln.Count - j)
      while (true)
        ln = f.ReadElt
        if (ln = ")") then
          break
        elseif (ln = nil) then
          break
        else
          source = source + ln
        end
      end
      for each x in escape.ReturnKeys
        source = source.Substitute(x,escape.Get(x))
      end
      srcDict.Add(n,source.Middle(1,source.Count-2))
      continue
    end
' ***Get the name of the SEd.***
    if ((flag = 1) and ln.Contains("Name:")) then
      n = ln.asTokens("""").Get(1)
      n.Substitute("\\","\")
      flag = 2
      continue
    end
' ***Start of an SEd section.***
    if (ln.Contains("(SEd.") or ln.Contains("(Script.")) then
      flag = 1
      continue
    end
```

```
end
```

```
key = MsgBox.ListAsString(srcDict.ReturnKeys,"Pick a
Script",fname.asString)
```

```
if (key = nil) then exit end
```

```
source = srcDict.Get(key)
```

```
newSEd = SEd.Make
```

```
newSEd.SetName(n)
```

```
newSEd.SetSource(source)
```

```
newSEd.GetWin.Open
```

## ◆ *ProjectInit*

Opens a specific project and initializes it with a maximized specific view. The script initializes the state of the themes, and sets the default tool.

**Topics**: Project, View, Theme, Tools

**Search Keys**: System.GetEnvVar, Project.Open, View.SetTOC, Theme.SetVisible, Theme.SetActive, Theme.GetHotScriptName, AV.GetActiveGui

**Requirements**: The DOS environment variable *SAMPLEDATA* must be set to the location of the ArcView sample data location.

```
' Variable Initialization
t = "ProjectInit"
DOSEnvVar = "sampledata"
NewProj = "\world.apr"
SpecificView = "World Map"
' Get DOS environment variable (DOSEnvVar).
_AVData = System.GetEnvVar(DOSEnvVar)
if (_AVData = nil) then
  _AVData = av.Run("SetEnvironmentVariable",DOSEnvVar)
  if (_AVData = nil) then
```

```
    MsgBox.Error("DOS Environment Variable" ++ DOSEnvVar ++ "Not
Set" + nl +
                "Please Set Environment Variable in DOS and Restart
Windows",
                "DOS Environment Variable")

    exit
  end
end
' Verifies the existence of a specified project file, NewProj.
theFName = _AVData + NewProj
fileexists = File.Exists(theFName.asFilename)
if (Fileexists = false) then
   MsgBox.Error("Project not found!" + nl +theFName,theFName)
   exit
end
' Save the current project if modified since last save.
theProject = av.GetProject
if (nil <> theProject) then
   if (theProject.IsModified) then
      res = MsgBox.SaveChanges("Do you want to save changes to " +
theProject.GetName + "?", "ArcView", true)
      if (nil = res) then exit end
      if (res) then
        av.Run("Project.Save", nil)
        if (theProject.IsModified) then exit end
      end
   end
end
' Close the current project and open the new project.
```

```
  if (nil <> theFName) then
    if (nil <> theProject) then
      theProject.Close
    end
' Set the current working directory to the DOS environment
' variable so that relative paths can be used.
    _AVData.asFileName.SetCWD
    Project.Open(theFName.asFileName)
end
' New project is open.
' Initialize the new project environment.
' Find a specified view, shut down the size of the Table of Contents,
' make the TOC unresizable, open the view window,
' and maximize the view window.
theView = av.GetProject.FindDoc(SpecificView)
if (theView.Is(View).Not) then
  MsgBox.Error("View Name" ++ theView.asString ++ "is Invalid","")
  exit
end
theView.SetTOCWidth(0)
theView.SetTOCUnReSizable(true)
theView.GetWin.Open
theView.GetWin.Maximize
' Get a list of the themes in the view, and if theme has
' hot link properties set up, make theme active and visible.
theThemeList = theView.GetThemes
for each theTheme in theThemeList
  if (theTheme.is(FTheme) and (theTheme.GetHotScriptName <> ""))
```

```
then
      theTheme.SetVisible(TRUE)
      theTheme.SetActive(TRUE)
   else
      theTheme.setVisible(FALSE)
      theTheme.SetActive(FALSE)
   end
end
' If any of the themes have hot link property set up,
' set active tool to hot link tool.
hotlinkcnt = 0
theThemeList = theView.GetThemes
for each theTheme in theThemeList
   if (theTheme.is(FTheme) and (theTheme.GetHotScriptName <> ""))
then
      hotlinkcnt = (hotlinkcnt + 1)
   end
end
if (hotlinkcnt <> 0) then
   theToolList = av.GetActiveGUI.GetToolbar.GetControls
   for each x in theToolList
     if (x.GetApply = "View.Hotlink") then
       x.Select
     end
   end
end
' Make specific themes visible.
theThemeList = theView.GetThemes
for each theTheme in theThemeList
```

```
    if ((theTheme.GetName = "Countries ('94)") or (theTheme.GetName
= "Major Rivers") or (theTheme.GetName = "Major Lakes")) then
        theTheme.SetVisible(TRUE)
        else
            the Theme.setVisible(FALSE)
    end
end
```

# ◆ *ProjectSave*

Ensures against corruption of the active project. This script saves every project to two files: the normal *project.apr* and the other file specified in the tag for the project. (The tag is a string stored with the project.) The ObjectTag is an object stored with the project. Use the tag to store a name and a number; the number is incremented with each save. This procedure results in a family history of projects. Be sure to occasionally execute a cleanup of the family history.

**Topics**: Project, Project.Save

**Search Keys**: Project, Project.Save

**Requirements**: To override the system *Project.Save,* the script must be named *Project.Save.* Alternatively, the script can be installed with a different name for the click event of a save button.

**Comments**: The script should be installed as the click event for the diskette button on the button bar of all GUIs.

```
' Variable Initialization
t = "Project.Save"
theProject = av.GetProject
theFileName = theProject.GetFileName
backupFlag = 0
objTag = theProject.GetObjectTag
if (objTag = nil) then
    backupFlag = 0
```

```
end
if ((objTag.Is(List)) and (objTag.Count = 2)) then
   tag1 = objTag.Get(0)
   tag2 = objTag.Get(1)
   if (tag1.Is(String).Not) then
     backupFlag = 0
   elseif (tag2.Is(Number).Not or (tag2 > 999) or (tag2 < 0)) then
     backupFlag = 0
   else
     backupFlag = 1
   end
end
if (backupFlag = 1) then
   backupName = objTag.Get(0)
   backupNum  = objTag.Get(1)
   backupNum = backupNum + 1
   theProject.SetObjectTag({backupName,backupNum})
   bakFileName = (backupName + backupNum.asString).asFileName
   theProject.SetFileName(bakFileName)
   theProject.Save
end
theProject.SetFileName(theFileName)
theProject.Save
av.ShowMsg("Project saved to '" + theFileName.GetBaseName + "'")
```

## ◆ *ProjectSaveReset*

Resets the counter to zero (0) in the current project.

**Topics**: ProjectSave, System Script

**Search Keys**: Project, Save, System Script

**Requirements**: The object tag for the project must be initialized with a backup name and counter value.

```
' Variable Initialization
t = "ProjectSaveReset"
objTag = av.GetProject.GetObjectTag
if (objTag = nil) then
' Initialize the object tag.
' The save program assumes that the stem for the backup name
' is only 5 characters long. It allows backups to include
' a suffix of up to 999 (e.g., "hmpbk999.apr").
  objTag = {"jpabk",0}
  av.GetProject.SetObjectTag(objTag)
  exit
end
if ((objTag.Is(List)) and (objTag.Count = 2)) then
  tag1 = objTag.Get(0)
  if (tag1.Is(String).Not) then
    exit
  end
  tag2 = objTag.Get(1)
  if (tag2.Is(Number).Not or (tag2 > 999) or (tag2 < 0)) then
    exit
  end
  av.GetProject.SetObjectTag({tag1,0})
end
```

# SEds

## ◆ *AddComment*

Adds a comment (or any text string) to an SEd at the last cursor insertion point. This script should be called from a button on the script button bar.

**Topics**: Script, SEd, Insert

**Search Keys**: Script, SEd, Insert

```
' Variable Initialization
t = "AddComment"
theString = "' Checked: 3.0 " + date.Now.asstring
theScript = av.GetActiveDoc
if (theScript.Is(SEd).Not) then exit end
theScript.Insert (theString)
```

## ◆ *AddScriptToEMail*

Adds an SEd to the end of the ScriptExportAsEMail message. The user is presented with a list of SEds. When an SEd is picked, it is formatted and added to the bottom of the email message. Consult *ScriptExportAsEMail* for more information. The email message is actually a script that installs the mailed scripts.

**Topics**: Script, SEd

**Search Keys**: EMail, Script, SEd, Package

```
' Variable Initialization
t = "AddScriptToEMail"
defaultEmailMessage = "EMail.004"
pro = av.GetProject
```

```
actDoc = av.GetActiveDoc
docList = pro.GetDocs
sedList = List.Make
for each d in docList
    if (d.Is(SEd)) then
      sedList.Add(d)
    end
end
' Ask for the name of the e-mail message.
email = MsgBox.Input("EMail Message: ",t,defaultEMailMessage)
if (email = nil) then
    MsgBox.Warning("No Name Specified: Process Will Stop",t)
    exit
end
d = pro.FindDoc(email)
if (d = nil) then
    MsgBox.Warning("Specified EMail Message" ++ email ++ "Does Not
Exist",t)
    exit
end
source = d.GetSource
' If the active document is an SEd, ask if it should be added.
if (actDoc.Is(SEd)) then
    ch = MsgBox.YesNo("Add the active document:" ++
actDoc.GetName,t,true)
    if (ch = true) then
      newSource = actDoc.GetSource
      tmpSource = "'Start:" ++ actDoc.GetName + nl
      for each x in newSource.asTokens(nl)
```

```
            tmpSource = tmpSource + "'" + x + nl
      end
      tmpSource = tmpSource + "'Finish:" ++ actDoc.GetName + nl
      source = source + tmpSource
    end
end
' Use the multi-select list box and
' add SEds at the end of the email message.
ch = MsgBox.MultiList(sedList,"Choose an SEd: ",t)
if (ch = nil) then exit end
for each d in ch
  newSource = d.GetSource
  tmpSource = "'Start:" + d.GetName + nl
  for each x in newSource.asTokens(nl)
    tmpSource = tmpSource + "'" + x + nl
  end
  tmpSource = tmpSource + "'Finish:" ++ d.GetName + nl
  source = source + tmpSource
end
d.SetSource(source)
```

# ◆ *ConvertScript*

Converts a script extracted from a project file into an SEd. Scripts are stored in project files as C-style strings. This means that special escape sequences handle newlines, quotes, and backslashes. This script handles most of the conversion required to obtain a compilable script.

**Topics**: SEd, Project

**Search Keys**: SEd, Project, Source, Compile, asTokens, SetInsertPos, Insert, Search

**Requirements**: The active document can be an SEd or the SEd can be passed to the script by a calling program.

```
' Start Here
' Variable Initialization
t = "ConvertScript"
newline = "\n"
quote    = "\""
qu       = """"
bslash   = "\\"
bl       = "\"
if (SELF.Is(SEd)) then
    d = SELF
else
    d = av.GetActiveDoc
    if (d.Is(SEd).Not) then
      MsgBox.Error("Active Document" ++ d.GetName ++ "is not an
SEd",t)
       exit
    end
end
' Convert the script installed from a project to a normal
' script format.
' This involves changing some of the character strings.
' Newlines
d.SetInsertPos(0)
while (d.Search(newline))
    d.Insert(nl)
end
' Quotes
```

```
d.SetInsertPos(0)
while (d.Search(quote))
    d.Insert(qu)
end
' Backslashes
d.SetInsertPos(0)
while (d.Search(bslash))
    d.Insert(bl)
end
```

## ◆ *DeleteEmbeddedScripts*

Deletes embedded scripts by collecting a list of embedded scripts and asking the user which ones to delete. Uses MultiListAsString MsgBox to collect scripts to delete.

**Topics**: Embedded Script, Dictionary

**Search Keys**: Dictionary, Embedded Script, SEd, GetScripts

```
' Start Here
' Variable Initialization
t = "DeleteEmbeddedScripts"
' Get the project and create a list of
' embedded scripts matching this pattern.
pro = av.GetProject
embedDict = pro.GetScripts
if (embedDict.Count = 0) then
    MsgBox.Warning("No Embedded Scripts in this Project",t)
    exit
end
```

```
aList = embedDict.ReturnKeys
aMsg = "Select Embedded Scripts to Delete"
aTitle = t
theDeleteList = MsgBox.MultiListAsString (aList, aMsg, aTitle)
if (theDeleteList.Count = nil) then
    MsgBox.Info("Leaving" ++ t,t)
    exit
else
  for each ES in theDeleteList
    pro.RemoveScript(ES)
  end
end
```

# ◆ *DeleteSEdsFromProject*

The user enters a search pattern, and the SEds starting with this pattern are deleted from the project. The user is asked if s/he wishes to verify each deletion.

**Topics**: SEd, Pattern

**Search Keys**: SEd, Pattern, IndexOf, RemoveDoc

```
' Variable Initialization
t = "DeleteSEdsFromProject"
' Get a search pattern.
pat = MsgBox.Input("Enter a Search Pattern",t,"a:\")
if (pat = nil) then
    MsgBox.Error("No Pattern Entered",t)
    exit
end
' Verify the deletion of each script.
```

```
verify = MsgBox.YesNo("Do you want to verify the deletion of each
script?",t,true)
' Get the project and create a list of documents matching
' this pattern.
pro = av.GetProject
docList = pro.GetDocs
for each i in (docList.count-1) .. 0
    d = docList.Get(i)
    ch = true
    if (d.GetName.IndexOf(pat) = 0 and d.Is(SEd)) then
      if (verify = true) then
          ch = MsgBox.YesNo(d.GetName,"Delete Script?",true)
      end
      if (ch = true) then
          pro.RemoveDoc(d)
      end
    end
end
```

## ◆ *ExecuteComment*

Executes the current set of highlighted lines in an SEd. The lines must be comments. The SEd is created as a temporary object and deleted after the SEd is compiled and executed.

**Topics**: SEd, av

**Search Keys**: av.Run, Source, Compile, Testing, Package

```
' Variable Initialization
t = "ExecuteComment"
d = av.GetActiveDoc
```

```
if (d.Is(SEd).Not) then
   MsgBox.Info("Active document" ++ d.GetName ++ "is not an SEd.",t)
    exit
end
pro = av.GetProject
' Get the source for this script and remove the initial
' comment symbol.
source = d.GetSelected.asTokens(nl)
tmpSource = ""
newLine = ""
for each x in source
    tmpSource = tmpSource + newLine + x.Right(x.Count-1)
    newLine = nl
end
' Make the temporary SEd.
temp = SEd.Make
temp.SetSource(tmpSource)
temp.Compile
av.Run(temp.GetName,nil)
' Remove the temporary SEd.
pro.RemoveDoc(temp)
```

## ◆ *LineEditingDriver*

Combines a set of operations on scripts. Many such operations can be carried out with the mouse, such as marking a line, moving to the beginning or end of a line, and marking a block. They are collected here to show how to programmatically manipulate various elements of an SEd. Marking a block is probably the most useful. First, position the cursor on the first line of the block and select *MarkBlockStart,* and then position the cursor on the last line of the block and select *MarkBlockEnd*. The block is then highlighted. This procedure is useful for marking a large section of a long script when dragging the cursor is not efficient. This script should be installed as a button on the script button bar.

**Topic**: SEd

**Search Keys**: Mark, Delete, SetInsertPos, GetInsertPos, Source, SEd

```
' Variable Initialization
t = "LineEditingDriver"
d = av.GetActiveDoc
if (d.Is(SEd).Not) then
    MsgBox.Error("Active document," ++ d.GetName + ", is not an
SEd.",t)
    exit
end
funcList = {"EndOfLine",
            "BeginOfLine",
            "MarkLine",
            "MarkBlockStart",
            "MarkBlockEnd",
            "DeleteLine"
            }
' Call function and branch to routine.
ch = MsgBox.ListasString(funcList,"Action: ",t)
if (ch = "EndOfLine") then
    d.Search(nl)
    d.SetInsertPos(d.ReturnInsertPos-1)
elseif (ch = "BeginOfLine") then
    pos0 = d.ReturnInsertPos
    posNew = pos0
```

```
     d.Search(nl)
     posNL = d.ReturnInsertPos
     while ( posNL > pos0)
       posNew = posNew - 1
       d.SetInsertPos(posNew)
       d.Search(nl)
       posNL = d.ReturnInsertPos
     end
     d.SetInsertPos(posNL)
elseif (ch = "MarkLine") then
     pos0 = d.ReturnInsertPos
     d.Search(nl)
     posEnd = d.ReturnInsertPos
     posNew = pos0
     posNL = d.ReturnInsertPos
     while ( posNL > pos0)
       posNew = posNew - 1
       d.SetInsertPos(posNew)
       d.Search(nl)
       posNL = d.ReturnInsertPos
     end
     posOffset = posNL
     posLength = posEnd - posOffset
     ln = d.GetSource.Middle(posOffset,posLength)
     d.SetInsertPos(posOffset - 1)
     d.Search(ln)
elseif (ch = "MarkBlockStart") then
     _posOffset = 0
```

```
    pos0 = d.ReturnInsertPos
    posNew = pos0
    d.Search(nl)
    posNL = d.ReturnInsertPos
    while ( posNL > pos0)
      posNew = posNew - 1
      d.SetInsertPos(posNew)
      d.Search(nl)
      posNL = d.ReturnInsertPos
    end
    _posOffset = posNL
elseif (ch = "MarkBlockEnd") then
    d.Search(nl)
    posEnd = d.ReturnInsertPos - 1
    posLength = posEnd - _posOffset
    ln = d.GetSource.Middle(_posOffset,posLength)
    d.SetInsertPos(_posOffset - 1)
    d.Search(ln)
elseif (ch = "DeleteLine") then
    pos0 = d.ReturnInsertPos
    d.Search(nl)
    posEnd = d.ReturnInsertPos
    posNew = pos0
    posNL = d.ReturnInsertPos
    while ( posNL > pos0)
      posNew = posNew - 1
      d.SetInsertPos(posNew)
      d.Search(nl)
```

```
        posNL = d.ReturnInsertPos
    end
    posOffset = posNL
    posLength = posEnd - posNL
    ln = d.GetSource.Middle(posOffset,posLength)
    d.SetInsertPos(posOffset - 1)
    d.Search(ln)
    d.CutSelected
else
    MsgBox.Warning("Nothing Picked: Line Editing Ended",t)
    exit
end
```

## ◆ *OpenScript*

Install this script as a button. When text in an SEd is highlighted and the button associated with this script is clicked on, the document referenced by the highlighted text is opened. Only SEds can be accessed using this script, but it is possible to open any document in the current project with this technique. Only SEds can be safely referenced in other projects. The text can be anywhere in the script, including in a comment. If an SEd in another project is referenced, use the following convention: *<SEd Name>@<Project Name>*. For example, the current script may be referenced as *OpenScript@bmp.apr*.

**Topic**: SEd

**Search Keys**: SEd, Select

**Requirements**: Script will not run properly without a compiled script named *ConvertScript* called by av.Run.

```
' Variable Initialization
t = "OpenScript"
```

```
' ***Critical Resource Test***
if ((av.GetProject.FindDoc("ConvertScript").is(SEd)) = false) then
   MsgBox.Error("Critical Resource NOT Available" + NL +
                "Program Cannot Continue" + NL +
                "Press OK to EXIT",t)
   exit
end
' ***      ***      ***      ***
pro = av.GetProject
d = av.GetActiveDoc
if (d.Is(SEd).Not) then
    MsgBox.Info("The active document," ++ d.GetName + ", is not an
SEd",t)
    exit
end
' Get the highlighted text and break it up at the @ symbol.
sel = d.GetSelected.AsTokens("@")
if (sel.Count = 1) then
    s = pro.FindDoc(sel.Get(0))
    if (s = nil) then
      s = SEd.Make
      s.SetName(sel.Get(0))
    end
    s.GetWin.Open
elseif (sel.Count = 2) then
    s = sel.Get(0)
    p = sel.Get(1).asFileName
    f = LineFile.Make(p,#FILE_PERM_READ)
    scrDict = Dictionary.Make(20)
```

```
firstline = f.ReadElt
if (firstline = "/2.0") then
  scriptFlag = "(SEd."
elseif (firstline = "/2.1") then
  scriptFlag = "(SEd."
end
while (f.IsAtEnd.Not)
  ln = f.ReadElt
  if (ln.Contains("(SEd.") or ln.Contains("(Script.")) then
     if (s = f.ReadElt.AsTokens("""").Get(1)) then
        y         = f.ReadElt
        while (true)
          if (y.Contains("Source:")) then break end
          if (y.Contains("SourceCode:")) then break end
          y = f.ReadElt
        end
        while (true)
          yy = f.ReadElt
          if (yy <> ")") then
             y = y + yy
          else
             break
          end
        end
        break
     end
  end
end
x = y.AsTokens("""")
```

```
      source = x.Get(1)
      if (x.Count > 2) then
        for each i in 2 .. (x.Count - 1)
          source = source + """" + x.Get(i)
        end
      end
      newSEd = SEd.Make
      newSEd.SetName(s)
      newSEd.SetSource(source)
      newSEd.GetWin.Open
      av.Run("ConvertScript",newSEd)
else
      MsgBox.Error("Name must have the form <script>@<project>",t)
      exit
end
```

# ◆ *ReadScriptsFromDiskette*

Prompts the user for a search pattern, loads scripts off the diskette into the current project, and compiles the scripts as they are loaded. The name given to a created SEd is the path to its location on the disk, which makes deletion easy.

**Topics**: SEd, TextFile, Pattern, TextWin, FileName

**Search Keys**: SEd, GetWin, ReadElt, ReadFiles

```
' Variable Initialization
t = "ReadScriptsFromDiskette"
pat = MsgBox.Input("Enter a Search Pattern","Load Scripts from
Diskette","a:\*.ave")
if (pat = nil) then
    MsgBox.Error("No Pattern Entered",t)
```

```
        exit
end

' Break pattern into directory and wildcard portions.
dir = pat.asFileName.ReturnDir
wildcard = pat.asFileName.GetBaseName
' Build list of scripts matching the pattern.
diskette = dir.ReadFiles(wildcard)
if (diskette.Count = 0) then
    MsgBox.Error("No Avenue Scripts on this Diskette","Load From
Diskette")
    exit
end
' Loads scripts into the project using complete path name and
' compiles them.
for each file_name in diskette
    aSEd = SEd.Make
    aSEd.GetWin.Open
    f = TextFile.Make(file_name, #FILE_PERM_READ)
    aSEd.Insert(f.Read(f.GetSize))
    f.Close
    aSEd.SetName(f.GetName)
    aSEd.Compile
    aSEd.GetWin.Close
end
```

## ◆ *ReplaceOldWithNew*

Can be called by another program, and takes three parameters. The first parameter is an SEd; the second, a string in the SEd; and the third, a new string to replace every occurrence of the existing string.

**Topics**: SEd, Search, Insert

**Search Keys**: SEd, Search, Insert

```
' Variable Initialization
t = "ReplaceOldWithNew"
p = av.GetProject
d = p.GetDocs
dLst = List.Make
if (SELF.Is(List).Not) then
    for each x in d
      if (x.Is(SEd)) then
         dLst.Add(x)
      end
    end
    ch = MsgBox.List(dLst,"Pick an SEd:",t)
    if (ch = nil) then exit end
    change = MsgBox.MultiInput("Enter New/Old Strings",t,{"Search
for:","Replace with:"},{"",""})
    if (change.Count <> 2) then exit end
    parmList = {ch,change.Get(0),change.Get(1)}
else
    parmList = SELF
end
if (parmList.Is(List).Not) then
```

```
        MsgBox.Error("Usage: av.Run(""ReplaceOldWithNew"",{aSEd,
    oldstring,newstring})",t)

        exit

    end

    if (parmList.Count <> 3) then

        MsgBox.Error("Usage: av.Run(""ReplaceOldWithNew"",{aSEd,
    oldstring,newstring})",t)

        exit

    end

    aSEd = parmList.Get(0)

    if (aSEd.Is(SEd).Not) then

        MsgBox.Error("First parameter must be an SEd",t)

        exit

    end

    old = parmList.Get(1)

    if (old.Is(String).Not) then

        MsgBox.Error("Second parameter must be the old string",t)

        exit

    end

    new = parmList.Get(2)

    if (new.Is(String).Not) then

        MsgBox.Error("Third parameter must be the new string",t)

        exit

    end

    aSEd.GetWin.Open

    d = aSEd

    ' Go to the top of the SEd.

    d.SetInsertPos(0)

    ' Loop and change old string to new string. Count number of changes.
```

```
changes = 0
while (d.Search(old))
    d.Insert(new)
    changes = changes + 1
end
' Write number of changes to message area at bottom left of
' the application frame.
av.ShowMsg(changes.asString ++ "occurrences of" ++ old ++ "changed
to" ++ new)
return changes
```

# ◆ *ScriptExportAsEMail*

This script is a package: the header is intended to be an email message; the next section is a script; and the third and final section is a collection of scripts embedded as comments. When the script is run, it extracts the embedded scripts, correctly sets their names, compiles them, and installs them in the current project. This script provides a convenient way to transfer packages of scripts, as well as to simplify installation. Two very simple scripts called *foo* and *foo1* are embedded in this script.

**Topics**: SEd, String

**Search Keys**: Compile, SetName, Left, Middle, Right, asTokens

**Requirements**: To run the script, the variable *t* must have the value of the script name. If you install it in a new project, the name *Script1* in a new script will be correct.

```
' E-mail message here
' Variable Initialization
t = "ScriptExportAsEMail"
d = av.GetProject.FindDoc(t)        ' Current SEd.
if (d.Is(SEd).Not) then
```

```
      MsgBox.Error("Active Document" ++ d.GetName ++ "is not an SEd",t)
        exit
  end
  source = d.GetSource.AsTokens(nl) ' Source as a list of lines.
  sedFlag = 0
  newSEdFlag = 0
  lineCount = 0
  ' Read through the source until the SEd section is found.
    for each x in source
       if (x.Contains("'SEd Section")) then
  ' Read to start of SEd section contained in this file.
         sedFlag = 1
       end
       if ((x.Left(7) = "'Start:") and (sedFlag = 1)) then
  ' Initialize newSource and SetName for the SEd; the name is
  ' after the colon.
         newSEd = SEd.Make
         n = x.asTokens(":").Get(1)
         newSEd.SetName(n)
         newSEdFlag = 1
         newSource = "'!" + n + nl
         lineCount = 0
       end
       if ((x.Left(8) = "'Finish:") and (sedFlag = 1)) then
  ' Close SEd, install, and compile.
         newSEd.SetSource(newSource)
         newSEd.Compile
         newSEdFlag = 0
```

```
      end
    if ((sedFlag = 1) and (newSEdFlag = 1)) then
' Strip out comment and add to source for this SEd.
        if (lineCount > 0) then
            newSource = newSource + nl + x.Right(x.Count - 1)
        end
        lineCount = lineCount + 1
    end
end
' ***The scripts are inserted here.***
'SEd Section
'Start: foo
'MsgBox.Info("First Script","Test1")
'Finish: foo
'Start: foo1
'l = av.GetProject.GetDocs
'ch = MsgBox.List(l,"Pick a Doc","foo1")
'ch.GetWin.Open
'Finish: foo1
```

# ◆  *ScriptFileLoad*

Rewrite of the *ScriptFileLoad* system script. Rename it if you do not want the script to interfere with the default system script. The only change is support for the script working directory. The value for this directory is set in the *CurrentWorkingDirectoryInitialize* script.

**Topics**: SEd, System Script

**Search Keys**: SEd, System Script, ScriptFileLoad

```
' Variable Initialization
t = "ScriptFileLoad"
theSEd = av.GetActiveDoc
' Read the global dictionary that sets the working directory.
_dirDict.Get("Scripts").SetCWD
file_name = FileDialog.Show("*", "Text File", "Load Script")
if (nil = file_name) then exit end
f = TextFile.Make(file_name, #FILE_PERM_READ)
theSEd.Insert(f.Read(f.GetSize))
f.Close
```

## ◆ *ScriptManagerDriver*

Puts a front end on the the script manager and makes it possible to select standard subsets of the installed scripts. The chosen script is installed as an SEd, given the same name as the system script, and opened.

**Topics**: SEd, ScriptMgr

**Search Keys**: ScriptMgr, SEd

**Requirements**: Script will not run properly without the compiled script named *ScriptManagerSubset* called by av.Run.

```
' Variable Initialization
t = "ScriptManagerDriver"
' ***Critical Resource Test***
if ((av.GetProject.FindDoc("ScriptManagerSubset").is(SEd)) =
false) then
  MsgBox.Error("Critical Resource NOT Available" + NL +
               "Program Cannot Continue" + NL +
               "Press OK to EXIT",t)
  exit
```

```
end
' ***      ***      ***      ***
d = Dictionary.Make(10)
d.Add("...",nil)
d.Add("All","*".asPattern)
d.Add("Project","Project*".asPattern)
d.Add("View","View*".asPattern)
d.Add("wptc","wptc*".asPattern)
d.Add("Add","*Add*".asPattern)
d.Add("Close","*Close*".asPattern)
d.Add("Open","*Open*".asPattern)
d.Add("Update","*Update*".asPattern)
' Choose a subclass of scripts.
sub = MsgBox.ListasString(d.ReturnKeys,"Pick a Class: ",t)
if (sub = nil) then
    MsgBox.Warning("No subclass picked: Exiting",t)
    exit
end
if (sub = "...") then
    sub = MsgBox.Input("Enter a SubClass:",t,"wptc*")
    if (sub = nil) then exit end
    scr = av.Run("ScriptManagerSubset",sub.asPattern)
else
    scr = av.Run("ScriptManagerSubset",d.Get(sub))
end
if (scr = nil) then
    MsgBox.Warning("No script picked: Exiting",t)
    exit
```

```
end

s = SEd.Make

if (scr.Is(Script)) then

    s.SetSource(scr.asString)

elseif (scr.Is(SEd)) then

    s.SetSource(scr.GetSource)

end

s.SetName(scr.GetName)

s.GetWin.Open
```

## ◆ *ScriptManagerSubset*

Returns the selected script name to the calling program. This script uses a pattern to select a subset of the script library. The default subset is set in the variable initialization section.

**Topics**: ScriptMgr, Pattern, GetFromSubset

**Search Key**: ScriptMgr

```
' Variable Initialization

t = "ScriptManagerSubset"

pat = "*"

if (SELF.is(Pattern)) then

  parm = SELF

else

  parm = pat.AsPattern

end

return(ScriptMgr.GetFromSubset(parm))
```

## ◆ SEd2TextFile

Extracts all SEds in a project and writes them to a single ASCII text file.

```
t =  Script.The.GetName
p = av.GetProject
' Substitute a file name that works on your system here.
fname = "d:\temp\script.txt".AsFileName
' Use the file name to make a line file.
lf = LineFile.Make(fname,#FILE_PERM_WRITE)
if (lf = NIL) then
  MsgBox.Error("The file," + fname.AsString +
              + ", could not be made.",t)
  exit
end
count = 0
for each d in p.GetDocsWithGUI(p.FindGUI("Script"))
  source = d.GetSource.AsTokens(nl)
  first = source.Get(0)
  if (first.Contains("Checked: 3.0")) then
    source.Remove(0)
    source.Add(nl+nl)
    lf.Write(source,source.Count)
    count = count + 1
  else
    continue
  end
end
lf.Close
MsgBox.Info(count.AsString,t)
```

# ◆ *SEdSetName*

Takes a highlighted phrase and sets the name of the active SEd to that phrase. This is very useful when importing scripts from another source. The name is usually in the header information. If this script is installed as a tool, then highlighting the name and pressing the button will set the correct name for the imported script. If there is no selected text, the user is prompted to enter a new name.

**Topic**: SEd

**Search Keys**: SEd, SetName, Input, GetSelected

```
' Variable Initialization
t = "SEdSetName"
' Get and check the active document.
d = av.GetActiveDoc
if (d.Is(SEd).Not) then
    MsgBox.Error("Active document is not an SEd!",t)
    exit
end
' The name will be changed to the selected text.
s = d.GetSelected
if (s = "") then
    s = MsgBox.Input("Enter a new name: ",t,"")
    if (s = nil) then
      MsgBox.Warning("No name entered: Name not changed",t)
      exit
    end
end
' Set the name of the SEd.
d.SetName(s)
```

# ◆ *WriteScriptsToDisk*

Writes a collection of SEds or embedded scripts to either a hard or floppy disk. It uses names such as *sed0.ave*, *sed1.ave*, *embed0.ave*, and *embed1.ave* for the scripts. Next, a log file is written in the same directory (*script.log*) that matches the names of the scripts to their file names. A comment containing the name of the script is also added to the beginning of each script.

**Topics**: SEd, FileName, LineFile, System, Pattern

**Search Keys**: SEd, FileName, System, LineFile, WriteElt, Execute, ReadFiles, Write

```
' Variable Initialization
t = "WriteScriptsToDisk"
sedN = 0
embedN = 0
defPath = "a:\"
pro = av.GetProject
mkdir = "command.com /c mkdir"
' Prompt for a path; path must include the final \.
path = MsgBox.Input("Enter Path (with final \): ",t,defPath)
if (path = nil) then
    MsgBox.Error("No Path Entered",t)
    exit
end
dirPath = path.AsFileName
if (dirPath.IsDir.Not) then
    ch = MsgBox.YesNo("Make a new directory?","Path" ++ path ++ "is
not a directory!",true)
    if (ch = false) then
      exit
    end
```

```
    newDir = path.asFileName.ReturnDir.GetName
    System.Execute(mkdir ++ newDir)
end

' Scripts and/or SEds.
embedYN = MsgBox.YesNo("Do you want to write out Embedded
Scripts?",t,true)
sedYN    = MsgBox.YesNo("Do you want to write out SEd's?",t,true)
if ((embedYN = false) and (sedYN = false)) then exit end
' Set a pattern for embedded scripts and SEds.
pat = MsgBox.Input("Starting with ...",t,"")
if (pat = nil) then
    ch = MsgBox.YesNo("No Pattern Entered: Do you want everything
to be selected?",t,true)
    if (ch = false) then exit end
    pat = ""
end
pat = pat.asPattern
' Open the log file.
log       = (path + "script.log").asFileName
logFile  = TextFile.Make(log,#FILE_PERM_WRITE)
' Write out embedded scripts.
if (embedYN = true) then
    embedList = pro.GetScripts
    for each s in embedList
      if (s.GetName.IndexOf(pat) = 0) then
        f = (path+"embed"+embedN.asString+".ave").asFileName
        embedN = embedN + 1
        out = TextFile.Make(f,#FILE_PERM_WRITE)
        source = s.asString
```

```
' Check for the magic cookie at the start.
        if (source.IndexOf("'!") <> 0) then
            source = "'!"+s.GetName + nl + source
        else
            lines = source.asTokens(nl)
            if ((line.Get(0) = s.GetName).Not) then
               source = "'!" + s.GetName + nl
               for each i in 1 .. (lines.Count-1)
                   source = source+ lines.Get(i) + nl
               end
            end
        end
        out.Write(source,source.Count)
        out.Close
        logRec = f.GetBaseName + "," + s.GetName + nl + cr
        logFile.Write(logRec,logRec.Count)
      end
    end
end
' Write out SEds.
if (sedYN = true) then
    docList = pro.GetDocs
    for each s in docList
        if (s.Is(SEd) and (s.GetName.IndexOf(pat) = 0)) then
            f = (path+"sed"+sedN.asString+".ave").asFileName
            sedN = sedN + 1
            out = TextFile.Make(f,#FILE_PERM_WRITE)
            source = s.GetSource
```

```
' Check for the magic cookie at the start.
        if (source.IndexOf("'!") <> 0) then
            source = "'!"+s.GetName + nl + source
        else
            lines = source.asTokens(nl)
            if ((lines.Get(0) = s.GetName).Not) then
              source = "'!" + s.GetName + nl
              for each i in 1 .. (lines.Count-1)
                  source = source+ lines.Get(i) + nl
              end
            end
        end
        out.Write(source,source.Count)
        out.Close
        logRec = f.GetBaseName + "," + s.GetName + nl + cr
        logFile.Write(logRec,logRec.Count)
      end
    end
end
logFile.Close
```

# Tables

## ◆ *ApplySelection*

Applies the bitmap selection set from a theme in one view to a theme in a second view. The FTabs for both themes must be identical in size. For example, you could set up two views, named *View1* and *View2*. In both views, you could add the *states.shp* polygon theme from the *USA* directory of the ArcView sample data set and make a selection in *View1*. The script could be installed as a button on the view button bar.

**Topics**: Table, FTab, Bitmap, Selection

**Search Keys**: FTab.GetSelection, FTab.SetSelection

**Requirements**: Script will not run properly without the following critical resources: view named *View1* referenced by variable *ViewName1*; view named *View2* referenced by variable *ViewName2*; and an active theme named *states.shp* in both views. To change the name of the views and themes, change the value of the string variables *ViewName1*, *ViewName2*, *ThemeName1*, and *ThemeName2* in the variable initialization section.

```
' Variable Initialization
t = "ApplySelection"
ViewName1 = "View1"
ThemeName1 = "states.shp"
ViewName2 = "View2"
ThemeName2 = "states.shp"
'***Critical Resource Test***
if ((av.GetProject.FindDoc(ViewName1).is(View)) = false) then
  MsgBox.Error("Critical Resource NOT Available" + NL +
               "Program Cannot Continue" + NL +
```

```
                        "Press OK to EXIT",t)

   exit

 elseif ((av.GetProject.FindDoc(ViewName2).is(View)) = false) then

     MsgBox.Error("Critical Resource NOT Available" + NL +
                  "Program Cannot Continue" + NL +
                  "Press OK to EXIT",t)

     exit

elseif (av.GetProject.FindDoc(ViewName1).FindTheme(ThemeName1) =
nil) then

     MsgBox.Error("Critical Resource NOT Available" + NL +
                  "Program Cannot Continue" + NL +
                  "Press OK to EXIT",t)

     exit

elseif (av.GetProject.FindDoc(ViewName2).FindTheme(ThemeName2) =
nil) then

     MsgBox.Error("Critical Resource NOT Available" + NL +
                  "Program Cannot Continue" + NL +
                  "Press OK to EXIT",t)

     exit

end

'***     ***      ***       ***

aFtab = av.GetProject.FindDoc(ViewName1)
.FindTheme(ThemeName1).GetFTab

SelectB = aFtab.GetSelection

bFtab = av.GetProject.FindDoc(ViewName2)
.FindTheme(ThemeName2).GetFTab

bFtab.SetSelection(SelectB)
```

## ◆ *BitmapClearSelection*

Demonstrates how to clear the selection set from a bitmap.

**Topics**: Table, VTab, Bitmap, Selection

**Search Keys**: VTab.GetSelection, Bitmap.ClearAll, VTab.UpdateSelection

```
' Variable Initialization
t = "BitmapClearSelection"
' Get the project.
theProj = av.GetProject
' Select the view.
theDocList = theProj.GetDocs
theDict = Dictionary.Make(theDocList.Count)
for each i in theDocList
    if (i.is(View)) then
       theDict.Add(i.GetName,i)
    end
end
theList = theDict.ReturnKeys
theView = theDict.Get(MsgBox.ListAsString(theList,"Choose a
View","VIEWS"))
if (theView = nil) then exit end
' Select the theme.
theThemes = theView.GetThemes
theDict = Dictionary.Make(theThemes.Count)
for each i in theThemes
    if (i.is(FTheme)) then
       theDict.Add(i.GetName,i)
    end
```

```
end
theList = theDict.ReturnKeys
theTheme = theDict.Get(MsgBox.ListAsString(theList,"Choose a
Feature Theme","FEATURE THEMES"))
if (theTheme = nil) then exit end
' Get the theme's FTab (VTab).
theVTab = theTheme.GetFTab
' Get the theme's selection set.
theBitMap = theVTab.GetSelection
' Clear the selection.
theBitMap.ClearAll
' Update the selection set drawn in the view.
theVTab.UpdateSelection
```

## ◆ *BitmapCountSelected*

Demonstrates how to obtain the number of selected records in a VTab bitmap.

**Topics**: Table, VTab, BitMap, Selection

**Search Keys**: VTab.GetNumRecords, BitMap.Count

```
' Variable Initialization
t = "BitmapCountSelected"
' Get the project.
theProj = av.GetProject
' Select the view.
theDocList = theProj.GetDocs
vList = {}
for each i in theDocList
    if (i.is(View)) then
```

```
        vList.Add(i)
    end
end
theView = MsgBox.List(vList,"Choose a View","VIEWS")
if (theView = nil) then exit end
' Select the theme.
theThemes = theView.GetThemes
theTheme = MsgBox.List(theThemes,"Choose a Feature Theme","FEATURE
THEMES")
if (theTheme = nil) then exit end
' Get the theme's FTab (VTab).
theVTab = theTheme.GetFTab
' Get the theme's selection set.
theBitMap = theVTab.GetSelection
' Count the number of records in the VTab.
theVTabCount = theVTab.GetNumRecords
' Count the number of selected records.
theSelCount = theBitMap.Count
msgbox.info(theSelCount.asstring ++ "Selected of" ++
theVTabCount.asstring ++ "Records","Selected Records")
```

## ◆ *BitmapQuery*

Shows the use of the AND, OR, and XOR set functions that can be used on a bit-map to run queries. Default values for selections are from the *cities.shp* data set in the *USA* directory of the ArcView sample data. The setup would be a view with the *cities.shp* theme from the ArcView sample data.

**Topics**: Table, VTab, Bitmap, Query

**Search Keys**: VTab.Query, Bitmap.And, Bitmap.Or, Bitmap.XOr

```
' Variable Initialization

t = "BitmapQuery"

p = av.GetProject

theQueryType = MsgBox.ListAsString({"AND","OR","XOR"},"Select
Complex Query Type:","Complex Bitmap Query")

if (theQueryType = nil) then exit end

' Get the project.

theProj = av.GetProject

' Select the view.

vList = theProj.GetDocsWithGUI(p.FindGUI("View"))

theView = MsgBox.List(vList,"Choose a View","VIEWS")

if (theView = nil) then exit end

' Select the theme.

theThemes = theView.GetThemes

theTheme = MsgBox.List(theThemes,"Choose a Feature Theme","FEATURE
THEMES")

if (theTheme = nil) then exit end

' Get the theme's FTab (VTab).

theVTab = theTheme.GetFTab

' Get the theme's selection set.

theBitMap = theVTab.GetSelection

' Create two working bitmaps of the same size as the bitmap
' for the VTab.

AWorkingBitMap = BitMap.Make(theVTab.GetNumRecords)

BWorkingBitMap = BitMap.Make(theVTab.GetNumRecords)

' Build query string.

' Query A.

fldList = theVTab.GetFields

theAField = MsgBox.List(fldList,"Choose a field to build first
```

```
Query","First Query Field-City_Name")

if (theAField = nil) then exit end

theAFieldType = theAField.GetType

firstQueryValue = msgBox.Input("Value in" ++ theAField.GetName ++
"Field equals:","First Query","Albany")

if (theAField.IsTypeString) then

' First Query String - "[City-Name] = "Albany""

    AQueryString = "[" ++ theAField.GetName ++ "] = " ++
firstQueryValue.quote

else

    AQueryString = "[" ++ theAField.GetName ++ "] = " ++
firstQueryValue

end

' Query B.

theBField = MsgBox.List(fldList,"Choose a field to build second
Query","SECOND QUERY FIELD-State_Name")

if (theBField = nil) then exit end

theBFieldType = theBField.GetType

secondQueryValue = msgBox.Input("Value in" ++ theBField.GetName ++
"Field equals:","First Query","California")

if (theBField.IsTypeString) then

' Second Query String - "[State_Name] = "California""

    BQueryString = "[" ++ theBField.GetName ++ "] = " ++
secondQueryValue.quote

else

    BQueryString = "[" ++ theBField.GetName ++ "] = " ++
secondQueryValue

end

AWorkingBitmap.Copy(theBitmap)

AWorkingBitmap.ClearAll
```

```
QueryA = theVTab.Query(AQueryString,AWorkingBitmap,
#VTAB_SELTYPE_NEW)

if (QueryA = False) then

    msgbox.info("QUERY A FAILED","")

end

BWorkingBitmap.Copy(theBitmap)

BWorkingBitmap.ClearAll

QueryB = theVTab.Query(BQueryString,BWorkingBitmap,
#VTAB_SELTYPE_NEW)

if (QueryB = False) then

    msgbox.info("QUERY B FAILED","")

end

if (theQueryType = "AND") then

    AWorkingBitmap.And(BWorkingBitmap)

else

    if (theQueryType = "OR") then

      AWorkingBitmap.Or(BWorkingBitmap)

    else

      if (theQueryType = "XOR") then

          AWorkingBitmap.XOr(BWorkingBitmap)

      end

    end

end

' Apply the selection to the VTab.

theVTab.SetSelection(AWorkingBitmap)

' Update the selection set drawn in the view.

theVTab.UpdateSelection
```

# ◆ *EditAttributes*

Integrates GetViewTheme and GetVTabFields into a single unit. The script dynamically selects a view, theme, and fields from the FTab in preparation for editing. It opens and maximizes the view, makes the selected theme active and visible, and makes all other themes inactive and invisible. Variables are saved as global values so that they can be used by the SelectFeatures tool.

**Topics**: View, Theme, Application

**Search Keys**: av.Run, Global Variable

**Requirements**: Must be run to set up the global variables for the SelectFeatures tool. This script will not run properly without the following critical resources: a compiled script named *GetViewTheme* called by av.Run, and a compiled script named *GetVTabFields* called by av.Run.

```
' Variable Initialization
t = "EditAttributes"
' ***Critical Resource Test***
if ((av.GetProject.FindDoc("GetViewTheme").is(SEd)) = false) then
  MsgBox.Error("Critical Resource NOT Available" + NL +
               "Program Cannot Continue" + NL +
               "Press OK to EXIT",t)
  exit
end
if ((av.GetProject.FindDoc("GetVTabFields").is(SEd)) = false) then
  MsgBox.Error("Critical Resource NOT Available" + NL +
               "Program Cannot Continue" + NL +
               "Press OK to EXIT",t)
  exit
end
' ***     ***      ***       ***
```

```
pro = av.GetProject

' Get and check the return values.

vtfList = av.Run("GetViewTheme",nil)

if ((vtfList = nil) or (vtfList.Is(List).not) or (vtfList.Count <>
3)) then
    MsgBox.Error("Could not properly initialize the view, theme, and
FTab: exiting the script",t)
        exit
end

_v  = vtfList.Get(0)

_th = vtfList.Get(1)

if ((_th = nil) or (_th.Is(FTheme).Not)) then
    MsgBox.Error("No theme with an editable FTab was picked: exiting
script",t)
        exit
end

_ft = vtfList.Get(2)

' Get and check the fields.

_editList = av.Run("GetVTabFields",_ft)

if ((_editList = nil) or (_editList.Is(List).Not) or
(_editList.Count = 0)) then
    MsgBox.Error("No Fields to Edit: Exiting the Program",t)
        exit
end

' Clear the screen, open view, and zoom to theme.

pro.CloseAll

for each x in _v.GetThemes
        if (x <> _th) then
                x.SetActive(FALSE)
```

```
            x.SetVisible(FALSE)
      else
            x.SetActive(TRUE)
            x.SetVisible(TRUE)
      end
end
_v.GetDisplay.ZoomToRect(_th.ReturnExtent)
tbl = Table.Make(_ft)
tbl.GetWin.Open
_v.GetWin.Open
```

## ◆ *ExportSortedTable*

Tables are viewers for VTabs. As such, the order of the rows in the table and the order of the records in the VTab are not the same. To export a table in row order, the script must convert the row number to the record number and contain a general purpose mechanism for selecting and writing out fields. A sort must be performed on the table before running *ExportSortedTable*.

**Topics**: Table, VTab, Export

**Search Keys**: Table, VTab, Export, Field, Sort

**Requirements**: A table in the current project. This script will not run properly without the following critical resources: a valid search path for file output referenced by *dirpath* variable, and a table in a project sorted in the order you wish to export.

```
' Variable Initialization
t = "ExportSortedTable"
dirpath = "c:\arcview"
' ***Critical Resource Test***
if (dirpath.asfilename.IsDir.Not) then
```

```
    MsgBox.Error("Critical Resource NOT Available" + NL +
              "Program Cannot Continue" + NL +
              "Press OK to EXIT",t)
  exit
end
' Test for table in project.
testlist = Av.GetProject.GetDocs
count = 0
for each t in testlist
  if (t.Is(Table)) then
    count = count + 1
  end
end
if (count = 0) then
  MsgBox.Error("Critical Resource NOT Available" + NL +
              "Program Cannot Continue" + NL +
              "Press OK to EXIT",t)
  exit
end
'***      ***      ***      ***
' ***Create the output file.***
dir = dirpath.asFileName
dir.SetCWD
f = FileDialog.Put((dirpath + "\out.txt").asFileName,"*.txt",t)
if (f = nil) then exit end
outFile = LineFile.Make(f,#file_perm_write)
if (outFile = nil) then exit end
' ***Select the table.***
```

```
if (av.GetActiveDoc.Is(Table)) then
    tbl = av.GetActiveDoc
else
    p = av.GetProject
    d = p.GetDocs
    tList = List.Make
    for each x in d
      if (x.Is(Table)) then
          tList.Add(x)
      end
    end
    tbl = MsgBox.List(tList,"Select a table:",t)
    if (tbl = nil) then exit end
    tbl.GetWin.Open
end
vt = tbl.GetVTab
' ***Select the fields to export.***
chList = vt.GetFields.Clone
fldList = MsgBox.MultiList(chList,"Choose fields to export:",t)
' ***Export records to output file.***
for each i in (0..(vt.GetNumRecords - 1))
    r = tbl.ConvertRowtoRecord(i)
    outLine = ""
    sep = ""
    for each x in fldList
      outLine = outLine + sep + vt.ReturnValue(x,r)
      sep = ","
    end
```

```
        outFile.WriteElt(outLine)
end
' ***Close the output file.***
outFile.Close
```

## ◆ *GetSelectionAddField*

Shows how to retrieve a selection set that was saved in the field of a table using the *SaveSelectionAddField* script.

**Topics**: Table, VTab, Field

**Search Keys**: VTab.SetSelection, VTab.UpdateSelection

```
' Variable Initialization
t = "GetSelectionAddField"
' Get the project.
theProj = av.GetProject
' Select the view.
theDocList = theProj.GetDocs
theDict = Dictionary.Make(theDocList.Count)
for each i in theDocList
    if (i.is(View)) then
        theDict.Add(i.GetName,i)
    end
end
theList = theDict.ReturnKeys
theView = theDict.Get(MsgBox.ListAsString(theList,"Choose a
view","VIEWS"))
if (theView = nil) then exit end
' Select the theme.
```

```
theThemes = theView.GetThemes
theDict = Dictionary.Make(theThemes.Count)
for each i in theThemes
    if (i.is(FTheme)) then
       theDict.Add(i.GetName,i)
    end
end
theList = theDict.ReturnKeys
theTheme = theDict.Get(MsgBox.ListAsString(theList,
                            "Choose a feature theme",
                            "FEATURE THEMES"))
if (theTheme = nil) then exit end
' Get the theme's FTab (VTab).
theVTab = theTheme.GetFTab
' Clear the theme's selection set.
theBitMap = theVTab.GetSelection
theBitMap.ClearAll
theVTab.UpdateSelection
theField = MsgBox.ListAsString(theVTab.GetFields,
            "Select Field to Use to Apply Selection",
            "Selection Field Name")
if (theField = nil) then exit end
for each i in 0 .. (theVTab.GetNumRecords - 1)
    if (theVTAB.ReturnValue(theField,i) = "T") then
       theBitmap.Set(i)
    end
end
theVtab.SetSelection(theBitmap)
theVTab.UpdateSelection
```

# ◆ *GetSelectionFromLookupTable*

Shows how to retrieve a selection from a look-up table that was created with the *SaveSelectionCreateLookupTable* script.

**Topics**: Table, VTab, Selection

**Search Keys**: VTab.FindField, VTab.ReturnValueNumber

**Requirements**: *SaveSelectionCreateLookupTable* output file. This script will not run properly without a dBase format file referenced by the *thefilename* variable.

```
' Variable Initialization
t = "GetSelectionFromLookupTable"
thefilename = "C:\foo.dbf"
' ***Critical Resource Test***
' Test for the existence of look-up table file.
theFile = thefilename.asfilename
filetest = File.Exists(theFile)
if (Filetest = False) then
  MsgBox.Error("Critical Resource NOT Available" + NL +
               "Program Cannot Continue" + NL +
               "Press OK to EXIT",t)
  exit
end
' ***     ***     ***     ***
' Get the project.
theProj = av.GetProject
' Select the view.
theDocList = theProj.GetDocs
theDict = Dictionary.Make(theDocList.Count)
for each i in theDocList
    if (i.is(View)) then
```

```
        theDict.Add(i.GetName,i)
    end
end
theList = theDict.ReturnKeys
theView = theDict.Get(MsgBox.ListAsString(theList,
                        "Choose a view",
                        "VIEWS"))
if (theView = nil) then exit end
' Select the theme.
theThemes = theView.GetThemes
theDict = Dictionary.Make(theThemes.Count)
for each i in theThemes
    if (i.is(FTheme)) then
        theDict.Add(i.GetName,i)
    end
end
theList = theDict.ReturnKeys
theTheme = theDict.Get(MsgBox.ListAsString(theList,
                        "Choose a feature theme",
                        "FEATURE THEMES"))
if (theTheme = nil) then exit end
' Get the theme's FTab (VTab).
theVTab = theTheme.GetFTab
' Clear the theme's selection set.
theBitMap = theVTab.GetSelection
theBitMap.ClearAll
theVTab.UpdateSelection
theNewVTab = VTab.Make(theFile,False,False)
```

```
theField = theNewVTab.FindField("VTabRecd")
for each r in theNewVTab
    theValue = theNewVTab.ReturnValueNumber(theField,r)
    theBitmap.Set(theValue)
end

theVTab.UpdateSelection
```

# ◆ *GetSelectionTableObjectTag*

Shows how to retrieve a bitmap that was saved as the table object tag using *Save-SelectionTableObjectTag*. The script applies the bitmap as the selection set to the table.

**Topics**: Table, VTab, Selection, ObjectTag

**Search Keys**: GetObjectTag

```
' Variable Initialization
t = "GetSelectionTableObjectTag"
' Get the project.
theProj = av.GetProject
' Select the view.
theDocList = theProj.GetDocs
theViewDict = Dictionary.Make(theDocList.Count)
theTableDict = Dictionary.Make(theDocList.Count)
for each i in theDocList
    if (i.is(View)) then
        theViewDict.Add(i.GetName,i)
    end
    if (i.is(Table)) then
        theTableDict.Add(i.GetName,i)
```

```
        end
end
theList = theViewDict.ReturnKeys
theView = theViewDict.Get(MsgBox.ListAsString(theList,
                "Choose a view",
                "VIEWS"))
if (theView = nil) then exit end
' Select the theme.
theThemes = theView.GetThemes
theDict = Dictionary.Make(theThemes.Count)
for each i in theThemes
    if (i.is(FTheme)) then
        theDict.Add(i.GetName,i)
    end
end
theList = theDict.ReturnKeys
theTheme = theDict.Get(MsgBox.ListAsString(theList,
                    "Choose a feature theme",
                    "FEATURE THEMES"))
if (theTheme = nil) then exit end
' Get the theme's FTab (VTab).
theVTab = theTheme.GetFTab
' Get the theme's selection set.
theBitMap = theVTab.GetSelection
' Clear the current selection.
theBitMap.Clearall
theVTab.UpdateSelection
' Check to see if the table exists.
```

```
theList = theTableDict.ReturnKeys

theList = theList.Add("NOT ON THE LIST")

theTable = MsgBox.ListAsString(theList,

          "Choose a Table to Get the Selection",

          "TABLE")

if ((theTable = "NOT ON THE LIST") or (theTable = nil)) then

    theTable = Table.Make(theVTab)

else

    theTable = theTableDict.Get(theTable)

end

theBitMap = theTable.GetObjectTag

theVtab.SetSelection(theBitmap)

theVTab.UpdateSelection
```

## ◆ *GetSelectionToODB*

Shows how to retrieve a selection set by getting a bitmap from an ODB (object database) file (*C:\bitmap.odb*) created with the *SaveSelectionToODB* script. The selection set is applied as the current selection set.

**Topics**: Table, VTab, Selection, ODB Files

**Search Keys**: ODB.Open, ODB.Get

**Requirements**: *SaveSelectionToODB* output ODB file. This script will not run properly without the ODB format file referenced by the *thefilename* variable.

```
' Variable Initialization

t = "GetSelectionToODB"

thefilename = "C:\bitmap.odb"

' ***Critical Resource Test***

' Test for the existence of ODB file.
```

```
theFile = thefilename.asfilename
filetest = File.Exists(theFile)
if (Filetest = FALSE) then
  MsgBox.Error("Critical Resource NOT Available" + NL +
               "Program Cannot Continue" + NL +
               "Press OK to EXIT",t)
  exit
end
'***      ***      ***       ***
' Open and read the ODB.
theODB = ODB.Open(theFile)
' Pull variables from SELF.
theProjName = theODB.Get(0).Get(0)
theViewName = theODB.Get(0).Get(1)
theThemeName = theODB.Get(0).Get(2)
theODBBitmap = theODB.Get(0).Get(3)
' Verify existence of the correct project, view, and theme.
theProj = av.GetProject
if ((theProj.GetName = theProjName).Not) then
    msgBox.Info("Wrong PROJECT","")
    exit
end
' Verify the existence of the view.
theDocList = theProj.GetDocs
for each i in theDocList
    if (i.is(View)) then
      if (i.GetName = theViewName) then
        ViewOK = True
```

```
            theView = i
        end
      end
end
if (viewOK.Not) then
    MsgBox.Info("Correct View Does Not Exist")
    exit
end
' Verify whether theme exists.
theThemeList = theView.GetThemes
for each i in theThemeList
    if (i.is(FTheme)) then
      if (i.GetName = theThemeName) then
          ThemeOK = True
          theTheme = i
      end
    end
end
if (ThemeOK.Not) then
    MsgBox.Info("Correct Theme Does Not Exist")
    exit
end
' Get the theme's FTab (VTab).
theVTab = theTheme.GetFTab
if (theVTab = nil) then exit end
' Get the theme's selection set.
theBitMap = theVTab.GetSelection
' Clear the theme's current selection set.
```

```
theBitMap.ClearAll

theVTab.UpdateSelection

theVtab.SetSelection(theODBBitmap)

theVTab.UpdateSelection
```

# ◆ *GetVTabFields*

Implements a dialog to select a list of fields from a VTab. The script returns a list of fields to the calling program.

**Topics**: VTab, FTab, Field, List, MsgBox, Table

**Search Keys**: VTab, FTab, Field, List, ListAsString, DeepClone

**Requirements**: Script expects to be called by another program that will pass it a VTab as a variable value. To run the script as a standalone, delete the comment symbol before the line that initializes the parm variable (i.e., *parm = av.Run("Get-ViewTheme",nil).Get(2)*). The Test Drive implements a simple calling program. Use the ExecuteComment script to explore how the calling works.

**Test Drive**: The following test drive calls the script and shows the results with minimal checking of the return value. The *GetViewTheme* script returns a view, theme, and FTab. See that script for more detail.

```
' lst = av.Run("GetViewTheme",nil)

' ft = lst.Get(2)

' if (ft = nil) then exit end

' fldList = av.Run("GetVTabFields",ft)

' if ((fldList = nil) or (fldList.Count = 0)) then exit end

' MsgBox.List(fldList,"List of Fields Selected","Test Drive")

' Variable Initialization

t = "GetVTabFields"

parm = SELF

' The following line is for testing purposes only.
```

```
' parm = av.Run("GetViewTheme",nil).Get(2)
if (parm.Is(VTab).Not) then
    MsgBox.Error("This script requires a VTab/FTab passed as a
parameter: returning",t)
    return (nil)
end
fldList = parm.GetFields
if (fldList = nil) then
    MsgBox.Info("No field list for this VTab: Returning",t)
    return (nil)
end
editList = MsgBox.MultiList(fldList,"Pick the fields: ","Press
cancel to end")
if ((editList = nil) or (editList.Count = 0)) then
    MsgBox.Warning("No fields picked: returning",t)
    return (nil)
end
return (editList)
```

## ◆ *SaveSelectionAddField*

Shows how to save a selection set by adding a field to the table and using a True/False value to designate TRUE = selected, or FALSE = unselected for the record. Use the *GetSelectionAddField* script to retrieve the selection from the field values.

**Topics**: Table, VTab, Field

**Search Keys**: VTab.SetValue, VTab.SetEditable, VTab.AddFields, Field.Make

**Requirements**: Unless the theme is a shape theme, ArcView cannot edit the table.

```
' Variable Initialization
t = "SaveSelectionAddField"
```

```
' Get the project.
theProj = av.GetProject
' Select the view.
theDocList = theProj.GetDocs
theDict = Dictionary.Make(theDocList.Count)
for each i in theDocList
   if (i.is(View)) then
      theDict.Add(i.GetName,i)
   end
end
theList = theDict.ReturnKeys
theView = theDict.Get(MsgBox.ListAsString(theList,
                        "Choose a view",
                        "VIEWS"))
if (theView = nil) then exit end
' Select the theme.
theThemes = theView.GetThemes
theDict = Dictionary.Make(theThemes.Count)
for each i in theThemes
   if (i.is(FTheme)) then
      theDict.Add(i.GetName,i)
   end
end
theList = theDict.ReturnKeys
theTheme = theDict.Get(MsgBox.ListAsString(theList,
                  "Choose a feature theme",
                  "FEATURE THEMES"))
if (theTheme = nil) then exit end
' Get the theme's FTab (VTab).
```

```
theVTab = theTheme.GetFTab
' Get the theme's selection set.
theBitMap = theVTab.GetSelection
theFldname = MsgBox.Input("Input Selection Field Name",
              "Selection Field Name",
                "Select")
if (theFldName = nil) then exit end
theField = field.make(theFldName,#Field_Char,1,0)
theVTab.SetEditable(True)
test = theVTAB.IsEditable
if (test = false) then
  msgbox.info("Table Not Editable","")
  exit
end
' Add the select field.
theVTab.AddFields({theField})
theObject = true
for each i in 0 .. (theVTab.GetNumRecords - 1)
  x = theBitmap.Get(i)
  if (x = True) then
    theVTAB.SetValue(theField,i,theobject)
  else
    theVTAB.SetValue(theField,i,theobject.not)
  end
end
theVTab.SetEditable(FALSE)
```

## ◆ *SaveSelectionCreateLookupTable*

Shows how to save a selection by creating a look-up table, which consists of a record number for the look-up table and the record number of the selected items in the bitmap for the table. Use the *GetSelectionCreateLookupTable* script to retrieve the selection set. The look-up table will be created at *C:\foo.dbf*. If you wish to change the look-up table name, change the value of the *thefname* variable.

**Topics**: Tables, VTab, Selection

**Search Keys**: VTab.MakeNew, Field.Make, VTab.AddFields, VTab.AddRecord, VTab.Flush, VTab.SetValueNumber

```
' Variable Initialization
t = "SaveSelectionCreateLookupTable"
' Define the look-up table name and location.
thefname = ("c:\foo.dbf").asfilename
' Get the project.
theProj = av.GetProject
' Select the view.
theDocList = theProj.GetDocs
theDict = Dictionary.Make(theDocList.Count)
for each i in theDocList
  if (i.is(View)) then
    theDict.Add(i.GetName,i)
  end
end
theList = theDict.ReturnKeys
theView = theDict.Get(MsgBox.ListAsString(theList,"Choose a
view","VIEWS"))
if (theView = nil) then exit end
' Select the theme.
theThemes = theView.GetThemes
```

```
theDict = Dictionary.Make(theThemes.Count)
for each i in theThemes
  if (i.is(FTheme)) then
    theDict.Add(i.GetName,i)
  end
end
theList = theDict.ReturnKeys
theTheme = theDict.Get(MsgBox.ListAsString(theList,
                "Choose a feature theme",
                "FEATURE THEMES"))
if (theTheme = nil) then exit end
' Get the theme's FTab (VTab).
theVTab = theTheme.GetFTab
if (theVTab = nil) then exit end
' Get the theme's selection set.
theBitMap = theVTab.GetSelection
' Create new dBase table.
filetest = File.Exists(thefname)
if (Filetest = true) then
  file.delete(theFname)
end
' Define fields.
field1 = Field.Make("RecordID",#Field_Short,5,0)
field2 = Field.Make("VTabRecd",#Field_Short,5,0)
' Add fields to field list.
fieldlist = {field1,field2}
' Create dBase file.
theNewVTab = VTab.MakeNew(theFname,dBASE)
' Add fields to the VTab.
```

```
theNewVTab.AddFields(fieldlist)
' Find selected records in the VTab and add
' those record numbers to the NEW VTab.
for each record in theVTab
  bittest = theBitmap.Get(record)
  if (bittest = true) then
    RecordNo = theNewVTab.AddRecord
    theNewVTab.SetValueNumber(field1,RecordNo,RecordNo)
    theNewVTab.SetValueNumber(field2,RecordNo,record)
  end
end
theNewVTab.Flush
```

## ◆ *SaveSelectionTableObjectTag*

Shows how to save a selection set by saving the bitmap of the selection set in the table object tag. Use the *GetSelectionTableObjectTag* script to retrieve the saved selection set.

**Topics**: Table, VTab, Selection, ObjectTag

**Search Keys**: SetObjectTag

```
' Variable Initialization
t = "SaveSelectionTableObjectTag"
' Get the project.
theProj = av.GetProject
' Select the view.
theDocList = theProj.GetDocs
theViewDict = Dictionary.Make(theDocList.Count)
theTableDict = Dictionary.Make(theDocList.Count)
```

```
for each i in theDocList
    if (i.is(View)) then
       theViewDict.Add(i.GetName,i)
    end
    if (i.is(Table)) then
       theTableDict.Add(i.GetName,i)
    end
end
theList = theViewDict.ReturnKeys
theView = theViewDict.Get(MsgBox.ListAsString(theList,
                "Choose a view",
                "VIEWS"))
if (theView = nil) then exit end
' Select the theme.
theThemes = theView.GetThemes
theDict = Dictionary.Make(theThemes.Count)
for each i in theThemes
    if (i.is(FTheme)) then
       theDict.Add(i.GetName,i)
    end
end
theList = theDict.ReturnKeys
theTheme = theDict.Get(MsgBox.ListAsString(theList,
                    "Choose a feature theme",
                    "FEATURE THEMES"))
if (theTheme = nil) then exit end
' Get the theme's FTab (VTab).
theVTab = theTheme.GetFTab
```

```
' Get the theme's selection set.
theBitMap = theVTab.GetSelection
' Check to see if the table exists.
theList = theTableDict.ReturnKeys
theList = theList.Add("NOT ON THE LIST")
theTable = MsgBox.ListAsString(theList,
          "Choose a Table to Save the Selection",
          "TABLE")
if ((theTable = "NOT ON THE LIST") or (theTable = nil)) then
    theTable = Table.Make(theVTab)
else
    theTable = theTableDict.Get(theTable)
end
theTable.SetObjectTag(theBitmap)
```

## ◆ *SaveSelectionToODB*

Shows how to save a selection set by writing the selection as a bitmap to an ODB (object database) file (*C:\bitmap.odb*). To change the name of the file where the ODB is stored, change the value of the *theFileName* variable. Use the *GetSelectionFromODB* script to retrieve the selection from an ODB file.

**Topics**: Table, VTab, Selection, ODB Files

**Search Keys**: ODB.Make, ODB.Add, ODB.Commit

```
' Variable Initialization
t = "SaveSelectionToODB"
theFileName = "C:\bitmap.odb"
theFile = theFileName.asfilename
filetest = File.Exists(theFile)
if (Filetest = true) then
```

```
        file.delete(theFile)
end
' Get the project.
theProj = av.GetProject
' Select the view.
theDocList = theProj.GetDocs
theDict = Dictionary.Make(theDocList.Count)
for each i in theDocList
    if (i.is(View)) then
       theDict.Add(i.GetName,i)
    end
end
theList = theDict.ReturnKeys
theView = theDict.Get(MsgBox.ListAsString(theList,
                        "Choose a view","VIEWS"))
if (theView = nil) then exit end
' Select the theme.
theThemes = theView.GetThemes
theDict = Dictionary.Make(theThemes.Count)
for each i in theThemes
    if (i.is(FTheme)) then
       theDict.Add(i.GetName,i)
    end
end
theList = theDict.ReturnKeys
theTheme = theDict.Get(MsgBox.ListAsString(theList,
                       "Choose a feature theme",
                       "FEATURE THEMES"))
if (theTheme = nil) then exit end
```

```
' Get the theme's FTab (VTab).
theVTab = theTheme.GetFTab
if (theVTab = nil) then exit end
' Get the theme's selection set.
theBitMap = theVTab.GetSelection
' Create the new ODB.
theODB = ODB.Make(theFile)
theList = List.Make
theList.Add(theProj.GetName)
theList.Add(theView.GetName)
theList.Add(theTheme.GetName)
theList.Add(theBitmap)
theODB.Add(theList)
theODB.Commit
```

# ◆ *SelectFeatures*

Uses a Theme.SelectByPoint to return a selection set. Upon using the view, theme, and field values initialized in *EditAttributes*, a MultiInput MsgBox is displayed to allow editing of the attribute values for the selected feature.

**Topics**: FTheme, MsgBox

**Search Keys**: FTheme, MsgBox, MultiInput

**Requirements**: Must be installed as a tool. Script will not run properly unless the compiled script named *EditAttributes* is run beforehand to initialize the global variables used in this script ( *_v*, *_th*, *_ft*, and *_editList*).

```
' Variable Initialization
t = "SelectFeatures"
' ***Critical Resource Test***
```

```
if (_v = nil) then
  MsgBox.Error("Critical Resource 1 NOT Available" + NL +
               "Program Cannot Continue" + NL +
               "Press OK to EXIT",t)
  exit
end
if (_th = nil) then
  MsgBox.Error("Critical Resource 2 NOT Available" + NL +
               "Program Cannot Continue" + NL +
               "Press OK to EXIT",t)
  exit
end
if (_ft = nil) then
  MsgBox.Error("Critical Resource 3 NOT Available" + NL +
               "Program Cannot Continue" + NL +
               "Press OK to EXIT",t)
  exit
end
if (_editList = nil) then
  MsgBox.Error("Critical Resource 4 NOT Available" + NL +
               "Program Cannot Continue" + NL +
               "Press OK to EXIT",t)
  exit
end
' ***      ***      ***      ***
p = _v.GetDisplay.ReturnUserPoint
if (p = nil) then
    MsgBox.Error("No point returned: no apply action taken",t)
```

```
     exit
end

labels = List.Make
for each x in _editList
     labels.Add(x.GetName)
end
_th.SelectByPoint(p,#VTAB_SELTYPE_NEW)
if (_ft.GetSelection.Count = 0) then
    MsgBox.Warning("No records selected: no apply action taken.",t)
    exit
end
for each r in _ft.GetSelection
     defaults = List.Make
' Get the current values for these fields.
     for each x in _editList
          defaults.Add(_ft.ReturnValue(x,r).asString)
     end
' Edit the values.
     newValues = MsgBox.MultiInput("Edit these fields: ","Update
the attributes",labels,defaults)
     if (newValues.Count = 0) then
        MsgBox.Info("Cancel selected: ending edit session",
"Editing FTab")
        exit
     end
' Start editing.
     _ft.SetEditable(TRUE)
     if (_ft.IsEditable = FALSE) then
```

```
            MsgBox.Info("FTab Is Not Editable: Ending Program",t)
            exit
        end
' Update the FTab.
        for each x in _editList
            _ft.SetValueString(x,r,newValues.Get(_editList.Find(x)))
        end
' Stop editing.
        _ft.SetEditable(FALSE)
end
```

# ◆ *SelectFieldsFromVTAB*

Returns a list of user-selected numeric, string, date, Boolean, or mixed field types from a table (SELF).

**Topics**: Tables, VTab, Field

**Search Keys**: VTab.GetFields, Field.GetType, List.Sort

**Requirements**: If this script is called from another script, SELF is a table. If not called from another script, the script expects the *codemog.def* file from the *USA\Tables* directory of the ArcView sample data to be in the table list.

```
' Variable Initialization
t = "SelectFieldFromVTAB"
if (SELF = nil) then
' Hard code variable values.
    wptcTable = "codemog.dbf"
else
' Set variables equal to SELF.
    wptcTable = SELF
```

```
end
' Verify existence of wptcTable.
theTable = av.GetProject.FindDoc(wptcTable)
if (theTable = nil) then
    MsgBox.Error(wptcTable + nl + "Table Not Found","Attribute
Table")
    exit
end
theVTab = theTable.GetVTab
fldList = theVTab.GetFields
numDict = Dictionary.Make(fldList.Count)
typeList = {"Number","String","Boolean","Date","All Types"}
typeswitch = MsgBox.ListAsString(typeList,"Pick a field type:
","Field Type")
for each x in fldList
    if (typeswitch = "Date") then
      if (x.GetType = #Field_Date) then
         numDict.Add(x.GetName,x)
      end
    else
      if (typeswitch = "Number") then
         if ((x.IsTypeNumber) and (x.GetType <> #Field_Date)) then
            numDict.Add(x.GetName,x)
         end
      else
         if (typeswitch = "String") then
            if ((x.IsTypeString) and (x.GetType <> #Field_logical))
then
               numDict.Add(x.GetName,x)
```

```
                    end
              else
                 if (typeswitch = "Boolean") then
                   if (x.GetType = #Field_Logical) then
                       numDict.Add(x.GetName,x)
                   end
                 else
                   if (typeswitch = "All Types") then
                       numDict.Add(x.GetName,x)
                   else
                       exit
                   end
                 end
              end
          end
      end
end
kList = numDict.ReturnKeys
kList.Sort(True)
if (kList.count = 0) then
    MsgBox.Error("No" ++ typeswitch ++ "fields in" ++ wptcTable ++
"Table","Exit")
    exit
end
rList = MsgBox.MultiListasString(kList,"Choose the" ++ typeswitch
++ "fields",typeswitch ++ "Fields")
if (rList = nil) then
    MsgBox.Error("No" ++ typeswitch ++ "Field Chosen","Exit")
    exit
```

```
end
```

```
' To see the list of chosen fields, "uncomment" the next line.
' MsgBox.ListAsString(rList,"RLIST","")
' At this point you have a list (rList) of the fields you
' have chosen.
Return rList
```

## ◆ *TableChange*

Allows a user to edit a record (row) in a table by clicking on the record in the table by using a tool. Install this script as a tool on the table GUI. Use Examine as the icon for the tool.

**Topics**: Tables, Editing, ConvertRowToRecord, Set Editable

```
' Variable Initialization
t = "TableChange"
tbl = av.GetActiveDoc
if (tbl.Is(Table).Not) then exit end
vt = tbl.GetVTab
' When you click the tool on the table, the record number in
' the VTab vt is returned.
r = tbl.ConvertRowToRecord(tbl.GetUserRow)
' Make a list of the field names.
fldList = List.Make
for each f in vt.GetFields
  fldList.Add(f.GetName)
end
' Make a list of the values for each field for the chosen record.
' Note the default value, "-".
```

```
defList = {}
for each fld in fldList
  val = vt.ReturnValue(vt.FindField(fld),r).AsString
  if (val <> "") then
    defList.Add(val)
  else
    deflist.Add("-")
  end
end
p = av.GetProject
' Give the user a chance to edit the record.
changeList = MsgBox.MultiInput("Change this information:",t,
fldList,defList)
if (changeList.Count = 0) then exit end
' Turn on editing.
vt.SetEditable(TRUE)
' Now change the field values.
if (vt.IsEditable.Not) then
  MsgBox.Warning("Cannot edit: " + tbl.GetName,t)
  exit
end
for each i in (0 .. (fldList.Count - 1))
  f = vt.FindField(fldList.Get(i))
  if (f.IsTypeString) then
    if (changeList.Get(i) <> "-") then
      vt.SetValue(f,r,changeList.Get(i))
    end
  elseif (f.IsTypeNumber) then
```

```
    if (changeList.Get(i) <> "-") then
      vt.SetValue(f,r,changeList.Get(i).AsNumber)
    end
  end
end
' Turn editing off.
vt.SetEditable(FALSE)
' Refresh the in-memory and the disk versions of the table.
vt.Refresh
```

# Themes

## ◆ *ImageMove*

Tool for moving the extents of an image theme by means of the user drawing a rectangle in a view. The extents of the image theme are changed to the extents of the user-drawn rectangle.

**Topics**: Theme, ImageTheme

**Search Keys**: View.ReturnUserRect, Theme.Is(ITheme), Theme.GetExtent, ISrc.SetMapExtent, ITheme.Make, View.AddTheme

**Requirements**: Script must be installed as a tool on the view menu. The script will not run properly unless the image theme in the current view is the first active theme in the view.

```
' Variable Initialization
t = "ImageMove"
' ***Critical Resource Test***
```

```
if (av.GetActiveDoc.is(View).Not) then
  MsgBox.Error("Critical Resource 1 NOT Available" + NL +
               "Program Cannot Continue" + NL +
               "Press OK to EXIT",t)
  exit
end
' Get first active image theme.
aTheme = av.GetActiveDoc.GetActiveThemes.Get(0)
if (aTheme.Is(ITheme).Not) then
  MsgBox.Error("Critical Resource 2 NOT Available" + NL +
               "Program Cannot Continue" + NL +
               "Press OK to EXIT",t)
  exit
end
' ***     ***      ***       ***
aView = av.GetActiveDoc
av.ShowMsg("Draw Rectangle")
xRect = aView.ReturnUserRect
if (xRect = nil) then
  MsgBox.Error("no rectangle drawn","error")
  exit
end
fromRect = aTheme.ReturnExtent
aView.DeleteTheme(aTheme)
anISrc = ISrc.Make(aTheme.GetSrcName)
anISrc.SetMapExtent(xRect)
aTheme = ITheme.Make(anISrc)
aTheme.SetVisible(true)
```

```
aTheme.SetActive(true)

aView.AddTheme(aTheme)

aTheme.Invalidate(TRUE)
```

# ◆ *InstallImage*

The purpose of this script is to move the extents of an image theme by passing values for both upper right and lower left image tic mark values and upper right and lower left real world coordinate tic values. The image is moved from the image tic marks to the geolocated tic marks.

**Topics**: Theme, ImageTheme

**Search Keys**: List.Get, Rect.Make, ISrc.SetMapExtent, ISrc.Make, GetWin.Open, GetWin.Maximize, Theme.SetActive

**Requirements**: An example calling statement appears below.

```
' av.Run(t.GetHotScriptName, ---InstallImage-from HotScriptName
' from Active Theme
' {ImageName,  ---Name of Image -- c:\botanic.tif
' theImage,  ---ISource - Botanic.tif
' pImgELLx@pImgELLy,  ---Point - Lower Left Extent of Image Theme
' pImgEURx@pImgEURy,  ---Point - Upper Right Extent of Image Theme
' pImgTLLx@pImgTLLy,  ---Point - Lower Left Tic Value of Image Theme
' pImgTURx@pImgTURy,  ---Point - Upper Right Tic Value of Image
' Theme
' pMapTllx@pMapTLLy,  ---Point - Lower Left Tic Value in Map Units
' pMapTURx@pMapTURy,  ---Point - Upper Right Tic Value in Map Units
' VName})  ---String - Name for Created View
' Example calling statement ends.
' Variable Initialization
```

```
t = "InstallImage"
theVal = SELF
if (SELF.count <> 9) then
  MsgBox.Info("Wrong Number of Parameters in List","")
  exit end

if (Self.Get(1).Is(ITheme).Not) then
  MsgBox.Info("Parameter #2 is not an ISource","")
  exit end

for each i in (2..7)
  if (SELF.Get(i).Is(Point).Not) then
    MsgBox.Info("Invalid Point" ++ i.asstring,"")
    exit end
end
ImgELLx = theVal.Get(2).GetX
ImgELLy = theVal.Get(2).GetY
ImgEURx = theVal.Get(3).GetX
ImgEURy = theVal.Get(3).GetY
ImgTLLx = theVal.Get(4).GetX
ImgTLLy = theVal.Get(4).GetY
ImgTURx = theVal.Get(5).GetX
ImgTURy = theVal.Get(5).GetY
MapTLLx = theVal.Get(6).GetX
MapTLLy = theVal.Get(6).GetY
MapTURx = theVal.Get(7).GetX
MapTURy = theVal.Get(7).GetY
' Compute corresponding image and map tic rectangle width and
```

```
' height.
FromWidth = ((ImgTURx) - (ImgTLLx)).abs
FromHeight = ((ImgTURy) - (ImgTLLy)).abs
' Map tic width and height.
ToWidth = ((MapTURx) - (MapTLLx)).abs
ToHeight = ((MapTURy) - (MapTLLy)).abs
' Compute X and Y direction scale factors.
ScaleX = (ToWidth/FromWidth)
ScaleY = (ToHeight/FromHeight)
' Compute offset of image tic values from image extents and get
' absolute values.
ImgToffx = ((ImgTLLx) - (ImgELLx)).abs
ImgToffy = ((ImgTLLy) - (ImgELLy)).abs
' Compute map X and Y origin values.
MapELLx = (MapTLLx) - ((ImgToffx * ScaleX))
MapELLy = (MapTLLy) - ((ImgToffy * ScaleY))
' Compute map origin as point.
MapOrg = MapEllx@MapElly
' Compute map width and height using image width and height * scaleX
' and scaleY.
MapWidth = ((ImgEURx) - (ImgELLx)) * ScaleX
MapHeight = ((ImgEURy) - (ImgELLy)) * ScaleY
' Compute map size.
MapSize = MapWidth@MapHeight
' Create a rectangle that will be the extents of the map.
MapRect = Rect.Make(MapOrg,MapSize)
' Get the image theme and reset the map extents of the ISrc.
theImageName = SELF.Get(0)
```

```
theISrc = ISrc.Make(SrcName.Make(theImageName))

theISrc.SetMapExtent(MapRect)

theITheme = ITheme.Make(theIsrc)

theITheme.SetVisible(true)

theITheme.SetActive(false)

theView = View.Make

' View name.

VName = (SELF.Get(8).asstring)

theView.SetName(VName.asstring)

' Add Itheme to new view.

theView.AddTheme(theITheme)

theITheme.SetActive(True)

theView.SetTOCWidth(0)

theView.SetTOCUnResizable(True)

theView.GetWin.Open

theView.GetWin.Maximize
```

## ◆ *LegendMakeSimple*

Allows you to make a theme legend into a simple legend, and to assign a color to the legend.

**Topics**: Theme, Legend

**Search Keys**: Theme.GetLegend, Legend.GetLegendType, Legend.GetSymbolsSymlist.UniformColor

**Requirements**: Script will not run properly without the compiled script named *ColorDict* called by av.Run.

```
' Variable Initialization

t = "LegendMakeSimple"
```

```
'***Critical Resource Test***
if ((av.GetProject.FindDoc("ColorDict").is(SEd)) = false) then
  MsgBox.Error("Critical Resource NOT Available" + NL +
               "Program Cannot Continue" + NL +
               "Press OK to EXIT",t)
  exit
end
' ***      ***       ***       ***
' Get the project.
theProj = av.GetProject
' Select the view.
theDocList = theProj.GetDocs
theDict = Dictionary.Make(theDocList.Count)
for each i in theDocList
    if (i.is(View)) then
      theDict.Add(i.GetName,i)
    end
end
theList = theDict.ReturnKeys
theView = theDict.Get(MsgBox.ListAsString(theList,
                "Choose a View",
                "VIEWS"))
if (theView = nil) then exit end
' Select the theme.
theThemes = theView.GetThemes
theDict = Dictionary.Make(theThemes.Count)
for each i in theThemes
    if (i.is(FTheme)) then
```

```
        theDict.Add(i.GetName,i)
      end
  end
  theList = theDict.ReturnKeys
  theTheme = theDict.Get(MsgBox.ListAsString(theList,
        "Choose a Feature Theme",
        "FEATURE THEMES"))
  if (theTheme = nil) then exit end
  ' Get the theme's legend.
  theLegend = theTheme.GetLegend
  if (theLegend.GetLegendType <> #Legend_Type_Simple) then
      ChangeTheme = MsgBox.YesNo("The Feature Theme" ++
                theTheme.asstring ++
                "Already has a legend built!" + NL +
                "Do you wish to make it a simple legend?",
                "LEGEND",True)
      if (ChangeTheme = False) then exit end
  end
  ' Get the FTtab for the theme.
  theFTab = theTheme.GetFTab
  theColor = av.run("ColorDict",
                 {"Pick a Color for the Theme",
                  "THEME COLOR"})
  if (theColor = NIL) then
      MsgBox.Error("NO color chosen - NO changes will be made!",
                "NO COLOR")
      exit
  else
```

```
    theLegend.SingleSymbol

    mysymlist = symbollist.fromlist(theLegend.GetSymbols)

    mysymlist.UniformColor(theColor)

end

theTheme.InvalidateLegend

theTheme.UpdateLegend
```

## ◆ *LegendSimple*

Allows you to make color changes to a simple theme legend by making selections from message boxes.

**Topics**: Theme, Legend

**Search Keys**: Legend.IsSimple, Legend.Default, Legend.GetSymbols

**Requirements**: Script will not run properly without the following critical resource: compiled script named *ColorDict* called by av.Run.

```
' Variable Initialization

t = "LegendSimple"

' ***Critical Resource Test***

if ((av.GetProject.FindDoc("ColorDict").is(SEd)) = false) then

  MsgBox.Error("Critical Resource NOT Available" + NL +

               "Program Cannot Continue" + NL +

               "Press OK to EXIT",t)

  exit

end

' ***     ***     ***      ***

' Get the project.

theProj = av.GetProject

' Select the view.
```

```
theDocList = theProj.GetDocs
theDict = Dictionary.Make(theDocList.Count)
for each i in theDocList
    if (i.is(View)) then
       theDict.Add(i.GetName,i)
    end
end
theList = theDict.ReturnKeys
theView = theDict.Get(MsgBox.ListAsString(theList,
      "Choose a view",
      "VIEWS"))
if (theView = nil) then exit end
' Select the theme.
theThemes = theView.GetThemes
theDict = Dictionary.Make(theThemes.Count)
for each i in theThemes
    if (i.is(FTheme)) then
       theDict.Add(i.GetName,i)
    end
end
theList = theDict.ReturnKeys
theTheme = theDict.Get(MsgBox.ListAsString(theList,
      "Choose a feature theme",
      "FEATURE THEMES"))
if (theTheme = nil) then exit end
' Get the theme's legend.
theLegend = theTheme.GetLegend
if (theLegend.GetLegendType <> #Legend_Type_Simple) then
```

```
    MsgBox.Info("The legend is NOT simple for the feature theme" ++
            theTheme.asstring,"")
    exit
end
' Get the FTab for the theme.
theFTab = theTheme.GetFTab
' Select color for legend if no classification
' field exists; use single symbol legend.
theColor = av.run("ColorDict",
        {"Pick a Color for the Theme",
         "THEME COLOR"})
if (theColor = NIL) then
    MsgBox.Error("NO Color Chosen - NO Changes Will Be Made!",
            "NO COLOR")
    exit
else
    theLegend.SingleSymbol
    mysymlist = symbollist.fromlist(theLegend.GetSymbols)
    mysymlist.UniformColor(theColor)
end
theTheme.InvalidateLegend
theTheme.UpdateLegend
```

# ◆ *LegendUnique*

Allows you to create a thematic legend for a theme using the unique classification scheme by making selections from message boxes. Functionality is similar to the legend editor, but gives you examples of how to manipulate the legend properties, among other things.

**Topics**: Theme, Legend

**Search Keys**: Legend.Unique, Legend.GetClassifications, Legend.GetSymbols, Symbol.SetColor

**Requirements**: Script will not run properly without the following critical resources: compiled script named *ColorDict* called by av.Run.

```
' Variable Initialization
t = "LegendUnique"
' ***Critical Resource Test***
if ((av.GetProject.FindDoc("ColorDict").is(SEd)) = false) then
  MsgBox.Error("Critical Resource NOT Available" + NL +
               "Program Cannot Continue" + NL +
               "Press OK to EXIT",t)
  exit
end
' ***     ***      ***       ***
' Get the project.
theProj = av.GetProject
' Select the view.
theDocList = theProj.GetDocs
theDict = Dictionary.Make(theDocList.Count)
for each i in theDocList
    if (i.is(View)) then
      theDict.Add(i.GetName,i)
    end
end
theList = theDict.ReturnKeys
theView = theDict.Get(MsgBox.ListAsString(theList,
    "Choose a view",
    "VIEWS"))
if (theView = nil) then exit end
```

```
' Select the theme.
theThemes = theView.GetThemes
theDict = Dictionary.Make(theThemes.Count)
for each i in theThemes
    if (i.is(FTheme)) then
       theDict.Add(i.GetName,i)
    end
end
theList = theDict.ReturnKeys
theTheme = theDict.Get(MsgBox.ListAsString(theList,
     "Choose a feature theme",
     "FEATURE THEMES"))
if (theTheme = nil) then exit end
' Get the theme's legend.
theLegend = theTheme.GetLegend
' Get the FTab for the theme.
theFTab = theTheme.GetFTab
' Select the classification field.
theFieldList = theFTab.GetFields
theDict = Dictionary.Make(theFieldList.Count)
for each i in theFieldList
  theDict.Add(i.GetName,i)
end
theList = theDict.ReturnKeys
theField = theDict.Get(MsgBox.ListAsString(theList,
      "Choose a classification field",
      "CLASSIFICATION FIELD"))
```

```
' Classification type is unique value.
if (theField <> Nil) then
  theLegend.Unique(theTheme,theField.GETName)


  theClassType = "Unique Value"
' Ability to select individual colors for each classification, ramp
' colors, or assign random colors.
  theChangeList = {"Specific Colors",
        "Random Colors"}
  theChangeType = MsgBox.ChoiceAsString(theChangeList,
        "Assign Colors",
        "COLORS")
  if (theChangeType = nil) then exit end
  if (theChangeType = "Specific Colors") then
      theLegClass = theLegend.GetClassifications
      theLegSym = theLegend.GetSymbols
      for each i in 0..(theLegClass.Count - 1)
          foo = theLegClass.Get(i).GetLabel
' Select color for this classification.
          theColor = av.run("ColorDict",
                {"Pick a Color for the " + NL + foo + NL +
                "Classification",
                "CLASSIFICATION COLOR"})
          if (theColor = nil) then
            theTheme.InvalidateLegend
            theTheme.UpdateLegend
            EXIT
          end
```

```
        foo2 = theLegSym.get(i).setcolor(theColor)
    end
    theTheme.InvalidateLegend
    theTheme.UpdateLegend
    exit
 else
   if (theChangeType = "Random Colors") then
       theTheme.InvalidateLegend
       theTheme.UpdateLegend
       exit
   end
 end
end
theTheme.InvalidateLegend
theTheme.UpdateLegend
```

## ◆ *ScatterDiagram*

Uses values from two user-selected numeric fields in a VTab to create an xyEvent (scatter diagram). The selection set from the original VTab is applied to the xyEvent. A view is made and the xyEvent installed. X and Y axes are created dynamically and are proportional to the values used for the xyEvent. Labels are created and added to the view.

**Topics**: Theme, XYEvent

**Search Keys**: VTab.GetSelection, VTab.SetSelection, Theme.Make, XYEvent.Get-Extent, View.Make, GraphicShape.Make, View.GetGraphics.Add

**Requirements**: The following sample data setup is required: (1) create a view; (2) add a theme *Cntry94.shp* in *World* subdirectory of sample data; (3) make theme active and visible; (4) join *Attributes of Cntry94.shp* and *demog.dbf* tables (found in *tables* subdirectory of *World* directory) on *Fips_Code* field, with *Attributes of Cntry94.shp* table as the destination table; (5) make view the active document and then run this script.

```
' Variable Initialization
t = "ScatterDiagram"
wptcTable = "Attributes of Cntry94.shp"
thePro = av.GetProject
theView = av.GetActiveDoc
if (theView.Is(View).Not) then
  MsgBox.Error(theView.asString + nl + "Active document is not a
view","Active Document")
  exit
end
theThemes = theView.GetActiveThemes
if (theThemes.Count <> 1) then
  MsgBox.Error("Set Only One Theme Active - Cntry94.shp","Active
Theme")
  exit
end
' Verify existence of wptcTable.
TableA = av.GetProject.FindDoc(wptcTable)
if (TableA = nil) then
  MsgBox.Error(wptcTable + nl + "Table Not Found","Attribute Table")
  exit
end
AvTab = TableA.GetVTab
SelectA = AvTab.getSelection
BvTab = AvTab
' Set selection set of the second VTab equal to the first VTab.
```

```
BvTab.SetSelection(SelectA)

fldList = BvTab.GetFields

numDict = Dictionary.Make(50)

for each x in fldList

  if (x.IsTypeNumber) then

    numDict.Add(x.GetName,x)

  end

end

kList = numDict.ReturnKeys

kList.Sort(True)

chx = MsgBox.ListasString(kList,"Choose a field for X
coordinate","X Coordinate")

if (chx = nil) then

  MsgBox.Error("No X Coordinate","Exit")

  exit

end

xEvent = numDict.Get(chx)

chy = MsgBox.ListasString(kList,"Choose a field for Y
coordinate","Y Coordinate")

if (chy = nil) then

  MsgBox.Error("No Y Coordinate","Exit")

  exit

end

yEvent = numDict.Get(chy)

xyEvent = Theme.Make(XYName.Make(BvTab,xEvent,yEvent))

if (xyEvent = nil) then

  MsgBox.Error("No XYName","Exit")

  exit

end
```

```
aRect = xyEvent.ReturnExtent
h = aRect.GetHeight
w = aRect.GetWidth
o = aRect.ReturnOrigin
org = aRect.ReturnOrigin - ((0.1*w)@(0.1*h))
pty = org + (0@(1.2*h))
ptx = org + ((1.2*w)@0)
v = View.Make
xAx = GraphicShape.Make(Line.Make(org,pty))
yAx = GraphicShape.Make(Line.Make(org,ptx))
lpty = org + (0@(0.5*h))
lpty = lpty - ((0.2* w)@0)
lptx = org + ((0.5*w)@0)
lptx = lptx - (0@(0.2*h))
txtsym8 = textsymbol.make
afont = font.make("Times New Roman","Bold")
txtsym8.SetFont(afont)
txtsym8.SetSize(8)
txtsym10 = textsymbol.make
afont = font.make("Times New Roman","Bold Italic")
txtsym10.SetFont(afont)
txtsym10.SetSize(10)
xLabel = GraphicText.Make(xEvent.GetName,lptx)
xLabel.SetAlignment(#TEXTCOMPOSER_JUST_CENTER)
xLabel.SetSymbols({txtsym8})
yLabel = GraphicText.Make(yEvent.GetName,lpty)
yLabel.SetAlignment(#TEXTCOMPOSER_JUST_CENTER)
yLabel.SetAngle(90)
```

```
yLabel.SetSymbols({txtsym8})
' Max labels
xmax = o + (w@0)
xmaxval = xmax.GetX
ymax = o + (0@h)
ymaxval = ymax.GetY
xmax = xmax - (0@(0.2*h))
ymax = ymax - ((0.2*w)@0)
' Min labels
xminval = o.GetX
yminval = o.GetY
xmin = o - (0@(0.2*h))
ymin = o - ((0.2*w)@0)
pty = org + (0@(1.2*h))
ptx = org + ((1.2*w)@0)
xorglabel = GraphicText.Make(xminval.asstring,xmin)
xorglabel.SetSymbols({txtsym8})
yorglabel = GraphicText.Make(yminval.asstring,ymin)
yorglabel.SetSymbols({txtsym8})
yorglabel.SetAngle(90)
ymaxlab = GraphicText.Make(ymaxval.asstring,ymax)
ymaxlab.SetAngle(90)
ymaxlab.SetSymbols({txtsym8})
xmaxlab = GraphicText.Make(xmaxval.asstring,xmax)
xmaxlab.SetSymbols({txtsym8})
' Tic marks
xtic1 = GraphicShape.Make(Line.Make((o.getx@(o.gety -
    (0.05*h))),(o.getx@(o.gety - (0.15*h)))))
xtic5 = GraphicShape.Make(Line.Make(((o.getx + (w * 0.5))@(o.gety -
```

```
(0.05*h))),((o.getx + (w * 0.5)) @(o.gety - (0.15*h)))))
xtic10 = GraphicShape.Make(Line.Make(((o.getx + w)@(o.gety -
(0.05*h))),((o.getx + w) @(o.gety - (0.15*h)))))
xtic5label = GraphicText.Make((o.getx +
(w * 0.5)).asstring,((o.getx + (w * 0.5))@(o.gety - (0.2*h))))
xtic5label.SetSymbols({txtsym8})
ytic1 = GraphicShape.Make(Line.Make(((o.getx -
(0.05*w))@o.gety),((o.getx - (0.15*w))@o.gety)))
ytic5 = GraphicShape.Make(Line.Make(((o.getx - (0.05*w))@(o.gety +
(h * 0.5))),((o.getx - (0.15*w))@(o.gety + (h * 0.5)))))
ytic10 = GraphicShape.Make(Line.Make(((o.getx - (0.05*w))@(o.gety +
h)),((o.getx - (0.15*w))@(o.gety + h))))
ytic5label = GraphicText.Make((o.gety +
(h * 0.5)).asstring,((o.getx - (0.2*w))@(o.gety + (h * 0.5))))
ytic5label.SetSymbols({txtsym8})
ytic5label.SetAngle(90)
v.GetGraphics.UnselectAll
v.GetGraphics.Add(xtic1)
v.GetGraphics.Add(xtic5)
v.GetGraphics.Add(xtic5label)
v.GetGraphics.Add(xtic10)
v.GetGraphics.Add(ytic1)
v.GetGraphics.Add(ytic5label)
v.GetGraphics.Add(ytic5)
v.GetGraphics.Add(ytic10)
v.GetGraphics.Add(xAx)
v.GetGraphics.Add(yAx)
v.GetGraphics.Add(xLabel)
v.GetGraphics.Add(yLabel)
v.GetGraphics.Add(xmaxlab)
v.GetGraphics.Add(ymaxlab)
```

```
v.GetGraphics.Add(xorglabel)

v.GetGraphics.Add(yorglabel)

v.AddTheme(xyEvent)

xyEvent.SetActive(True)

xyEvent.SetVisible(True)

v.SetName(xEvent.asString ++ "by" ++ yEvent.asString ++ "Scatter
Diagram")

v.SetTOCWidth(0)

v.SetTOCUnresizable(true)

v.GetWin.Open
```

# Views

## ◆ ChooseViewTheme

Dialog that prompts the user to select a view and then a theme in that view.

**Topics**: View, Theme, FTheme, Project

**Search Keys**: View, Theme, Project, GetThemes, GetDocs, Is

**Requirements**: Script can be called from another script; it returns the chosen view, theme, and the FTab for the theme.

```
' Variable Initialization

t = "ChooseViewTheme"

pro = av.GetProject

' Get a list of views.

docList = pro.GetDocs

vList = List.Make
```

```
for each x in docList
  if (x.Is(View)) then
   vList.Add(x)
  end
end
' Select a view or exit.
v = MsgBox.List(vList,"Select a View:",t)
if (v = nil) then
  MsgBox.Warning("No View Chosen: Ending Process",t)
  exit
end
themeList = v.GetThemes
if (themeList.Count = 0) then
    MsgBox.Error("No themes for this view:" ++ v.GetName,t)
    exit
end
aTheme = MsgBox.List(themeList,"Select a theme:",t)
if (aTheme = nil) then
    MsgBox.Warning("No Theme Selected",t)
    exit
end
if (aTheme.Is(FTheme)) then
  fT = aTheme.GetFTab
else
  fT = nil
end
return {v,aTheme,fT}
```

# ◆ *GetDoc*

This script is intended to be called by another program. Use the test section and *ExecuteComment* to test it. The script tests to determine whether a document of a given class exists. If the document does not exist, the script presents a list selection dialog of all documents of that class. This procedure provides a certain level of error recovery for changed document names embedded in programs.

**Topic**: av

**Search Keys**: ShowMsg, Run

```
' t = "Test Driver for GetDoc"
' docName = "ExecuteComment"
' className = "Sed"
' rc = av.Run("GetDoc",{className,docName})
' if (rc = nil) then
' MsgBox.Warning("Document Not Found and No Replacement Chosen",t)
' exit
' end
' MsgBox.Info(rc.GetName,t)
' Test Driver ends
' Variable Initialization
t = "GetDoc"
parmList = SELF
' The following line is included for testing purposes.
' parmList = {"SEd","ToolDocumentation"}
if (parmList.Is(List).Not) then exit end
if (parmList.Count <> 2) then exit end
' The first parameter is the class type.
cl = parmList.Get(0)
if (cl.Is(String).Not) then
```

```
    av.ShowMsg("First parameter must be a Doc Class")
    return nil
  end
  ' The second parameter is the document name.
  docName = parmList.Get(1)
  if (docName.Is(String).Not) then
    av.ShowMsg("Second parameter must be the name of Doc in this
    project.")
    return nil
  end
  ' If the document exists, return it.
  pro = av.GetProject
  d = pro.FindScript(docName)
  if (d <> nil) then
    return d
  else
    MsgBox.Warning(docName ++ "not found",t)
  end
  ' If the document does not exist, return a selection of documents
  ' of this class.
  docList = pro.GetDocs
  cList = List.Make
  for each x in docList
    if (x.GetClass.GetClassName = cl) then
     cList.Add(x)
    end
  end
  MsgBox.Info(cList.Count.asString,t)
```

```
' Check for documents of this class.
if (cList.count = 0) then
  av.ShowMsg("No documents of class" ++ cl ++ "in this project")
  return nil
end
' Present the user with a selection list.
d = MsgBox.List(cList,"Select a Document: ","Doc Class:" ++ cl)
' Return the selection, which may be nil, and must be handled by the
' calling program.
return d
```

## ◆ GetViewTheme

Presents the user with a dialog to select a view from a project, and then to select a theme. If the theme is an FTheme, the FTab is also returned. The result is packaged in a list and returned to the calling program.

**Topics**: View, Theme, FTab, MsgBox

**Search Keys**: View, Theme, FTheme, FTab, GetThemes, Is, GetFields, List, ListAsString

**Requirements**: This script can be run as a standalone or called by another program. The calling program must expect the returned values in the following format: (1) the returned value is a list; (2) the first element of the list is a view; (3) the second element of the list is nil or a theme in that view; and (4) the third element of the list is nil or an FTab. The test drive implements a simple calling program. Use the *ExecuteComment* script to explore how the calling works.

**Test Drive**: If you use the following test drive and fail to select a view and theme, it will bomb. The v.GetName and th.GetName requests will generate a run-time error if *v* or *th* is nil.

```
' lst = av.Run("GetViewTheme",nil)
' v = lst.Get(0)
' th = lst.Get(1)
```

```
' ft = lst.Get(2)
' MsgBox.Info(v.Getname + "/" + th.GetName,"Test Drive")
' Variable Initialization
t = "GetViewTheme"
parmlist = SELF
pro = av.GetProject
' Create the list of views.
vList = List.Make
for each x in pro.GetDocs
  if (x.Is(View)) then
    vList.Add(x)
  end
end
' Present the user with a selection of views.
if (vList.Count = 0) then
  MsgBox.Warning("There are no views in this project: Returning",t)
  return(nil)
end
v = MsgBox.List(vList,"Pick a view:",t)
if (v = nil) then
  MsgBox.Error("No View Picked: Returning",t)
  return nil
end
' Present the user with a selection of themes from this view.
tList = v.GetThemes
if (tList.Count = 0) then
  MsgBox.Warning("No themes for this view: Returning",t)
  return ({v,nil,nil})
end
```

```
th = MsgBox.List(tList,"Pick a theme:",t)
if (th = nil) then
  MsgBox.Warning("No Theme Picked: Returning",t)
  return({v,nil,nil})
end
' If th is a feature theme or FTheme, then return its FTab.
if (th.Is(FTheme)) then
  return ({v,th,th.GetFTab})
else
  return ({v,th,nil})
end
```

## ◆ *HotlinkClickSwitch*

Shows how the Hot Link tool function can be changed to act differently depending on the theme that is active when the tool is invoked. This script should be installed as the click event for the Hot Link tool for views, and the *counties.shp* theme should be installed in the active view. If the only active theme is the *counties.shp* theme (county polygon coverage in ArcView sample data), the user is prompted to select the type of hot link s/he wishes to perform (link to image or link to text file) when the Hot Link tool is selected.

**Topics**: View, Hot link

**Search Keys**: Theme.SetHotField, Theme.SetHotScriptName

**Requirements**: You must run the *HotlinkTableAdd* script before running this script in order to set up the correct environment.

```
' Variable Initialization
t = "HotlinkClickSwitch"
theProj = av.getProject
theView = av.GetActiveDoc
```

```
if (theView.Is(View).Not) then exit end
theTheme = theView.FindTheme("counties.shp")
' ***Critical Resource Test***
if (theTheme.Is(Theme).Not) then
  MsgBox.Error("Critical Resource NOT Available" + NL +
               "Program Cannot Continue" + NL +
               "Press OK to EXIT",t)
  exit
end
' ***     ***     ***       ***
theActiveThemes = theView.GetActiveThemes
for each i in theActiveThemes
  if (i <> theTheme) then
    MsgBox.Info("To select the hot link options for counties.shp"
+ NL +
             "Make the counties.shp theme the only active theme","")
    exit
  end
end


HotChoice = MsgBox.ChoiceAsString({"Hot link to image",
           "Hot link to write document"},
           "Select counties.shp theme hot link option",
           "counties.shp hot link option")
if (HotChoice = nil) then
  MsgBox.Info("No hot link choice made -- " + nl +
             "Defaulted to link to image","")
  theTheme.SetHotField(theTheme.GetFTab.FindField("Image"))
```

```
    theTheme.SetHotScriptName("Link.ImageFile")
else
    if (HotChoice = "Hot link to image") then
        theTheme.SetHotField(theTheme.GetFTab.FindField("Image"))
        theTheme.SetHotScriptName("Link.ImageFile")
    else
        if (HotChoice = "Hot link to write document") then
            theTheme.SetHotField(theTheme.GetFTab.FindField
("Executable"))
            theTheme.SetHotScriptName("Link.TextFile")
        end
    end
end
```

# ◆ *HotlinkTableAdd*

Performs the setup for the *HotlinkClickSwitch* script. This script creates a new dBase table, and then joins the table to the ArcView sample data *counties.shp* theme based on the *Cnty_fips* field. The joined table has a field named *image,* with the path and name to an image file, and a field named *executable,* with the name of a Windows executable file. These two new fields are used in the *HotlinkClick-Switch* script. Create *View1* with the *counties.shp* feature theme prior to running this script. (County data derive from the ArcView USA sample data.)

**Topics**: View, Hot link

**Search Keys**: VTab.Join

**Requirements**: Script will not run properly without the following critical resources: compiled script named *CheckEnvironmentVariable* called by av.Run; a view in the project with its name referenced in the *Viewname* variable; and a theme in the view with its name referenced in the *ThemeName* variable.

```
' Variable Initialization
```

```
t = "HotlinkTableAdd"

Viewname = "View1"

ThemeName = "counties.shp"

' ***Critical Resource Test***

if ((av.GetProject.FindDoc("CheckEnvironmentVariable").is(SEd)) =
false) then

  MsgBox.Error("Critical Resource 1 NOT Available" + NL +
                "Program cannot continue" + NL +
                "Press OK to EXIT",t)

  exit

end

if (av.GetProject.FindDoc(Viewname) = nil) then

  MsgBox.Error("Critical Resource 2 NOT Available" + NL +
                "Program cannot continue" + NL +
                "Press OK to EXIT",t)

  exit

end

if (av.GetProject.FindDoc(Viewname).FindTheme(ThemeName) = nil)
then

  MsgBox.Error("Critical Resource 3 NOT Available" + NL +
                "Program cannot continue" + NL +
                "Press OK to EXIT",t)

  exit

end

' ***     ***     ***     ***

CntyField = "Cnty_fips"

theProj = av.getProject

theView = theProj.FindDoc(Viewname)

theTheme = theView.FindTheme(ThemeName)
```

```
userExt = av.Run("CheckEnvironmentVariable","USEREXT")
' Get VTab for county coverage and number of records in table.
countyVTab = theTheme.GetFTab
cntycount = countyVTAB.GetNumRecords
theField = countyVTAB.FindField(CntyField)
' Get user input for image file (tif) location and name.
Image = MsgBox.Input("Input TIF file location and name",
        "TIF file location and name",
        "c:\ESRI\AV_GIS30\ArcView\???.tif")
if (Image = nil) then exit end
' Get user input for location of a readme.wri file.
Executable = MsgBox.Input("Image readme.wri file location and
name",
        "readme.wri file location and name",
        "c:ESRI\AV_GIS30\ArcView\Etc\Readme.wri")
if (Executable = nil) then exit end
recordDict = Dictionary.Make(cntycount)
for each record in countyVTAB
    recordDict.Add(countyVTAB.ReturnValueString
(theField,record),{Image,Executable})
end
' Create field objects for new dBase file.
field1 = field.make("cnty_fips",#Field_Char,3,0)
field2 = field.make("Image",#Field_Char,64,0)
field3 = field.make("Executable",#Field_Char,64,0)
fieldlist = {field1,field2,field3}
' File name for new dBase file.
theFile = (userext + "\hotlink.dbf").asfilename
' Check for existence of file; if file exists, delete it.
```

```
filetest = File.Exists(theFile)
if (filetest = True) then
     file.delete(theFile)
end
' Create new dBase file and create VTab object for new file.
theVTab = VTab.makenew(theFile,dBASE)
' Add the fields to the VTab.
theVTab.AddFields(fieldlist)
theList = recordDict.ReturnKeys
for each key in theList
   RecordNo = theVTab.AddRecord
   theVTab.setvaluestring(field1,RecordNo,key)
   theVTab.setvaluestring(field2,RecordNo,
recordDict.Get(key).Get(0))
   theVTab.setvaluestring(field3,RecordNo,
recordDict.Get(key).Get(1))
end
theVTab.Flush
countyVTab.Join(theField,theVTab,theVTab.FindField(CntyField))
```

# ◆ *RotateWorld*

"Rotates" the world data set from the ArcView sample data by altering the central meridian of the view projection.

**Topics**: View, Projection

**Search Keys**: View.GetProjection, Projection.SetCentralMeridian, Display.Flush

**Requirements**: Create a view with the *World* polygon theme. Set the view with the following projection: Category=Projections of the World, Type=World From Space, Projection=Orthographic (Perspective).

```
' Variable Initialization
t = "RotateWorld"
' Make the view the active document.
' Run the script.
' Get project.
theproj = av.getproject
' Get active document (should be a view).
theView = av.getactiveDoc
if (theView.Is(View).Not) then
  msgbox.error("Active document must be a view",t)
  exit
end
' Get the display for the view.
theDisp = theView.GetDisplay
' Get the projection for the view.
theprj = theView.GetProjection
' Set the reference latitude.
theprj.setReferenceLatitude(10)
' Returns the current central meridian. You could start rotation
' from here.
' theCenMer = theprj.ReturnCentralMeridian
' Change central meridian by increments of 40 degrees
' for each i in {180,140,100,60,20,-20,-60,-100,-140,-180}.
for each i in {140,100,20,-60,-100,-140,-180}
    theprj.setCentralMeridian(i)
    theprj.recalculate
    theDisp.invalidate(true)
    theDisp.flush
```

```
' Process for slowing down the change in the central meridian
' so that the screen can re-draw.
  av.ShowStopButton
  for each i in 0..1000
    progress = (i/1000) * 100
    doMore = av.SetStatus(progress)
' If stop button is pressed on status bar, rotation stops.
    if (not doMore) then
      exit
    end
  end
end
```

## ◆ *ViewAdd*

A minor rewrite of the *View.Add* system script. Rename this script to *View.Add* if you wish to overwrite the *View.Add* system script. The only change is support for the working directory for feature themes. The value for this directory is set in the *CurrentWorkingDirectoryInitialize* script.

**Topics**: View, System Script

**Search Keys**: View, System Script

**Requirements**: The global variable *_dirDict* must be defined before this script will successfully execute. See the *ScriptFileLoad* and *CurrentWorkingDirectory-Initialize* scripts. Next, this script will not run properly without the global variable named *_dirDict* initialized in the *CurrentWorkingDirectoryInitialize* script.

```
' Variable Initialization
t = "ViewAdd"
' ***Critical Resource Test***
if (_dirDict = nil) then
```

```
  MsgBox.Error("Critical Resource NOT Available" + NL +
               "Program Cannot Continue" + NL +
               "Press OK to EXIT",t)
  exit
end
' ***      ***      ***      ***
theView = av.GetActiveDoc
if (theView.Is(View).Not) then
  msgbox.error("Active document Must Be a view",t)
  exit
end
ch = MsgBox.ListAsString({"Feature Themes","Image Themes"},"Pick a
Theme Type","Add a Theme")
if (ch = nil) then exit end
' The next line sets the current working directory (CWD) from the
' global values.
_dirDict.Get(ch).SetCWD
srcnames = SourceDialog.Show("")
zoom = (theView.GetThemes.Count = 0)
for each n in srcnames
  theView.AddTheme(Theme.Make(n))
end
if ((theView.GetActiveThemes.Count = 0) and (srcnames.Count > 0))
then
  theView.GetThemes.Get(0).SetActive(TRUE)
end
if (zoom) then theThemes = theView.GetThemes
  r = Rect.MakeEmpty
  for each t in theThemes
```

```
    r = r.UnionWith(t.GetExtent)
  end
  if (r.IsEmpty) then
   exit
  elseif ((r.ReturnSize) = (0@0)) then
   theView.GetDisplay.PanTo(r.ReturnOrigin)
  else
   theView.GetDisplay.SetExtent(r.Scale(1.1))
  end
end
if (theView.GetThemes.Count > 0) then
  av.GetProject.SetModified( TRUE )
end
```

# Index

**A**

ActiveGUI.GetButtonBar 285
Add 283, 290, 299, 304, 358
AddButton 302
AddChoice 302
AddComment 412
AddScriptToEMail 412
Applications 281–282, 450
Apply 400
Apply Event 348
ApplySelection 442
asPattern 359
asTokens 323, 414, 430
av 418, 504
av.GetActiveGUI 285, 405
av.Run 306, 318, 323, 327–328, 346, 348, 351, 418, 450

**B**

BasicMarkerStyleEnum 16
BasicPenCapEnum 17
BasicPenJoinEnum 17
Bitmap 442, 444–446
Bitmap.ClearAll 444
BitMap.Count 445
BitmapClearSelection 444
BitmapCountSelected 445
BitmapQuery 446
Button 283, 285–286, 288, 306, 308–309, 318
ButtonAdd 283, 297
ButtonBar 283, 285–286, 288, 306
ButtonBarDocumentation 306
ButtonDeleteEdit 288
ButtonDocumentation 308
ButtonReport 309

**C**

Chart 311
ChartDisplayEnum 22, 24
ChartDisplayLocEnum 25, 255
ChartDisplayMarkEnum 24
ChartDisplayViewEnum 24
ChartDocumentation 311

CheckEnvironmentVariable 281
Choice 292, 295, 297, 300, 312, 314, 318, 326, 328, 387
ChoiceAdd 290
ChoiceasString 387
ChoiceDeleteEdit 292
ChoiceDocumentation 312
ChoiceReport 314
ChooseViewTheme 502
Click Event 348
Clone 389–390
Color 368, 371
Color(GetBlack) 393
ColorDict 392
ColorMap 368, 371, 375
Colors 393
CommonScript 414
Compile 414, 418, 430
Control 295, 302, 304, 314, 318, 348
Control.GetClick 285, 308, 309
Control.GetHelp 308–309
Control.GetIcon 308–309
Control.GetLabel 312
Control.GetUpdate 308–309
ControlSets 283, 288, 290, 292, 299–300, 302, 304
ConvertRowToRecord 480
CurrentWorkingDirectoryInitialize 358

**D**

DeepClone 390, 464
Delete 420
DeleteEmbeddedScripts 416
DeleteSEdsFromProject 417
Dictionary 358, 368, 393–394, 416
Dictionary.Add 393–394
Dictionary.Get 393–394
Dictionary.Make 393
Dictionary.ReturnKeys 393
Display 369
Display.Flush 513
DisplayResEnum 54, 144, 177
DLLProcTypeEnum 55
Doc.Is(View) 351

DocGUI  285–286, 290, 292, 294–295, 297, 300, 306, 312, 326, 346, 348, 350, 400
Documentation  306, 308–309, 312, 316, 323, 325, 330, 338–339, 341, 344, 351, 353
DrawSomething  368
DrawText  369

**E**

EditAttributes  450
Editing  480
EMail  412
Embedded Script  416
EmbeddedScriptDocumentation  315
Enumerations
   BasicMarkerStyleEnum  16
   BasicPenCapEnum  17
   BasicPenJoinEnum  17
   ChartDisplayEnum  22, 24
   ChartDisplayLocEnum  25, 255
   ChartDisplayMarkEnum  24
   ChartDisplayViewEnum  24
   DisplayResEnum  54, 144, 177
   DLLProcTypeEnum  55
   FieldEnum  70
   FieldStatusEnum  71
   FileEolTypeEnum  72
   FilePermEnum  72, 138, 246
   FontEnum  77
   FontWgtEnum  78
   FontWidenessEnum  79
   FrameQualityEnum  79
   FrameRefreshEnum  80
   FTabRelTypeEnum  84, 88
   MatchPrefEnum  148
   PageManagerSizeEnum  178
   PaletteListEnum  180, 238
   PrinterFormatEnum  189
   PrjDisplayQualityEnum  190
   PrjEnum  191
   PrjRectQualityEnum  191
   RasterFillStyleEnum  196
   ScaleBarFrameStyleEnum  202
   SpheroidEnum  191, 229
   SymbolEnum  145, 237, 239
   SymbolWinPaneEnum  239
   SystemArchEnum  241
   SystemLookEnum  241
   SystemOSEnum  241
   SystemOSVariantEnum  241
   TextComposerJustEnum  101, 245

TextPositionerHAlignEnum  247
TextPositionerVAlignEnum  247
UnitsLinearEnum  54, 178, 202, 253, 259
VectorFillStyleEnum  261
VTabSelTypeEnum  84, 89, 273
VTabSummaryEnum  85, 273
Environment Variable  281–282
Execute  438
ExecuteComment  418
Export  452
ExportSortedTable  452
Extensions  283
ExtractScript  402

**F**

Field  452, 455, 464, 477
Field.GetType  477
Field.IsBase  341, 344
Field.IsTypeNumber/String/Shape  341, 344
Field.IsVisible  341, 344
Field.Make  468
FieldEnum  70
FieldStatusEnum  71
File  358–359, 362, 364
FileEolTypeEnum  72
FileExists  358
FileExistsInSearchPath  359
FileName  364, 426, 438
FilePermEnum  72, 138, 246
Files  358
Filter  359
Font  369
FontEnum  77
FontWgtEnum  78
FontWidenessEnum  79
FrameQualityEnum  79
FrameRefreshEnum  80
FTab  365, 442, 464, 506
FTab.GetSelection  442
FTab.SetSelection  442
FTabRelTypeEnum  84, 88
FTheme  316, 474, 502, 506
FThemeDocumentation  316

**G**

GetActiveGUI  292
GetDocs  502, 504
GetFields  506
GetFromSubset  435
GetGraphics  371
GetInsertPos  420

GetLabel 314
GetObjectTag 459
GetScripts 314, 416
GetSelected 437
GetSelectionAddField 455
GetSelectionFromLookupTable 457
GetSelectionTableObjectTag 459
GetSelectionToODB 461
GetThemes 502, 506
GetUpdate 314
GetViewTheme 506
GetVTabFields 464
GetWin 426
GetWin.Maximize 484
GetWin.Open 484
Global Variable 358, 450
Graphic 368–369, 371, 375
GraphicButtons 371
Graphics 368
GraphicShape 368, 371, 375
GraphicShape.Make 496
GraphicText 369
GUI.GetButtonBar.GetControls 308–309
GUI.GetMenuBar.GetControls 312
GUIReport 318

**H**

Hot link 508, 510
HotlinkClickSwitch 508
HotlinkTableAdd 510
HTML 306, 315, 323, 327, 330, 338, 346, 348

**I**

Icon 294, 302, 304, 308–309, 398
IconGet 398
IconMgr 294
ImageMove 482
ImageTheme 482, 484
Indent 323
IndexOf 417
Input 384, 437
Insert 412, 414, 428
Install 400
InstallaTool 294
InstallImage 484
Is 502, 506
Is(Layout) 325
Is(Tool) 350
IsDir 359, 364
ISrc.Make 484
ISrc.SetMapExtent 482, 484

ITheme 323
ITheme.Make 482
IThemeDocumentation 323

**L**

LabelButtons 399
Layout 325
LayoutDocumentation 325
Left 430
Legend 487, 490, 492
Legend.Default 490
Legend.GetClassifications 493
Legend.GetLegendType 487
Legend.GetSymbols 490, 493
Legend.GetSymbolsSymlist.UniformColor 487
Legend.IsSimple 490
Legend.Unique 493
LegendMadeSimple 487
LegendSimple 490
LegendUnique 492
LineEditingDriver 419
LineFile 362, 365, 438
Linefile.Make 402
List 387, 390, 464, 506
List.Get 484
List.Sort 477
ListAsString 387, 464, 506

**M**

Make 362, 365, 371
MakeColorWheel 368
Mark 420
Marker 144
MatchPrefEnum 148
Menu 286, 290, 292, 295, 297, 299, 300, 318, 326, 328
MenuAdd 297
MenuBar 295, 299, 326
MenuBarDocumentation 326
MenuCopy 299
MenuDelete 300
MenuDocumentation 328
Message Boxes 383
MessageBoxFlowControl 383
MessageBoxInput 384
MessageBoxReport 385
MessageBoxSelection 387
Metadata 311, 315, 323
Middle 430
MiniYesNo 383

Miscellaneous 389
MsgBox 383–385, 387, 464, 474, 506
MsgBox.Report 316, 325
MultiInput 384, 474
MultiList 387
MultlistAsString 387

**O**

ObjectTag 459, 470
ODB Files 461, 472
ODB.Add 472
ODB.Commit 472
ODB.Get 461
ODB.Make 472
ODB.Open 461
OpenScript 423

**P**

Package 412, 418
PageManagerSizeEnum 178
PaletteListEnum 180, 238
Pattern 417, 426, 435, 438
PLasXYEvent 362
Polygon 371
PolyLine 365
Pop 401
PrettyButtons 375
PrinterFormatEnum 189
PrjDisplayQualityEnum 190
PrjEnum 191
PrjRectQualityEnum 191
Project 330, 341, 344, 353, 405, 409, 411, 414, 502
Project.Open 405
Project.Save 409
ProjectDocumentation 330
ProjectInit 405
Projection 513
Projection.SetCentralMeridian 513
Projects 402
ProjectSave 409–410
ProjectSaveReset 410
Push 401

**Q**

Query 446

**R**

RasterFillStyleEnum 196
Read 362, 365
ReadElt 426

ReadFiles 426, 438
ReadScriptsFromDiskette 426
Rect.Make 484
Remove 283, 288, 290, 292, 304
RemoveDoc 417
ReplaceOldWithNew 428
Report 385
ReportCoordinates 400
Right 430
RotateWorld 513
Run 504

**S**

Save 411
SaveSelectionAddField 465
SaveSelectionCreateLookupTable 468
SaveSelectionTableObjectTag 470
SaveSelectionToODB 472
ScaleBarFrameStyleEnum 202
ScatterDiagram 496
Script 315, 412
ScriptExportAsEMail 430
ScriptFileLoad 432
ScriptManagerDriver 433
ScriptManagerSubset 435
ScriptMgr 433, 435
Search 414, 428
SEd 338, 412, 414, 416–418, 420, 423, 426, 428,
        430, 432–433, 437, 438
SEd2TextFile 436
SEdDocumentation 338
SEds 412
SEdSetName 437
Select 423
SelectFeatures 474
SelectFieldsFromVTab 477
Selection 442, 444–445, 457, 459, 461, 468, 470,
        472
Set Editable 480
SetClick 288
SetCWD 364
SetDirectories 364
SetEnvironmentVariable 282
SetFont 369
SetHelp 288
SetIcon 288
SetInsertPos 414, 420
SetName 430, 437
SetObjectTag 470
SetSize 369

Shape 365
ShowMsg 504
Sort 452
Source 414, 418, 420
SpheroidEnum 191, 229
Stack 401
StackTest 401
String 323, 430
Symbol 368, 375
Symbol.SetColor 493
SymbolEnum 145, 237, 239
SymbolWinPanelEnum 239
System 281, 282, 438
System Script 410–411, 432, 515
System.GetEnvVar 405
SystemArchEnum 241
SystemLookEnum 241
SystemOSEnum 241
SystemOSVariantEnum 241

**T**

Table 339, 341, 344, 442, 444–446, 452, 455, 457, 459, 461, 464–465, 470, 472
Table.GetObjectTag 341, 344
Table.GetVTab 341, 344
TableChange 480
TableDocumentation 339
TableReport 341
TableReportFile 344
Tables 442, 468, 477, 480
Testing 418
TextComposerJustEnum 101, 245
TextFile 362, 365, 426
TextPositionerHAlignEnum 247
TextPositionerVAlignEnum 247
TextSymbol 369
TextWin 426
TextWin.Make 354
Theme 362, 405, 450, 482, 484, 487, 490, 492, 496, 502, 506
Theme.GetComments 316
Theme.GetExtent 482
Theme.GetExtract 316
Theme.GetHotField 316
Theme.GetHotScriptName 405
Theme.GetHotScriptname 316
Theme.GetLegend 487
Theme.GetSrcName 316
Theme.Is(ITheme) 482
Theme.Make 496

Theme.SetActive 405, 484
Theme.SetHotField 508
Theme.SetHotScriptName 508
Theme.SetVisible 405
Themes 482
Tiger 365
Tool 302, 304, 318, 346, 348, 350, 400
Tool.GetApply 350
Tool.GetClick 350
Tool.GetHelp 350
Tool.GetIcon.GetName 350
Tool.GetUpdate 350
ToolAdd 297, 302
ToolBar 302, 304, 346, 348, 350
ToolBarDocumentation 346
ToolDeleteEdit 304
ToolDocumentation 348
ToolReport 350
Tools 405
tVariable 282

**U**

UnitsLinearEnum 54, 178, 202, 253, 259
Update Events 348, 399
UserExtVariable 283

**V**

VectorFillStyleEnum 261
View 316, 351, 353, 362, 400, 405, 450, 502, 506, 508, 510, 513, 515
View.AddTheme 482
View.GetDisplay.GetReportUnits 354
View.GetDisplay.GetUnits 354
View.GetGraphics.Add 496
View.GetGUI 354
View.GetProjection 513
View.GetThemes 351
View.Make 496
View.ReturnExtent 354
View.ReturnScale 354
View.ReturnUserRect 482
View.SetTOC 405
ViewAdd 515
ViewDocumentation 351
ViewReport 353
Views 502
VTab 339, 444–446, 452, 455, 457, 459, 461, 464–465, 468, 470, 472, 477
VTab.AddFields 468
VTab.AddRecord 468

VTab.FindField 457
VTab.Flush 468
VTab.GetFields 341, 344, 477
VTab.GetNumRecords 445
VTab.GetSelection 444, 496
VTab.Join 510
VTab.MakeNew 468
VTab.ReturnValueNumber 457
VTab.SetEditabl 465
VTab.SetSelection 455, 496
VTab.SetValue 465
VTab.SetValueNumber 468
VTab.UpdateSelection 444, 455
VTabSelTypeEnum 84, 89, 273

VTabSummaryEnum 85, 273

**W**

Write 365, 438
WriteElt 362, 438
WriteScriptsToDisk 438

**X**

XYEvent 362, 496
XYEvent.GetExtent 496

**Y**

YesNo 383
YesNoCancel 383

# More OnWord Press Titles

## Computing/Business

*Lotus Notes for Web Workgroups*
$34.95

*Mapping with Microsoft Office*
$29.95 Includes Disk

## Geographic Information Systems (GIS)

*GIS: A Visual Approach*
$39.95

*The GIS Book, 4E*
$39.95

*INSIDE MapInfo Professional*
$49.95 Includes CD-ROM

*MapBasic Developer's Guide*
$49.95 Includes Disk

*Raster Imagery in Geographic Information Systems*
$59.95

*INSIDE ArcView GIS, 2E*
$39.95 Includes CD-ROM

*ArcView GIS Exercise Book, 2E*
$49.95 Includes CD-ROM

*ArcView GIS/Avenue Developer's Guide, 2E*
$49.95

*ArcView GIS/Avenue Programmer's Reference, 2E*
$49.95

*101 ArcView/Avenue Scripts: The Disk*
Disk $101.00

*ArcView GIS/Avenue Scripts: The Disk, 2E*
Disk $99.00

*ARC/INFO Quick Reference*
$24.95

*INSIDE ARC/INFO*
$59.95 Includes CD-ROM

## MicroStation

*INSIDE MicroStation 95, 4E*
$39.95 Includes Disk

*MicroStation 95 Exercise Book*
$39.95 Includes Disk
Optional Instructor's Guide $14.95

*MicroStation 95 Quick Reference*
$24.95

*MicroStation 95 Productivity Book*
$49.95

*Adventures in MicroStation 3D*
$49.95  Includes CD-ROM

*MicroStation for AutoCAD Users, 2E*
$34.95

MicroStation Exercise Book 5.X
$34.95 Includes Disk
Optional Instructor's Guide $14.95

MicroStation Reference Guide 5.X
$18.95

Build Cell for 5.X
Software $69.95

101 MDL Commands (5.X and 95)
Executable Disk $101.00
Source Disks (6) $259.95

# Pro/ENGINEER and Pro/JR.

Automating Design in Pro/ENGINEER
with Pro/PROGRAM
$59.95 Includes CD-ROM

INSIDE Pro/ENGINEER, 3E
$49.95  Includes Disk

Pro/ENGINEER Exercise Book, 2E
$39.95  Includes Disk

Pro/ENGINEER Quick Reference, 2E
$24.95

Thinking Pro/ENGINEER
$49.95

Pro/ENGINEER Tips and Techniques
$59.95

INSIDE Pro/JR.
$49.95

# Softdesk

INSIDE Softdesk Architectural
$49.95 Includes Disk

Softdesk Architecture 1 Certified
Courseware
$34.95  Includes CD-ROM

Softdesk Architecture 2 Certified
Courseware
$34.95  Includes CD-ROM

INSIDE Softdesk Civil
$49.95  Includes Disk

Softdesk Civil 1 Certified Courseware
$34.95 Includes CD-ROM

Softdesk Civil 2 Certified Courseware
$34.95 Includes CD-ROM

# Other CAD

Manager's Guide to Computer-Aided
Engineering
$49.95

Fallingwater in 3D Studio
$39.95  Includes Disk

# Interleaf

*INSIDE Interleaf (v. 6)*
$49.95  Includes Disk

*Interleaf Quick Reference (v. 6)*
$24.95

*Interleaf Exercise Book (v. 5)*
$39.95 Includes Disk

*Interleaf Tips and Tricks (v. 5)*
$49.95  Includes Disk

*Adventurer's Guide to Interleaf LISP*
$49.95 Includes Disk

# Windows NT

*Windows NT for the Technical Professional*
$39.95

# SunSoft Solaris

*SunSoft Solaris 2.\* for Managers and Administrators*
$34.95

*SunSoft Solaris 2.\* User's Guide*
$29.95  Includes Disk

*SunSoft Solaris 2.\* Quick Reference*
$18.95

*Five Steps to SunSoft Solaris 2.\**
$24.95  Includes Disk

*SunSoft Solaris 2.\* for Windows Users*
$24.95

# HP-UX

*HP-UX User's Guide*
$29.95

*Five Steps to HP-UX*
$24.95  Includes Disk

# OnWord Press Distribution

## End Users/User Groups/Corporate Sales

OnWord Press books are available worldwide to end users, user groups, and corporate accounts from local booksellers or from Softstore/CADNEWS Bookstore: call 1-800-CADNEWS (1-800-223-6397) or 505-474-5120; fax 505-474-5020; write to SoftStore, Inc., 2530 Camino Entrada, Santa Fe, NM 87505-4835, USA or e-mail orders@hmp.com. SoftStore, Inc., is a High Mountain Press Company.

## Wholesale, Including Overseas Distribution

High Mountain Press distributes OnWord Press books internationally. For terms call 1-800-4-ONWORD (1-800-466-9673) or 505-474-5130; fax to 505-474-5030; e-mail orders@hmp.com; or write to High Mountain Press, 2530 Camino Entrada, Santa Fe, NM 87505-4835, USA.

## On the Internet: http://www.hmp.com

OnWord Press, 2530 Camino Entrada, Santa Fe, NM 87505-4835 USA